Origin and Evolution of Biodiversity

Pierre Pontarotti
Editor

Origin and Evolution
of Biodiversity

 Springer

Editor
Pierre Pontarotti
Aix-Marseille Université
  IRD, APHM, Microbe, Evolution,
  Phylogénie, Infection, IHU Méditerranée
  Infection
Marseille, France

and

CNRS
Marseille, France

ISBN 978-3-030-07118-9   ISBN 978-3-319-95954-2   (eBook)
https://doi.org/10.1007/978-3-319-95954-2

This Springer imprint is published by the registered company Springer Nature Switzerland AG
The registered company address is: Gewerbestrasse 11, 6330 Cham, Switzerland

Preface

For the eleventh year, we publish a book on evolutionary biology concept and application.

We try to really catch the evolution and progress of this field for this goal; we are really helped by the Evolutionary Biology Meeting in Marseille. The goal of this annual meeting is to allow scientists of different disciplines, who share a deep interest in evolutionary biology concepts, knowledge and applications, to meet and exchange and enhance interdisciplinary collaborations. The Evolutionary Biology Meeting in Marseille is now recognised internationally as an important exchange platform and a booster for the use of evolutionary-based approaches in biology and also in other scientific areas.

The chapters have been selected from the meeting presentations and from proposition born by the interaction of meeting participants.

The reader of the evolutionary biology books as well as the meeting participants would maybe like us witness years after years during the different meetings and book editions a shift on the evolutionary biology concepts. The fact that the chapters of the book are selected from a meeting enables the quick diffusion of the novelties.

We would like to underline that the eleven books are complementary one to another and should be considered as tomes.

The articles are organised in the following categories

Genome/Phenotype Evolution (Chapters "*Pmela* and *Tyrp1b* Contribute to Melanophore Variation in Mexican Cavefish"–"Mini-bioreactors as Tools for Adaptive Laboratory Evolution for Antibiotic Drug Resistance and Evolutionary Tuning of Bacterial Optogenetic Circuits")

Self/Nonself Evolution (Chapters "Deciphering the Evolution of Vertebrate Immune Cell Types with Single-Cell RNA-seq"–"Immunoglobulin-Like Domains Have an Evolutionarily Conserved Role During Gamete Fusion in *C. elegans* and Mouse")

Origin of Biodiversity (Chapters "Feralisation—The Understudied Counterpoint to Domestication"–"Metagenomic Approaches Highlight the Organization and Dynamics of Plankton at the Species Level")

Origin of Life (Chapters "Ion-Molecule Reactions as a Possible Synthetic Route for the Formation of Prebiotic Molecules in Space" and "Did Gene Expression Co-evolve with Gene Replication?")

Concepts (Chapters "Biological Dogmas in Relation to the Origin of Evolutionary Novelties"–"Natura Fecit Saltum: Punctuationalism Pervades the Natural Sciences")

Marseille, France   Pierre Pontarotti

Acknowledgements

We would like to thank all the authors and the reviewers of the different chapters.

We thank the sponsors of the meeting: Aix Marseille Université, CNRS, ECCOREV FEDERATION, Conseil Départemental 13, ITMO, Ville de Marseille.

We wish to thank the A.E.E.B team for the organisation of the meeting.

We also wish to thank the Springer's edition staff and, in particular, Andrea Schlitzberger for her competence and help.

Marseille, France
May 2018

Marie Hélène Rome
A.E.E.B director
Pierre Pontarotti
A.E.E.B and CNRS

Contents

Part I
Genome/Phenotype Evolution

Pmela and *Tyrp1b* Contribute to Melanophore Variation in Mexican Cavefish

Bethany A. Stahl, Connor R. Sears, Li Ma, Molly Perkins and Joshua B. Gross

Abstract Regressive evolution is a widespread phenomenon that affects every living organism, yet the mechanisms underlying trait loss remain largely unknown. Cave animals enable the study of degenerative disorders, owing to the frequent loss of eyes and pigmentation among lineages evolving in the subterranean habitat. Here, we utilize the blind Mexican cavefish, *Astyanax mexicanus*, to investigate regressive loss of pigmentation because "ancestral" surface-dwelling morphs allow direct comparisons with cave-dwelling forms. Two genes (*Oca2*-albinism and *Mc1r-brown*) have been linked to specific pigmentation alterations in several cavefish populations. Pigment cell (melanophore) number is a complex trait governed by multiple genes, and variation in this trait may contribute to pigmentation diversity in *Astyanax*. To uncover genes associated with this trait, we assembled a high-resolution linkage map and used automated phenotypic scoring to quantify melanophore number variation across seven body regions in a surface × Pachón cave F_2 pedigree. QTL mapping yielded several markers strongly associated with melanophore number variation in the dorsal mid-lateral stripe area and superior head region, which anchor to regions of the *Astyanax* genome and the zebrafish genome. Within these syntenic regions, we identified two candidate genes, *Tyrp1b* and *Pmela*, with known roles in pigmentation based on gene ontology annotation. Mutant forms of these candidate genes in other organisms cause global and regional pigmentation variation, respectively. In *Astyanax*, these genes harbor coding sequence mutations and demonstrate differential expression in Pachón cavefish compared to surface morphs. In sum, this work identifies genes involved with complex aspects of *Astyanax* pigmentation and provides insight into genetic mechanisms governing regressive phenotypic change.

B. A. Stahl · C. R. Sears · L. Ma · M. Perkins · J. B. Gross (✉)
Department of Biological Sciences, University of Cincinnati, Cincinnati, OH 45221, USA
e-mail: grossja@ucmail.uc.edu

Present Address
B. A. Stahl
Jupiter Life Science Initiative, Florida Atlantic University, Jupiter, FL 33458, USA

© Springer International Publishing AG, part of Springer Nature 2018
P. Pontarotti (ed.), *Origin and Evolution of Biodiversity*,
https://doi.org/10.1007/978-3-319-95954-2_1

3

1   Introduction

Pigmentation varies dramatically across the animal kingdom—from crypsis to colorful ornamental displays—suggesting that coloration serves dynamic and adaptive functions. In many animals, these roles vary from mate choice selection (Protas and Patel 2008), defense from predation (Linnen et al. 2009), UV protection, structural support, and thermoregulation (Hubbard et al. 2010). Pigmentation traits have also served as a powerful approach for linking specific genes to phenotypic characters (Hoekstra 2006).

In this study, we investigated the naturally occurring pigmentation loss in the blind cavefish *Astyanax mexicanus* (Jeffery 2001; Borowsky 2008). This species harbors two distinct morphotypes: a pigmented surface-dwelling form that populates the rivers of NE Mexico, and several depigmented (or albino) cave-dwelling morphs that reside in the subterranean environment. This species demonstrates recurrent loss since *Astyanax* cave morphs have repeatedly colonized the cave, providing natural biological "replicates" (Gross 2012a).

The gene underlying the absence of melanin (albinism) was identified and confirmed by CRISPR mutagenesis in surface fish, as *Oca2* in two independent cavefish lineages (Protas et al. 2005; Gross and Wilkens 2013; Ma et al. 2015; Klaassen et al. 2018). A second pigmentation phenotype, *brown,* is associated with the gene *Mc1r* in three cavefish populations (Gross et al. 2009). Although these studies have discovered the genetic basis for monogenic components of pigmentation loss, the genes contributing to complex pigmentation loss in cavefish have not been identified. A prior mapping study in *Astyanax mexicanus* did confirm that melanophore (pigment cell) numerical variation is indeed complex and linked to 18 QTL associated with pigment cell number, yet the identity of the genes underlying this trait still remains unknown (Protas et al. 2007).

The previous characterization of melanogenesis in other animals shows that melanophores are derived from a set of migratory cells that give rise to numerous cell types including cranial cartilage and bone, peripheral neurons, fat cells, and pigment-producing melanophores (Erickson and Perris 1993; Huang and Saint-Jeannet 2004). Due to the diversity of neural crest cell derivatives, it would be less likely to acquire mutations within genes of the neural crest pathway due to potentially lethal consequences (Jeffery 2009). Labeling experiments revealed normal neural crest migration during cavefish development (McCauley et al. 2004), and quantification of cell apoptosis after neural crest-derived precursor migration showed comparable numbers in both cavefish and surface (Jeffery 2006). These combined results suggest that evolutionary changes leading to pigment cell regression in cave morphs may be mediated by alterations late in melanogenesis (Jeffery 2009).

To identify pigmentation-related genes, we employed a second-generation linkage map (Carlson et al. 2015) inclusive of >3,000 genomic markers to perform high-resolution mapping of melanophore number diversity in a large cave × surface F_2 pedigree (Fig. 1a–e). A quantitative trait locus (QTL) mapping study yielded numerous significant associations linked with 20 different regions of our linkage

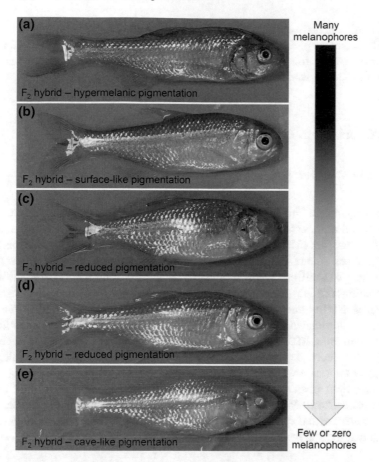

Fig. 1 Surface × Pachón cave F₂ sibling hybrids display a vast array of coloration. Hybrid offspring from a surface × Pachón cave cross reveal a varying degree of melanic-based pigmentation. The "dark" versus "light" appearance is associated with the number of melanophores (pigment cells) that an individual harbors. We observed levels of pigmentation that are darker (**a**) than normal surface fish (**b**). Some hybrid individuals showed dramatic reductions in pigmentation (**c, d**), while albino individuals (which still retain melanophores) produce no melanin, rendering melanophores invisible (**e**)

map. We then leveraged available genomics resources (McGaugh et al. 2014) to nominate candidate genes. Comparative genomics identified the critical syntenic region for each QTL in the *Astyanax* cavefish draft genome alongside conserved intervals in the distantly related zebrafish genome. We nominated candidate genes by screening the genes within these syntenic regions for gene ontology (GO) terms related to pigmentation. These analyses yielded two genes, *Tyrp1b* and *Pmela*, with well-characterized roles in melanin-based pigmentation in other animals. We further characterized the coding sequence and expression of candidate genes. Through these studies, we propose *Tyrp1b* and *Pmela* as genes that likely contribute to complex pigmentation in *Astyanax* cavefish. Identification of additional pigmentation genes provides a clearer picture of the mechanisms contributing to regressive evolution in *Astyanax* and informs our understanding of the broader principles governing trait loss in the natural world.

2 Materials and Methods

2.1 Melanophore Scoring

To quantify melanophore number, we analyzed numerous regions of the body where pigment cell number varied within our F_2 pedigree. These included seven regions: near the anal fin (MelAnalfinSquare), below the mid-lateral stripe (MelUnder-Stripe), dorsal square (MelDorsalSquare), area above the stripe (MelAboveStripe), above the eye (MelHeadSquare), full head (MelHead), and neighboring the anal fin (MelAnalfinTriangle). Regions were selected based on a consistent set of landmarks for reproducibility and were similar to the areas previously assayed (Protas et al. 2007). For automated counts, we employed ImageJ (v.1.6; National Institutes of Health, Bethesda, MD) by inverting the color in a selected area and then counting the lighter objects (e.g., pigment cells, now white) with a preset "noise tolerance." The noise tolerance was set so only markings (melanophores in this case) above the preset light pixel intensity were counted. Each image was reviewed and any melanophores "missed" in the automatic quantification were manually added. When appropriate, we transformed the melanophore counts to $\log^{10}$ values to generate a normal distribution for association studies.

2.2 Quantitative Trait Locus (QTL) Association Mapping

All QTL analyses were performed using a previously published linkage map (Carlson et al. 2015). We employed the software program R/qtl (v.1.30; Broman et al. 2003) for all association analyses. We analyzed each trait using four mapping methods: marker regression (MR; Kearsey and Hyne 1994), expectation maximization (EM; Xu and Hu 2010), Haley-Knott (HK; Haley and Knott 1992), and nonparametric (NP; Kruglyak and Lander 1995) as described in Gross et al. (2014). Significant linkages were set at a LOD score threshold of ≥ 4.0—as used in other QTL studies (Protas et al. 2007; Gross et al. 2009). To confirm associations, permutation tests involving 1000 iterations were performed to identify statistically significant QTL ($P < 0.05$). Effect plots for associations were generated using the closest linked genetic marker. QTL regions were then anchored (~6–8 cM on each side of the top marker) using the NCBI BLAST Toolkit (v.2.28+) to the *Astyanax* genome (Ensembl build v.75; McGaugh et al. 2014). We also determined the syntenic interval in the zebrafish genome, with previously demonstrated synteny with *Astyanax* (Gross et al. 2008; O'Quin et al. 2013; Carlson et al. 2015). Visual representations of synteny between our linkage map and the genomes were created with Circos (v.0.64; Krzywinski et al. 2009; Fig. 3a, d).

2.3 Gene Ontology (GO) Term Analysis

To nominate prospective candidate genes, we interrogated all genes within the syntenic interval in the *Astyanax* draft genome for gene ontology (GO) terms. From these analyses, we collected the GO terms for all genes located in the *Astyanax* syntenic interval using BioMart (v.0.8; Kasprzyk 2011). This approach yielded hundreds of GO terms for each significant QTL. We then narrowed our search to terms with potential involvement in pigmentation, such as "pigment", "pigmentation", "melanin", "eumelanin", "phaeomelanin", "melanophore", "melanocyte", "melanosome", "xanthophore", "iridophore", "chromatophore", and "carotene". This approach enabled the selection of genes based on annotation information from known functions in other organisms.

2.4 RNA-Seq, Qualitative, and QPCR Expression Analyses

We evaluated genes located within predicted critical genomic regions for expression differences using RNA-seq. Total RNA was isolated from pools of 50 surface or cave individuals using the RNeasy Plus Mini Kit (Qiagen) at each of five developmental stages. These included 10 hours post-fertilization (hpf), 24, 36, and 72 hpf, and from three individuals during juvenilehood (~4 months). Library preparation (TruSeq v.2 kit) and sequencing (Illumina 2500 Hi-Seq) was performed in triplicate (10–72 hpf) or duplicate (juvenile) at the DNA Sequencing Core (Cincinnati Children's Hospital and Medical Center). All samples were sequenced to a ~10 million read depth for 50-bp, single-end reads. Normalized gene expression was calculated using ArrayStar (DNAStar). All expressions were evaluated using comparative read counts between cave and surface fish with the RPKM normalization method (Mortazavi et al. 2008). Raw sequencing reads are deposited at the NCBI SRA (BioProject: PRJNA258661).

We validated expression profiles at the 72 hpf stage using qualitative and quantitative PCR analyses, described in Stahl and Gross (2017). Template cDNA from surface and cavefish RNA pools (n = 50 embryos each; RNeasy Plus Mini kit, Qiagen) was synthesized for both experiments using the Transcriptor RT kit (Roche). Quantitative PCR (qPCR) experiments were performed as described in Stahl and Gross (2017). All samples were analyzed in sextuplet and normalized expression values (C_q) and significant differences (two-tailed Student's t-test) were determined using the CFX Manager software program (v.3.1; BioRad).

Genes residing within the syntenic region in the *Astyanax* genome were analyzed for sequence alterations. Sequencing reads derived from surface fish and Pachón cavefish (~280 million reads total) were aligned to the draft *Astyanax* genome (Ensembl.v.75; McGaugh et al. 2014) using default parameters for the program Seq-Man NGen (DNASTAR) and evaluated for diverse sequence mutations (e.g., SNP, indels) segregating between surface and cave morphs.

2.5 Whole-Mount In Situ Hybridization

RNA probes for in situ hybridization were generated by PCR for the genes *Tyrp1b* (forward primer: 5′-GAAACAGCCCTCAGTTCGAG-3′, reverse primer: 5′-AGGTGGGCCAGATTGTGTAG-3′) and *Pmela* (forward primer: 5′-CTACTGATGCTGCCACTGGA-3′, reverse primer: 5′-AGAGCCGTAGCGGTAGATCA-3′). The resulting PCR products were cloned into the TOPO TA Dual Promoter cloning vector (Life Technologies) and confirmed by sequencing. Sense and antisense digoxygenin (DIG)-labeled RNA probes for *Astyanax Tyrp1b* and *Pmela* were transcribed with either SP6 or T7 RNA polymerase (Roche). Whole-mount in situ hybridizations were performed on embryos at stages <12 hours post-fertilization (hpf), 24, 36, and 72 hpf according to Ma et al. (2014). Embryos were washed in NTM, PBT, and TBST and fixed in 4% PFA/PBS for storage at 4 °C and visualized using Leica Microscope M205 FA (with LAS software v.3.8.0) montage imaging (Fig. 5a–d). The numbers of cells with positive expression for *Pmela* or *Tyrp1b*, respectively, were counted.

2.6 Functional Validation in Zebrafish with MO Knockdowns

The translational blocking morpholino oligonucleotides were targeted to the first 25 base pairs of the zebrafish ORF of *Pmela* (5′-GAGGAAGATGAGAGATGTCCACAT-3′) and *Tyrp1b* (5′-GCACTAAACACACACTCTTCCACAT-3′; Gene Tools, LLC.). Morpholinos were injected into one-cell stage zebrafish embryos in a 1 nl volume at a 0.2 mM concentration. Control individuals were administered a mock injection composed of phenol red and Danieaux's solution (Wingert et al. 2004), or a control oligo (5′-CCTCTTACCTCAGTTACAATTTATA-3′; Gene Tools, LLC). Imaging was performed using a Leica Microscope M205 FA (with LAS software v.3.8.0) stereoscope (Fig. 5e–g). Phenotypic analysis was performed on embryos fixed at 5dpf in 4% PFA. Counts were performed on the yolk sac of the number of light and dark melanophores, and the number of clustered and isolated melanophores, both relative to total melanophore number on the yolk sac (Fig. 5h, i).

3 Results

3.1 Cavefish × Surface Fish Hybrid Individuals Demonstrate Diversity in Melanophore Numbers in Distinct Regions Spanning the Body

Direct observations reveal a wide array of pigmentation variation among F_2 siblings (Fig. 1a–e), and substantial pigment cell variation has been reported in four different body regions (Protas et al. 2007). We quantified melanophore number in seven areas and found variation across our F_2 population (n = 170). Counts ranged within each region (reported as minimum to max number melanophore distribution for a given region among the entire F_2 pedigree): 1–99 melanophores in the square near the anal fin (MelAnalfinSquare), 1–113 pigment cells below the mid-lateral stripe (MelUnderStripe), 4–600 cells in the dorsal square (MelDorsalSquare), 1–2,762 melanophores in the area above the stripe (MelAboveStripe), 1–89 pigment cells above the eye (MelHeadSquare), 11–1,395 melanophores present in the full head region (MelHead), and 0–252 pigment cells by the anal fin (MelAnalfinTriangle). Numerical variation in melanophore numbers did not simply increase or decrease proportionately across all regions of the body, instead pigment cell numbers in some cases varied independently between different assayed regions.

3.2 QTL Analysis Revealed 19 Genomic Regions Associated with Complex Melanophore Variation

We scored melanophore variation in seven regions and detected numerous QTL (Fig. 2a). We did not discover QTL for melanophores on the entire head (MelHead). This was surprising since we detected QTL for pigment cell counts in the head square above the eye (MelHeadSquare). Additionally using four different mapping methods, our results yielded multiple significant (independent) associations (n = 41 markers) with the linkage map for the following pigment cell number traits: MelAnalfinSquare (n = 1 QTL), MelUnderStripe (n = 13 QTL; Fig. 2d, e), MelDorsalSquare (n = 3 QTL; Fig. 2b, c), MelAboveStripe (n = 1 QTL), MelHeadSquare (n = 3 QTL), and MelAnalfinTriangle (n = 6 QTL). In some instances, the different mapping methods identified significant associations with same top marker (e.g., MR, EM, HK, and NP methods all noted the marker ASTYANAX_414 at single map location (LG 16, 22.2 cM for MelUnderStripe), and in other cases, the different approaches yielded multiple significant markers within a small positional window—for example, three-markers for each MR, EM, and HK respective method for MelDorsalSquare on LG 20 (53–61 cM). Not surprisingly, some of these QTL co-localized near the same position in our linkage map. One "hotspot" in our map was on linkage group 1 from 75 to 93 cM, wherein associations with five traits (MelHeadSquare, MelUnder-

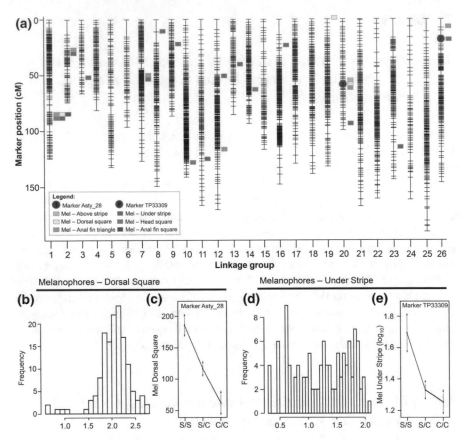

Fig. 2 High-resolution mapping identified multiple QTL associations with melanophore number variation. Our quantitative trait locus (QTL) study detected numerous associations (n = 40) with pigment cell variation (**a**). Many of these co-localized to similar positions on our linkage map. With this approach, we were able to capture the dramatic variation of pigment cell number, including the range of melanophores detected in the dorsal square (**b**) and under the mid-lateral stripe (**d**). The frequency denotes the number of individuals harboring a given number of melanophores. The markers underlying these two QTL demonstrated a significant effect of phenotype, where individuals harboring two copies of the surface allele "S/S" demonstrated increased numbers of melanophores, and hybrids with two cavefish alleles "C/C" exhibited fewer pigment cells. This effect is evident in both the dorsal (**c**) and stripe (**e**) regions assayed

Stripe, MelAnalfinTriangle, MelDorsalSquare, MelAnalfinSquare) yielded significant QTL (Fig. 2a). Moreover, three different regions assayed (MelDorsalSquare, MelAnalfinTriangle, MelAboveStripe) returned associations with linkage group (LG) 20 (53–61 cM), and two traits (MelAnalfinTriangle, MelUnderStripe) tended to co-localize together at several positions: LG 2 (29.8–33.6 cM), LG 7 (54.64–63.7 cM) and LG 26 (8.98–18 cM). In total, we discovered 19 distinct regions (QTLs) of the genome are associated with melanophore variation (Fig. 2a).

3.3 Comparative Analyses Narrowed Melanophore QTL Positions to Critical Genomic Regions in Astyanax and Danio

To further search for candidate genes residing near our top markers, we analyzed the positions of each QTL in the *Astyanax* cavefish draft genome (~10,000 scaffolds) and the current zebrafish genome (Ensembl.Zv9). First, we queried marker sequences to the *Astyanax* genome (Ensembl.v.75) using the standard nucleotide BLAST algorithm to find the position of the top genomic marker, and the locations of the markers within the QTL interval (immediately adjacent ~6–8 cM to the top marker), to identify regions of synteny. Since these scaffolds varied in length (876–9,823,298 bp), the number of scaffolds reported back from our BLAST search ranged from 4 to 27 scaffolds for each association (data not shown).

We also capitalized on the well-annotated genome of teleost fish *Danio rerio*. Three comparative studies previously identified large syntenic regions shared between *Astyanax* and *Danio*, which diverged ~120 MYa (Gross et al. 2008; Gross 2012b; O'Quin et al. 2013; Carlson et al. 2015). Our genomic comparisons revealed several syntenic blocks based on the positions of BLAST hits in *Danio*, yielding regions of synteny on different zebrafish chromosomes. We sought to further support the syntenic block from our direct comparisons of marker sequences to zebrafish, since usually <20 markers yielded direct BLAST hits for each QTL. Accordingly, we collected every predicted gene (often 100+ genes) from the respective *Astyanax* scaffolds associated with each QTL and queried the full-length gene sequences to the zebrafish genome.

3.4 Tyrp1b and Pmela Are Two Candidate Genes Associated with Numerical Melanophore Diversity

Nearly all of the assayed pigmentation traits yielded multiple QTL associated with multiple distinct regions of the linkage map (Fig. 2a). We characterized the wide range of melanophore number variation in our pedigree, including the dorsal square (Fig. 2b) and inferior to the mid-lateral stripe (Fig. 2d). Our association studies revealed multiple loci for these traits, including the markers ASTYANAX_28 on LG 20 (LOD^{MR} = 6.01, p = 0.015) and TP33309 on LG 26 (LOD^{MR} = 5.08, p = 0.041), respectively (see red circles in Fig. 2a). Corresponding effect plots demonstrated a significant effect of genotype for each marker: homozygous surface alleles "S/S" harbor more melanophores and homozygous cave alleles "C/C" have fewer melanophores (Fig. 2c, e). The heterozygous genotype for ASTYANAX_28 presented an intermediate number of melanophores at this locus, whereas the heterozygous genotype at TP33309 has melanophore numbers similar to the homozygous cave genotype, perhaps indicating the dominance of the cave allele in hybrid individuals.

To pursue candidate genes, we collected the ontology for genes within the syntenic blocks. For the QTL association on LG 20 (marker: ASTYANAX_28), we identified 951 terms affiliated with 157 genes on 10 *Astyanax* scaffolds, and 868 terms associated with 193 genes from the syntenic interval anchoring to chromosome 1 in zebrafish. Within this dataset, we identified two GO terms related to pigmentation: "melanosome membrane" (GO:0033162) and "melanin biosynthetic process" (GO:0042438) of which both terms are associated with two genes *Tyrp1b* and *Tyrp1a* in *Astyanax,* and "pigmentation" (GO:0043473) with the gene *Tyrp1b* in the zebrafish syntenic block. We note that the *Tyrp1a* paralog resides on a different chromosome (Chr7) in *Danio rerio,* whereas *Tyrp1b* resides on chromosome 1 in zebrafish, (described above). Additionally, we discovered 1,136 GO terms for 409 genes associated with 12 scaffolds for "MelUnderStripe" on LG 26 (marker: TP33309), and 3,017 terms for the 654 genes within the critical block on chromosome 11 in *Danio rerio.* This analysis revealed three pigmentation GO terms in *Astyanax* including "melanosome transport" (GO:0032402) linked to *Ippk* and "developmental pigmentation" (GO:0048066) and "eye pigmentation" (GO:0048069) assigned to *Pmela.* In *Danio*, we identified "developmental pigmentation" (GO:0048066) and "eye pigmentation" (GO:0048069) associated with *Pmela.* By co-analyzing gene ontology and positional information from both organisms, we nominated candidate genes for each QTL: *Tyrp1b* (LG20) and *Pmela* (LG26). These two genes reside in close proximity to the top two QTL markers, ~7 MB and ~0.05 MB, respectively (Fig. 3a, d).

Next, we interrogated the coding sequences by aligning surface- and cave-tagged RNA sequencing reads to the *Astyanax* draft genome. *Tyrp1b,* near a marker where multiple melanophore traits were mapped, harbored three mutations, including a G-to-A substitution in exon 2 at position 630 (Fig. 3b). This is predicted to be a synonymous change (F210F); however, it falls within a tyrosinase copper-binding domain (205–222 bp; Ensembl.v.75) which could affect interactions with tyrosinase—an enzyme catalyzing melanin biosynthesis (Oetting 2000). Synonymous mutations such as these can affect splicing, stability, structure, and protein folding (Hunt et al. 2009). For the gene *Pmela*, sequence analyses revealed numerous alterations, such as a G-to-A change at position 1738 in exon 7 (Fig. 3e). This non-synonymous mutation impacts the amino acid sequence, causing a change from a hydrophobic alanine residue to a hydrophilic threonine in cavefish (A580T). In addition, *Pmela* demonstrates several more mutations including two other non-synonymous changes, two silent SNPs, and three potential splice variants.

With respect to expression, *Tyrp1b* showed reduced expression in cave relative to surface beginning at 24 hpf (Fig. 3c). We observed expression differences of 7-fold down (36 hpf) and 5.5-fold down (72 hpf; Fig. 3c). Expression profiles of cave versus surface forms demonstrated significant differences at 10 hpf ($p = 0.0472$), 36 hpf ($p = 0.0312$), and 72 hpf ($p = 0.00167$). *Pmela* similarly revealed reduced expression at 24 hpf (Fig. 3f). *Pmela* demonstrated a 12-fold reduction in expression at 72 hpf. Levels of gene expression were significantly different between morphs at four stages assayed: 10 hpf ($p = 0.00607$), 24 hpf ($p = 0.00117$), 36 hpf ($p = 0.0311$), and 72 hpf ($p = 0.000239$). Using qualitative and quantitative PCR, we observed weaker bands

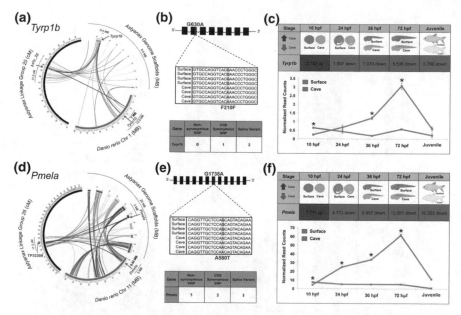

Fig. 3 Integrative analyses reveal two candidate genes for melanophore variation. The evaluation of melanophore variation using QTL analyses identified two top candidate genes that may be associated with pigmentation loss in *Astyanax mexicanus* cavefish. Pigment cell variation under the mid-dorsal stripe region showed a significant association with marker ASTYANAX_28 on LG 20 (LOD^{MR} = 6.01, p = 0.015), which anchors to genomic regions in *Astyanax* and *Danio* inclusive of *Tyrosinase-related protein 1b* (*Tyrp1b*; **a**). Distances between the top marker and candidate gene are in boldface. A second melanophore trait in the superior dorsal region demonstrates a significant association with marker TP33309 on linkage group 26 (LOD^{MR} = 5.08, p = 0.041; **d**). Regions of synteny were identified between *Astyanax* and *Danio rerio*. The syntenic region was mined for any genes with potential roles in the pigmentation pathway using gene ontology (GO) terms. This led to the discovery of the gene *Premelanosome protein a* (*Pmela*). These two candidates, and any other pigment-related genes that exist within the syntenic region, were evaluated for prospective coding mutations and expression alterations using RNA-seq technologies (**b–c, e–f**). This combined approach has led to discovery of candidate genes that may contribute to loss of pigmentation in cave-dwelling fish

in Pachón cavefish compared to the surface fish for *Tyrp1b* and *Pmela* (gel images not shown), and significantly reduced expression in Pachón cavefish for *Tyrp1b* (p = 0.005967; Fig. 4a, c) and *Pmela* (p < 0.000001; Fig. 4b, d).

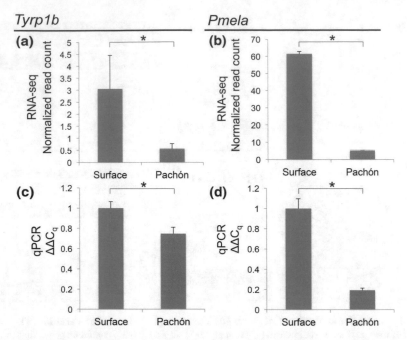

Fig. 4 Quantitative (qPCR) confirms reduced expression of _Tyrp1b_ and _Pmela_ in Pachón cavefish. To further validate reduced expression patterns from our RNA-seq expression profiling (**a, b**), we performed quantitative PCR on surface and Pachón cavefish at 72 hpf. These analyses confirmed dramatic reduction of expression in cave-dwelling morphotypes for both _Tyrp1b_ (p = 0.005967) and _Pmela_ (p < 0.000001)

3.5   **Tyrp1b** _and_ **Pmela** _Demonstrate Distinct Melanophore-Specific Expression Patterns Between Cave and Surface Morphotypes_

Both _Pmela_ and _Tyrp1b_ appear to influence both melanophore position and melanin density in _A. mexicanus_ and _D. rerio_ (Fig. 5). Analysis of the location of _Pmela_ (Fig. 5a, b) and _Tyrp1b_ (Fig. 5c, d) gene expression following in situ hybridization indicates a heterochronic delay in cavefish compared to surface fish. Early in development (<12 hpf), _Pmela_ and _Tyrp1b_ gene expressions are apparent in more cells in surface fish than in cavefish. Later in development, surface fish exhibits higher numbers of cells with positive _Pmela_ expression posteriorly than cavefish.

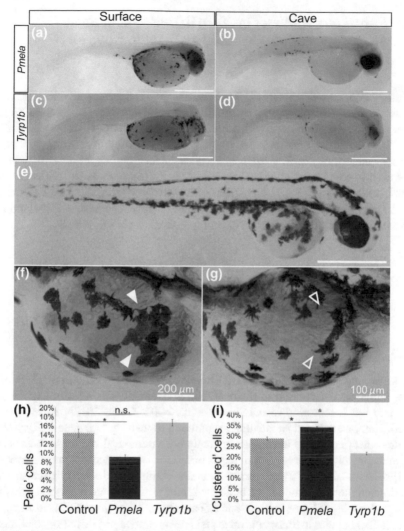

Fig. 5 Expression and functional analysis of *Pmela* and *Tyrp1b* candidate genes. In situ hybridization of *Pmela* (**a, b**) and *Tyrp1b* (**c, d**) of cave (**b, d**) and surface fish (**a, c**) embryos. At this stage, surface fish embryos have begun to develop melanin pigmentation across the yolk sac and putative eye regions, which appears to coincide with *Tyrp1b* and *Pmela* gene expression. In individuals stained for *Pmela*, we see a difference in spatial distribution of gene expression between cave and surface embryos. Morpholino knockdown of 5dpf *D. rerio* embryos (**e–g**). The control-injected (**e**) embryos display a "normative" melanophore distribution and density in comparison to *Pmela* (**f**) and *Tyrp1b* (**g**) knockdowns. **f** Filled arrowheads show clustering of melanophores in *Pmela* morphants. (**g**) Open arrowheads show a higher dispersal of melanophores in *Tyrp1b* morphants. A total number of melanophores on the yolk sac were counted to determine the ratio of pale (**h**) or clustered (**i**) melanophores. *Pmela* knockdowns exhibit a clustering of melanophores across the yolk sac (p < 0.05) and a lesser ratio of "pale" melanophores (p < 0.05) than control-injected embryos. (**i**) *Tyrp1b* knockdowns have a higher distribution of melanophores across the yolk sac (p < 0.05) and a higher number of "pale" melanophores (n.s.) than control. Scale bars at 500 μm, unless otherwise noted

3.6 Functional Analyses of Tyrp1b and Pmela Reveal Altered Melanophore Dispersion and Structure in Morphants

We utilized the closely related (120 MYa diverged), and laboratory-established fish species *Danio rerio,* to carry out morpholino experiments (Fig. 5e–g). *Pmela* knockdown via MO injection resulted in clustering of melanophores across the yolk sac (Fig. 5f, i), coincident with a lesser number of pale melanophores (Fig. 5h). In contrast to *Pmela* morphants, *Tyrp1b* morphants have a higher number of pale cells compared to control individuals (Fig. 5h). This corresponds to the wider dispersal of melanophores (Fig. 5g) and a higher number of "pale" melanophores on the yolk sac of *Tyrp1b* morphants. Both *Pmela* and *Tyrp1b* appear to influence melanophore darkness and position, based upon phenotypes of *D. rerio* morphants.

4 Discussion

4.1 Tyrp1b and Pmela Contribute to Melanophore Number Variation in Cavefish

Cave-dwelling *Astyanax* has evolved diverse phenotypes upon colonization of the cave, including a reduction in pigmentation and eyes, and character expansion of taste and touch sensation (Montgomery et al. 2001; Gross 2012a; Kowalko et al. 2013). Here, we utilized an integrative approach to identify two genes—*Tyrp1b* and *Pmela*—that play a role in complex melanophore number diversity in *Astyanax*. In *Astyanax*, deeply conserved sets of pigmentation-related processes may be governed by *Tyrp1b* and *Pmela*. *Tyrp1* serves as a stabilizing protein for tyrosinase which functions in melanin production, and when the *Tyrp1b* protein is absent, tyrosinase rapidly degrades (Müller et al. 1988; Kobayashi and Hearing 2007). In mice, two distinct *Tyrp1* alleles in transgenic mice yield degenerative pigmentation: one causing brown coat coloration or nearly white fur relative to the normally black mice (*b* allele), and another form that could induce albinism (*c* allele; Kwon et al. 1989). Similarly, variants of *Tyrp1* are the cause of "chocolate" and "cinnamon" coat colors in the domestic cat (Lyons et al. 2005; Schmidt-Küntzel et al. 2005). Numerous *Tyrp* alleles have also been identified in humans as a cause of oculocutaneous albinism type III, which are often caused by small deletions or single base pair changes impacting the amino acid sequence (Forshew et al. 2005; Rooryck et al. 2006; Chiang et al. 2008; Kenny et al. 2012).

While most mutations in *Tyrp1* are the result of non-synonymous changes, we discovered a synonymous alteration in the coding sequence (F210F; Fig. 3b). Some studies of *Tyrp1b* also include synonymous mutations (e.g., six silent SNPs in the domestic cat) for which the precise impact has not yet been explained (Lyons et al. 2005).

Our RNA-seq and qPCR studies indicated significantly reduced *Tyrp1b* expression in Pachón cavefish compared to the surface-dwelling form. The reduced expression in cavefish may contribute to instability of tyrosinase and may be responsible, in part, for the reduced numbers of melanophores observed in Pachón cave morphotypes.

The gene *Pmela* which is also known as "*Silv*" or "*Pmel17*" is a similarly well-described gene associated with pigmentation (Theos et al. 2005). This gene is responsible for recessive dilution of coat color that deteriorates with age in inbred mouse strains (Dunn and Thigpen 1930). Variants of *Pmela* cause dilution or hypotrichosis of coat color in the domestic yak (Zhang et al. 2014), cattle (Jolly et al. 2011; Schmutz and Dreger 2013), and the "silver" phenotypes in horse (Brunberg et al. 2006) and zebrafish (Schonthaler et al. 2005).

We discovered three non-synonymous changes to the *Pmela* coding sequence, including G580A in cavefish, which causes an alanine-to-threonine substitution. This is intriguing since *Pmela* undergoes posttranslational modification and processing imperative for proper functioning (Theos et al. 2013). The change from a hydrophobic alanine (nonpolar side chains) to a hydrophilic threonine (polar side chains) may alter the Pmela protein. Ala-to-Thr residue changes in diverse proteins have been shown to induce self-aggregation into amyloids (i.e., an insoluble β-pleated sheet formed by the alteration in secondary structure). Although some amyloids are native to Pmel, additional changes in this structure-sensitive protein may impact normal functioning (Fowler et al. 2005).

Pmela has been described as a melanocyte-specific type 1 transmembrane encoded protein enriched in melanosomes (i.e., the pigment-producing organelles within the melanocytes/melanophores; reviewed in Theos et al. 2005). The gene *Pmela* plays a critical role in the premelanosome "fibril" ultrastructure, and Pmel-targeted antibodies are evident in fibrous, stage II melanosomes as Pmel is thought to polymerize fibrillar arrays that ultimately form the backbone of eumelanosomes (Spanakis et al. 1992; Solano et al. 2000; Raposo et al. 2001; Berson et al. 2001; Raposo and Marks 2002). Moreover, high levels of *Pmela* expression in non-agouti (solid dark) mice are necessary to construct the fibrils associated with the shape of eumelanosomes (Theos et al. 2005).

We detected the substantial reduction in *Pmela* expression across early development in Pachón cavefish relative to surface fish. Reduced *Pmela* expression could impede fibril formation, leading to the aberrant melanophore morphology observed in our knockdown studies. Other reports similarly describe changes in melanosome shape but modest effects on overall body pigmentation in mice (Hellström et al. 2011). *Pmela* mutants often have reduced or scattered pigment granules in individual hairs (Dunn and Thigpen 1930). Furthermore, the "merle" phenotype in domestic dogs (e.g., Australian shepherds) is caused by codominance of the merle (M) and non-merle (m) *Pmel* alleles in heterozygous individuals that collectively yield patches of dark and light-colored fur due to spatially random protein instability (Clark et al. 2006; Schmutz and Berryere 2007). These degenerative pigmentation phenotypes, combined with the structural changes and severely reduced expression in colorless cavefish, suggest that *Tyrp1b* and *Pmela* contribute to the complex trait of melanophore variation in *Astyanax mexicanus*.

4.2   Analysis of Complex Pigmentation Informs the Genetic Basis for Regressive Evolution in Cavefish

Three hypotheses seeking to explain regressive evolution include natural selection, neutral mutation/genetic drift, and pleiotropy (Culver 1982). A prior QTL study yielded 18 loci associated with numerical melanophore variation. In certain cases, the homozygous cave genotype was associated with *increased* numbers of melanophores (Protas et al. 2007). Our study revealed a similar result—genotypic effect plots varied in their polarity for members our experimental F_2 pedigree. This may provide evidence for neutral mutation/genetic drift since it would be unlikely that depigmented cavefish (if under selection) would harbor alleles that increase pigment cell number (Protas et al. 2007). However, one potential "selective" benefit for loss of pigmentation in cave-dwelling morphs may be activation of the catecholamine pathway to encourage foraging behavior in the nutrient-poor subterranean environment (Bilandž-ija et al. 2013). Although the precise evolutionary pressures governing regression remain unknown, further studies will reveal the genetic mechanisms accompanying trait loss in the wild.

4.3   Astyanax mexicanus *Enables Investigation of Degenerative Pigmentation Disorders*

Degenerative pigmentation can be found widely around the globe, including clinical diseases impacting humans such as albinism, vitiligo, and skin melanoma (Oetting et al. 1996; Agarwal 1998; Hocker and Tsao 2007). Often the etiologies for these disorders remain unknown. Animals that naturally demonstrate degenerative traits, such as the blind Mexican cavefish, can help understand pigmentation losses since they have recurrently evolved extreme pigmentation changes as a consequence of the extreme cave environment (Jeffery 2005; Gross 2012a). Since the functions of many pigmentation genes are shared broadly across taxa, we can make progress toward understanding the comprehensive role of novel genes in the regulation of animal pigmentation, including albinism (Protas et al. 2005) and *brown* (Valverde et al. 1995; Flanagan et al. 2000; Rees 2003).

Here, we suggest a role for *Tyrp1b* and *Pmela* in melanophore variation in cave-fish, and these genes are also critical for the normal production of melanin in humans. For instance, in humans, *Tyrp1* is responsible for oculocutaneous albinism type III due to coding sequence alterations. In cavefish, *Tyrp1* has been repeatedly mutated, including alleles associated with different geographic regions (Manga et al. 1997; Forshew et al. 2005; Rooryck et al. 2006). However, more common *Tyrp1* alleles contribute to normal hair, skin, and iris variation in humans (Frudakis et al. 2003; Sulem et al. 2008; Han et al. 2008; Liu and Fisher 2010; Eriksson et al. 2010). Our knowledge of *Pmela* variants in humans is more limited, and some of the disorders present in other animals (e.g., double merle dogs) also cause hearing and ocular deficiencies due to pigmentation losses in the ears and eyes (Clark et al. 2006). Evaluation of pigmentation genes in cavefish will continue to identify vulnerable genes, shared broadly across taxa that may improve our knowledge of human, pigmentation-related diseases.

Funding This study was funded by a grant from the National Science Foundation, Washington D.C., USA, to JBG (grant number DEB-1457630).

Ethical Statement All applicable international, national, and/or institutional guidelines for the care and use of animals were followed. All procedures performed in studies involving animals were in accordance with the ethical standards of the institution or practice at which the studies were conducted. The protocol was approved by the Institutional Animal Care and Use Committee (IACUC) of the University of Cincinnati (Protocol Number 10-01-21-01).

References

Agarwal GA (1998) Vitiligo: an under-estimated problem. Fam Pract 15(Suppl 1):S19–23

Berson JF, Harper DC, Tenza D, Raposo G, Marks MS (2001) *Pmel17* initiates premelanosome morphogenesis within multivesicular bodies. Mol Biol Cell 12:3451–3464

Bilandžija H, Ma L, Parkhurst A, Jeffery WR (2013) A potential benefit of albinism in *Astyanax* cavefish: Downregulation of the *oca2* gene increases tyrosine and catecholamine levels as an alternative to melanin synthesis. PLoS ONE 8:e80823-14

Borowsky R (2008) *Astyanax mexicanus*, the blind Mexican cave fish: a model for studies in development and morphology. Cold Spring Harbor Protocols 2008:pdb.emo107

Broman KW, Wu H, Sen Ś, Churchill GA (2003) R/qtl: QTL mapping in experimental crosses. Bioinformatics 19:889–890

Brunberg E, Andersson L, Cothran G, Sandberg K, Mikko S, Lindgren G (2006) A missense mutation in *PMEL17* is associated with the Silver coat color in the horse. BMC Genet 7:46

Carlson BM, Onusko SW, Gross JB (2015) A high-density linkage map for *Astyanax mexicanus* using genotyping-by-sequencing technology. G3: Genes Genomes Genet 5:241–251

Chiang PW, Fulton AB, Spector E, Hisama FM (2008) Synergistic interaction of the *OCA2* and *OCA3* genes in a family. Am J Med Genet Part A 146A:2427–2430

Clark LA, Wahl JM, Rees CA, Murphy KE (2006) Retrotransposon insertion in *SILV* is responsible for merle patterning of the domestic dog. Proc Natl Acad Sci 103:1376–1381

Culver DC (1982) Cave life: evolution and ecology. Harvard University Press, Cambridge, 189 pp

Dunn LC, Thigpen LW (1930) The silver mouse: a recessive color variation. J Hered 21:495–498

Erickson CA, Perris R (1993) The role of cell-cell and cell-matrix interactions in the morphogenesis of the neural crest. Dev Biol 159:60–74

Eriksson N, Macpherson JM, Tung JY, Hon LS, Naughton B, Saxonov S, Avey L, Wojcicki A, Pe'er I, Mountain J (2010) Web-based, participant-driven studies yield novel genetic associations for common traits. PLoS Genet 6:e1000993

Flanagan N, Healy E, Ray A, Philips S, Todd C, Jackson IJ, Birch-Machin MA, Rees JL (2000) Pleiotropic effects of the melanocortin 1 receptor (*Mc1r*) gene on human pigmentation. Hum Mol Genet 9:2531–2537

Forshew T, Khaliq S, Tee L, Smith U, Johnson CA, Mehdi SQ, Maker ER (2005) Identification of novel *TYR* and *TYRP1* mutations in oculocutaneous albinism. Clin Genet 68:182–184

Fowler DM, Koulov AV, Alory-Jost C, Marks MS, Balch WE, Kelly JW (2005) Functional amyloid formation within mammalian tissue. PLoS Biol 4:e6

Frudakis T, Thomas M, Gaskin Z, Venkateswarlu K, Chandra KS, Ginjupalli S, Gunturi S, Natrajan S, Ponnuswamy VK, Ponnuswamy KN (2003) Sequences associated with human iris pigmentation. Genetics 165:2071–2083

Gross JB (2012a) Cave evolution. In: Encyclopedia of life sciences, eLS. Wiley

Gross JB (2012b) The complex origin of *Astyanax* cavefish. BMC Evol Biol 12:105–122

Gross JB, Wilkens H (2013) Albinism in phylogenetically and geographically distinct populations of *Astyanax* cavefish arises through the same loss-of-function *Oca2* allele. Heredity 111:122

Gross JB, Borowsky R, Tabin CJ (2009) A novel role for *Mc1r* in the parallel evolution of depigmentation in independent populations of the cavefish *Astyanax mexicanus*. PLoS Genet 5:e1000326-14

Gross JB, Protas ME, Conrad M, Scheid PE, Vidal O, Jeffery WR, Borowsky R, Tabin CJ (2008) Synteny and candidate gene prediction using an anchored linkage map of *Astyanax mexicanus*. Proc Natl Acad Sci 105:20106–20111

Gross JB, Krutzler AJ, Carlson BM (2014) Complex craniofacial changes in blind cave-dwelling fish are mediated by genetically symmetric and asymmetric loci. Genetics 196:1303–1319

Haley CS, Knott SA (1992) A simple regression method for mapping quantitative trait loci in line crosses using flanking markers. Heredity 69:315–324

Han J, Kraft P, Nan H, Guo Q, Chen C, Qureshi A, Hankinson SE, Hu FB, Duffy DL, Zhao ZZ, Martin NG, Montgomery GW, Hayward NK, Thomas G, Hoover RN, Chanock S, Hunter DJ (2008) A genome-wide association study identified novel alleles associated with hair color and skin pigmentation. PLoS Genet 4:e1000074

Hellström AR, Watt B, Fard SS, Tenza D, Mannström P, Mannström K, Ekesten B, Ito S, Wakamatso K, Larsson J, Ulfendahl M, Kullander K, Raposo G, Kerje S, Hallböök F, Marks MS, Andersson L (2011) Inactivation of *Pmel* alters melanosome shape but has only a subtle effect on visible pigmentation. PLoS Genet 7:e1002285

Hocker T, Tsao H (2007) Ultraviolet radiation and melanoma: a systematic review and analysis of reported sequence variants. Hum Mutat 28:578–588

Hoekstra HE (2006) Genetics, development and evolution of adaptive pigmentation in vertebrates. Heredity 97:222–234

Huang X, Saint-Jeannet JP (2004) Induction of the neural crest and the opportunities of life on the edge. Dev Biol 275:1–11

Hubbard JK, Uy JAC, Hauber ME, Hoekstra HE, Safran RJ (2010) Vertebrate pigmentation: from underlying genes to adaptive function. Trends Genet 26:231–239

Hunt R, Sauna ZE, Ambudkar SV, Gottesman MM, Kimchi-Sarfaty C (2009) Silent (synonymous) SNPs: should we care about them? Single nucleotide polymorphisms. Humana Press, Totowa, NJ, pp 23–29

Jeffery WR (2001) Cavefish as a model system in evolutionary developmental biology. Dev Biol 231:1–12

Jeffery WR (2005) Adaptive evolution of eye degeneration in the Mexican blind cavefish. J Hered 96:185–196

Jeffery WR (2006) Regressive evolution of pigmentation in the cavefish *Astyanax*. Isr J Ecol Evolut 52:405–422

Jeffery WR (2009) Evolution and development in the cavefish *Astyanax*. Curr Top Dev Biol 86:191–221

Jolly RD, Wills JL, Kenny JE, Cahill JI, Howe L (2011) Coat-colour dilution and hypotrichosis in Hereford crossbred calves. N Z Vet J 56:74–77

Kasprzyk A (2011) BioMart: driving a paradigm change in biological data management. Database 2011:bar049

Kearsey MJ, Hyne V (1994) QTL analysis: a simple "marker-regression" approach. Theor Appl Genet 89:698–702

Kenny EE, Timpson NJ, Sikora M, Yee M-C, Moreno-Estrada A, Eng C, Huntsman S, González Burchard E, Stoneking M, Bustamante CD, Myles S (2012) Melanesian blond hair is caused by an amino acid change in *TYRP1*. Science 336:554–554

Klaassen H, Wang Y, Adamski K, Rohner N, Kowalko JE (2018) CRISPR mutagenesis confirms the role of *oca2* in melanin pigmentation in *Astyanax mexicanus*. Dev Biol (in press). https://doi.org/10.1016/j.ydbio.2018.03.014

Kobayashi T, Hearing VJ (2007) Direct interaction of tyrosinase with *Tyrp1* to form heterodimeric complexes *in vivo*. J Cell Sci 120:4261–4268

Kowalko JE, Rohner N, Linden TA, Rompani SB, Warren WC, Borowsky R, Tabin CJ, Jeffery WR, Yoshizawa M (2013) Convergence in feeding posture occurs through different genetic loci in independently evolved cave populations of *Astyanax mexicanus*. Proc Natl Acad Sci 110:16933–16938

Kruglyak L, Lander ES (1995) A nonparametric approach for mapping quantitative trait loci. Genetics 139:1421–1428

Krzywinski M, Schein J, Birol I, Connors J, Gascoyne R, Horsman D, Jones SJ, Marra MA (2009) Circos: an information aesthetic for comparative genomics. Genome Res 19:1639–1645

Kwon BS, Halaban R, Chintamaneni C (1989) Molecular basis of mouse Himalayan mutation. Biochem Biophys Res Commun 161:252–260

Linnen CR, Kingsley EP, Jensen JD, Hoekstra HE (2009) On the origin and spread of an adaptive allele in deer mice. Science 325:1095–1098

Liu JJ, Fisher DE (2010) Lighting a path to pigmentation: mechanisms of *MITF* induction by UV. Pigment Cell Melanoma Res 23:741–745

Lyons LA, Imes DL, Rah HC, Grahn RA (2005) Tyrosinase mutations associated with Siamese and Burmese patterns in the domestic cat (*Felis catus*). Anim Genet 36:119–126

Manga P, Kromberg JGR, Box NF, Sturm RA, Jenkins T, Ramsay M (1997) Rufous oculocutaneous albinism in Southern African Blacks is caused by mutations in the *TYRP1* gene. Am J Hum Genet 61:1095–1101

Ma L, Parkhurst A, Jeffery WR (2014) The role of a lens survival pathway including *sox2* and *aA-crystallin* in the evolution of cavefish eye degeneration. EvoDevo 5:28

Ma L, Jeffery WR, Essner JJ, Kowalko JE (2015) Genome editing using TALENs in blind Mexican cavefish, *Astynax mexicanus*. PLoS One 10:e0119370

McCauley DW, Hixon E, Jeffery WR (2004) Evolution of pigment cell regression in the cavefish *Astyanax*: a late step in melanogenesis. Evolut Dev 6:209–218

McGaugh SE, Gross JB, Aken B, Blin M, Borowsky R, Chalopin D, Hinaux H, Jeffery WR, Keene A, Ma L, Minx P, Murphy D, O'Quinn KE, Rétaux S, Rohner N, Searle SMJ, Stahl BA, Tabin C, Volff J, Yoshizawa M, Warren WC (2014) The cavefish genome reveals candidate genes for eye loss. Nat Commun 5:5307

Montgomery JC, Coombs S, Baker CF (2001) The mechanosensory lateral line system of the hypogean form of *Astyanax fasciatus*. The biology of hypogean fishes. Springer Netherlands, Dordrecht, pp 87–96

Mortazavi A, Williams BA, McCue K, Schaeffer L, Wold B (2008) Mapping and quantifying mammalian transcriptomes by RNA-seq. Nat Methods 5:621–628

Müller G, Ruppert S, Schmid E, Schütz G (1988) Functional analysis of alternatively spliced tyrosinase gene transcripts. EMBO J 7:2723

O'Quin KE, Yoshizawa M, Doshi P, Jeffery WR (2013) Quantitative genetic analysis of retinal degeneration in the blind cavefish *Astyanax mexicanus*. PLoS ONE 8:e57281-11

Oetting WS (2000) The tyrosinase gene and oculocutaneous albinism type 1 (*OCA1*): a model for understanding the molecular biology of melanin formation. Pigment Cell Res 13:320–325

Oetting WS, Brilliant MH, King RA (1996) The clinical spectrum of albinism in humans. Mol Med Today 2:330–335

Protas ME, Conrad M, Gross JB, Tabin C, Borowsky R (2007) Regressive evolution in the Mexican cave tetra, *Astyanax mexicanus*. Curr Biol 17:452–454

Protas ME, Patel NH (2008) Evolution of coloration patterns. Annu Rev Cell Dev Biol 24:425–446

Protas ME, Hersey C, Kochanek D, Zhou Y, Wilkens H, Jeffery WR, Zon LI, Borowsky R, Tabin CJ (2005) Genetic analysis of cavefish reveals molecular convergence in the evolution of albinism. Nat Genet 38:107–111

Raposo G, Marks MS (2002) The dark side of lysosome-related organelles: Specialization of the endocytic pathway for melanosome biogenesis. Traffic 3:237–248

Raposo G, Tenza D, Murphy DM, Berson JF, Marks MS (2001) Distinct protein sorting and localization to premelanosomes, melanosomes, and lysosomes in pigmented melanocytic cells. J Cell Biol 152:809–824

Rees JL (2003) Genetics of hair and skin color. Annu Rev Genet 37:67–90

Rooryck C, Roudaut C, Robine E, Müsebeck J, Arveiler B (2006) Oculocutaneous albinism with *TYRP1* gene mutations in a Caucasian patient. Pigment Cell Res 19:239–242

Schmidt-Küntzel A, Eizirik E, O'Brien SJ, Menotti-Raymond M (2005) *Tyrosinase* and *Tyrosinase Related Protein 1* alleles specify domestic cat coat color phenotypes of the *albino* and *brown* loci. J Hered 96:289–301

Schmutz SM, Berryere TG (2007) Genes affecting coat colour and pattern in domestic dogs: a review. Anim Genet 38:539–549

Schmutz SM, Dreger DL (2013) Interaction of *MC1R* and *PMEL* alleles on solid coat colors in Highland cattle. Anim Genet 44:9–13

Schonthaler HB, Lampert JM, von Lintig J, Schwarz H, Geisler R, Neuhauss SC (2005) A mutation in the *silver* gene leads to defects in melanosome biogenesis and alterations in the visual system in the zebrafish mutant *fading vision*. Dev Biol 284:421–436

Solano F, Martínez Esparza M, Jimenez-Cervantes C, Hill SP, Lozano JA, García-Borrón JC (2000) New insights on the structure of the mouse *silver* locus and on the function of the *silver* protein. Pigment Cell Res 13:118–124

Spanakis E, Lamina P, Bennett DC (1992) Effects of the developmental colour mutations silver and recessive spotting on proliferation of diploid and immortal mouse melanocytes in culture. Development 114:675–680

Stahl BA, Gross JB (2017) A comparative transcriptomic analysis of development in two *Astyanax* cavefish populations. J ExpZool Part B: Mol Dev Evolut 328:515–532

Sulem P, Gudbjartsson DF, Stacey SN, Helgason A, Rafnar T, Jakobsdottir M, Steinberg S, Gudjonsson SA, Palsson A, Thorleifsson G, Pálsson S, Sigurgeirsson B, Thorisdottir K, Ragnarsson R, Benediktsdottir KR, Aben KK, Vermeulen SH, Goldstein AM, Tucker MA, Kiemeney LA, Olafsson JH, Gulcher J, Kong A, Thorsteinsdottir U, Stefansson K (2008) Two newly identified genetic determinants of pigmentation in Europeans. Nat Genet 40:835–837

Theos AC, Truschel ST, Raposo G, Marks MS (2005) The *Silver* locus product *Pmel17/gp100/Silv/ME20*: controversial in name and in function. Pigment Cell Res 18:322–336

Theos AC, Watt B, Harper DC, Janczura KJ, Theos SC, Herman KE, Marks MS (2013) The PKD domain distinguishes the trafficking and amyloidogenic properties of the pigment cell protein *PMEL* and its homologue *GPNMB*. Pigment Cell Melanoma Res 26:470–486

Valverde P, Healy E, Jackson I, Rees JL, Thody AJ (1995) Variants of the *melanocyte-stimulating hormone receptor* gene are associated with red hair and fair skin in humans. Nat Genet 11:328–330

Wingert RA, Brownlie A, Galloway JL, Dooley K, Fraenkel P, Axe JL, Davidson AJ, Barut B, Noriega L, Sheng X, Zhou Y, Zon LI (2004) The chianti zebrafish mutant provides a model for erythroid-specific disruption of *transferrin receptor 1*. Development 131:6225–6235

Xu S, Hu Z (2010) Mapping quantitative trait loci using distorted markers. Int J Plant Genomics 2009:1–11

Zhang MQ, Xu X, Luo SJ (2014) The genetics of brown coat color and white spotting in domestic yaks (*Bos grunniens*). Anim Genet 45:652–659

Adaptive Evolution of Yeast Under Heat Stress and Genetic Reconstruction to Generate Thermotolerant Yeast

Kouichi Kuroda and Mitsuyoshi Ueda

Abstract *Saccharomyces cerevisiae* is a microorganism that is widely used for the bioproduction of useful substances, including ethanol. However, during bioproduction, yeast cells are subjected to various stresses. In particular, bioproduction via fermentation generates heat, leading to decreased rates of cell growth and fermentation. Improving the thermotolerance of yeast can help to maintain metabolic activity and bioproduction efficiency at high temperatures. In this chapter, we describe the improvement of thermotolerance in yeast through stepwise adaptive evolution under heat stress. The adaptation strategy, in which the cells were selected under two selective pressures (heat stress and growth rate), improved thermotolerance while maintaining growth rate. Through this adaptation strategy, a thermotolerant yeast strain, YK60-1, was successfully isolated after adaptation to 38 °C. Transcriptome and non-targeted metabolome analyses revealed that YK60-1 induced stress-responsive genes and accumulated more trehalose than the wild-type parent strain. Furthermore, comparative genomic analysis of the intermediate populations after adaptation to each elevated temperature revealed key mutations for improving thermotolerance in the *CDC25* gene. A thermotolerant yeast strain was also reconstructed by introducing *CDC25* mutations in the wild-type strain. *CDC25* mutation is thought to alter global transcriptional regulation through downregulation of the cAMP/PKA pathway, leading to improved stress tolerance.

1 Introduction

Living organisms encounter various environmental conditions, but must preserve the intracellular environment to maintain biological activities. When the extracellular

K. Kuroda · M. Ueda (✉)
Division of Applied Life Sciences, Graduate School of Agriculture, Kyoto University,
Kitashirakawa Oiwake-cho, Sakyo-ku, Kyoto 606-8502, Japan
e-mail: miueda@kais.kyoto-u.ac.jp

K. Kuroda
e-mail: k_kuro@kais.kyoto-u.ac.jp

© Springer International Publishing AG, part of Springer Nature 2018
P. Pontarotti (ed.), *Origin and Evolution of Biodiversity*,
https://doi.org/10.1007/978-3-319-95954-2_2

environment suddenly changes, their internal environment may be perturbed, causing stress. Even in severe environments, living organisms can restore their intracellular environment to survive and adapt. Under stress, some cells within a population may spontaneously acquire mutations that can improve the cellular response to stress, leading to successful adaptation to various environments.

Stress tolerance is an important consideration in bioindustrial processes that produce valuable substances by employing microorganisms as whole-cell biocatalysts and can reduce the increasing environmental burden. The environmental conditions in industrial processes are not necessarily favorable, and the cells are usually subject to various stresses, such as high or low pH, high temperature, hyperosmotic pressure, and organic solvents. Furthermore, in some cases, the target product, by-product, or intermediate product can be hydrophobic or even toxic; thus, the cells are subject to stress, leading to decreased efficiency. Therefore, improving cellular stress tolerance has the potential to promote bioindustrial reactions. Among the various microorganisms used in bioindustrial processes, the budding yeast *Saccharomyces cerevisiae* is one of the most commonly used microorganisms due to their advantages, including (i) their rapid growth in inexpensive media; (ii) aerobic or anaerobic growth under various conditions; and (iii) generally recognized as safe (GRAS) status. Various attempts have been made to improve the stress tolerance of yeast, including overexpression and deletion of individual stress-related genes (Pacheco et al. 2009; Yazawa et al. 2007), cell-surface engineering to display desired proteins (Kuroda and Ueda 2017; Satomura et al. 2014), and adaptive evolution (Dragosits and Mattanovich 2013; Cakar et al. 2012; Winkler and Kao 2014; Wallace-Salinas and Gorwa-Grauslund 2013). In this chapter, we focused on adaptive evolution of yeast under heat stress.

2 Heat Stress in *S. cerevisiae*

The environmental conditions in which yeast cells grow are constantly fluctuating; therefore, they have several strategies to deal with environmental stresses. When the cells detect stress, they first adjust the transcriptional levels of stress-related genes to modify cellular metabolism, strengthen cellular structures, and protect proteins. However, yeast cells cannot grow when these environmental fluctuations exceed certain limits, such as a sudden rise in temperature. Therefore, bioindustrial processes are limited to the environmental conditions under which yeast cells can survive and grow, because bioproduction is usually associated with cellular growth. For bioethanol production, fermentation heat and organic acids, such as acetic acid, are produced during biomass pretreatment, which can reduce survival rates and the efficiency of fermentation. Cooling down the fermentation apparatus is an effective method to relieve the stresses caused by fermentation heat. However, this is not cost-efficient, as a lot of water and energy are used for cooling. Furthermore, most of the commercial enzymes used for the saccharification of lignocellulose have optimal temperatures above 30 °C, and fermentation at high temperatures is also desirable for preventing microbial contamination. To save on production costs and promote more efficient bioproduction, stress-tolerant yeast strains are needed.

3 Mechanisms of Thermotolerance in *S. cerevisiae*

The heat-shock response is a well-known molecular mechanism that makes cells more thermotolerant. In this response, a specific transcription factor activates the expression of genes encoding chaperone proteins, called heat-shock proteins (HSPs), in response to high temperature (heat shock). These chaperons play important roles in restoring the normal structure of proteins that became abnormal due to the temperature change by refolding them. According to the *Saccharomyces* genome database (SGD), there are at least 19 HSPs and other chaperones in *S. cerevisiae*. The production of many of these proteins is induced by heat shock. However, some HSPs are produced under normal conditions and are increased by heat shock. As denatured proteins impair cellular functions and even their existence causes problems, it is also important to remove them from cells to improve thermotolerance. The transcription of many HSPs is activated by Hsf1p, which regulates the expression of 5% of the genes involved in protein folding, energy generation, carbohydrate metabolism, and cell wall organization in *S. cerevisiae* (Eastmond and Nelson 2006; Hahn et al. 2004). In the regulatory region of Hsf1p target genes, there is a highly conserved sequence called the heat-shock element (HSE), and a homotrimer of Hsf1p binds to the HSE to activate target gene transcription. Although Hsf1p is always phosphorylated, it becomes excessively phosphorylated by heat shock. This causes a structural change in Hsf1p, leading to the binding of Hsf1p to the HSE and subsequent transcriptional activation (Sakurai and Takemori 2007). At low temperatures, the Hsf1p-mediated expression levels of genes encoding HSPs are low; however, at high temperatures, these genes are strongly upregulated (Chen and Pederson 1993; Giardina and Lis 1995). Although Hsf1p regulates the expression of several stress-responsive genes, an Hsf1p mutant that hyperactivates downstream genes failed to sufficiently improve thermotolerance (Gibney et al. 2013). Therefore, these Hsf1p mutants could not fully activate all the genes required for thermotolerance, and an additional signaling pathway may be required to further induce these genes. Additionally, variations in HSE sequences and the gaps between HSEs are thought to lead to variations in induction levels by Hsf1p.

In addition to Hsf1p, thermotolerance in yeast is also controlled by the transcription factors Msn2p and Msn4p (Martinez-Pastor et al. 1996). These two factors activate more than 200 genes, including the HSP genes *HSP12*, *HSP104*, and *SSA4* and the trehalose synthesis genes *TPS1*, *TPS2*, and *TSL1* (Martinez-Pastor et al. 1996; Boy-Marcotte et al. 1998; Schmitt and McEntee 1996). Msn2p and Msn4p are regulated by the cAMP-dependent protein kinase (PKA) signaling pathway, which is a well-conserved pathway involving intracellular cAMP as a second messenger (Thevelein and de Winde 1999). Intracellular cAMP levels are regulated by adenylate cyclase (Cyr1p), which converts ATP to cAMP (Matsumoto et al. 1982). Increased intracellular cAMP is known to antagonize the activities of Msn2p and Msn4p through inhibition of Bcy1p and activation of the cAMP-dependent protein kinase PKA (Gorner et al. 1998).

Trehalose is also known to contribute to thermotolerance. In *S. cerevisiae*, trehalose-6-phosphate is synthesized from glucose-6-phosphate and uridine diphosphate glucose (UDP-glucose); then, the phosphate group of trehalose-6-phosphate is removed to produce trehalose. These reactions are catalyzed by Tps1p and Tps2p, and the expression of the respective trehalose synthase genes, *TPS1* and *TPS2*, is activated by heat shock. Trehalose is thought to protect the structure of proteins during heat shock, while suppressing the refolding of denatured proteins. Although the latter function is problematic when cells are released from heat shock, trehalose is almost completely decomposed within approximately 1 h when cells are returned to normal temperature.

4   Methods for Improving Stress Tolerance

To improve the tolerance of *S. cerevisiae* to stress, such as higher temperature, several approaches have been utilized. Random genetic modification, followed by selection or genetic modification of target genes, can improve stress tolerance. Targeted genetic modification mainly includes overexpression or deletion of individual stress-related genes that are selected based on known mechanisms of stress tolerance. However, this approach depends on the availability of knowledge regarding the key factors involved in stress tolerance. When less information is available on the genes whose overexpression, deletion, or specific mutation can improve stress tolerance, this approach is very limited. Stress tolerance is complicated, and multiple factors are necessary for improved tolerance. Therefore, with some exceptions (Gorsich et al. 2006; Yazawa et al. 2007), modification of single genes has limitations for improving stress tolerance. Cell-surface engineering (Kuroda and Ueda 2013; Ueda and Tanaka 2000) has also been employed to improve stress tolerance through artificial modification of the cell surface by displaying various functional proteins/peptides on the cell surface. The cell surface, including the cell wall and cell membrane, has an important function as the first barrier to the environment, and it is the cellular component that is involved in exchanging substances and transmitting environmental information to cells. Therefore, the cell surface is an attractive target for modification to improve stress tolerance. In fact, tolerance to acidic stress and organic solvent stress in yeasts have been successfully improved by the screening of random protein/peptide-display libraries (Zou et al. 2001, 2002; Matsui et al. 2009; Kuroda and Ueda 2017; Satomura et al. 2014). However, cell wall modification cannot be used to improve heat stress because heat directly affects cellular proteins and membrane fluidity.

Non-directed approaches, which include random mutagenesis, genome shuffling (Shi et al. 2009), protoplast fusion (Kida et al. 1992), and adaptive evolution (also called evolutionary engineering or whole-cell directed evolution) (Winkler and Kao 2014; Cakar et al. 2012), have been used to improve stress tolerance. In adaptive evolution, a population of yeast cells is continuously propagated under the targeted stress to provide selective pressure. Adapted mutants are generated either naturally through random mutation during DNA replication or by physical or chemical mutagenesis to increase genetic diversity. The percentage of adapted mutants in the population increases in proportion to fitness in the selective environment. In some cases, the cycle of generating genetic diversity and selecting under specific conditions is repeated. Adaptive evolution has been applied to improve substrate utilization ability and cellular tolerance to environmental stresses or toxic chemicals (Cakar et al. 2009; Kuyper et al. 2005; Wisselink et al. 2009; Almario et al. 2013; Wallace-Salinas and Gorwa-Grauslund 2013; Satomura et al. 2013).

As other non-directed approaches, global transcription machinery engineering (gTME) has been applied to improve complicated phenotypes, such as stress tolerance, through simultaneous modification of multiple gene expressions (Alper et al. 2006). In gTME, random mutations are introduced into a global transcription factor controlling many downstream genes. As a strong mutator, an exonuclease-deficient DNA polymerase δ variant showing defective proofreading ability has also been used to improve stress tolerance (Abe et al. 2009). Physical or chemical mutagenesis by UV or EMS, respectively, can damage the cells, and most of the cells treated with such strong mutagens are likely to die or show undesirable traits. However, using the approach with an exonuclease-deficient DNA polymerase δ variant does not damage the cells because the mutations occur during DNA replication.

Mutant strains showing improved stress tolerance generated by these non-directed approaches would be useful for clarifying the mechanisms of stress tolerance. Recent advances in multi-omics and whole-genome sequencing have enabled the characterization of isolated tolerant mutants and the identification of key mutations for improved stress tolerance, leading to a better understanding of the genetic determinants and molecular mechanisms underlying stress tolerance (Satomura et al. 2014; Kuroda and Ueda 2017). Accumulation of the obtained knowledge will be useful for future rational engineering approaches for improved stress tolerance.

5   Acquisition of Thermotolerance in Yeast by Adaptive Evolution

Among the various methods for improving stress tolerance described above, we chose to employ adaptive evolution to generate thermotolerant yeast (Satomura et al. 2013). However, a non-directed approach like adaptive evolution has a disadvantage: Repeated cycles of mutation and selection under stress can result in highly tolerant strains but they are inferior to the parent strain for other important pheno-

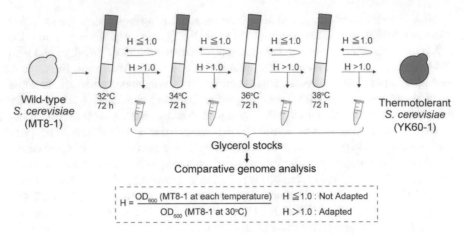

Fig. 1 Stepwise adaptive evolution under the selective pressures of heat stress and growth rate

type, namely crippled mutant strains (Cakar et al. 2005). Therefore, in our work, adaptation evolution was performed under two selective pressures, heat stress and growth rate, in a stepwise manner (Fig. 1). Specifically, a wild-type yeast strain was successively cultured for 72 h at 32, 34, 36, and 38 °C until the growth rate at each elevated temperature exceeded that of the wild-type strain at 30 °C. For example, in the first step of this stepwise adaptive evolution method, if the OD_{600} after cultivation at 32 °C for 72 h was higher than that of the wild-type strain at 30 °C for 72 h, the cultivated yeast cells were regarded as successfully adapted to 32 °C and were then cultivated at 34 °C in the next step. However, if the OD_{600} after cultivation at the 32 °C was not higher than that at 30 °C, the yeast cells were cultivated at 32 °C again, and cultivation at 32 °C was repeated until the cells adapted to 32 °C. Using this method, a thermotolerant yeast strain, YK60-1, was successfully obtained after 9, 8, 21, and 22 rounds of successive culture at 32 °C, 34 °C, 36 °C, and 38 °C, respectively. YK60-1 can grow better at 38 °C than the wild-type strain at 30 °C, and can grow at 40 °C, a temperature at which the wild-type strain cannot grow (Fig. 2). Furthermore, YK60-1 showed improved tolerance not only to prolonged heat stress at 40 °C but also to short sublethal heat shock (at 50 °C for 30 min), and these thermotolerant phenotypes were enhanced by mild heat shock (at 37 °C for 45 min; Fig. 3). Interestingly, YK60-1 showed better growth than the wild-type strain at 30 °C (Fig. 2). Therefore, adaptive evolution under the dual selective pressure of heat stress and growth rate is a promising strategy for improving stress tolerance without significantly decreasing the growth rate compared to the wild-type strain.

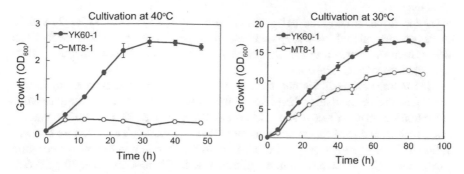

Fig. 2 Thermotolerance of strain YK60-1 isolated via adaptive evolution. YK60-1 and the wild-type parental strain MT8-1 were cultivated in YPD medium at 40 °C or 30 °C

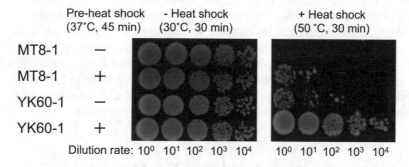

Fig. 3 Effect of mild heat shock and short sublethal heat shock on the viability of YK60-1 and MT8-1. Yeast cells were exposed to a short sublethal heat shock (at 50 °C for 30 min) after a mild heat shock (at 37 °C for 45 min), and then grown on a YPD plate at 30 °C for 72 h

6 Comprehensive Analysis of the Thermotolerant Yeast Strain YK60-1

In most previous studies aimed at improving thermotolerance through non-directed approaches, the molecular mechanisms underlying thermotolerance have not been elucidated. This limits the reconstruction of thermotolerance in other yeast strains. Although it is a very challenging task, knowledge about the mechanisms or key factors involved in the improved thermotolerance is important for future applications, including reconstruction of thermotolerant yeast. The recent development of comprehensive analysis technologies, such as transcriptomics, proteomics, metabolomics, and genomics, has enabled the discovery of clues to clarify tolerance mechanisms (Bro and Nielsen 2004). Thus, as a first attempt to clarify the mechanism of tolerance, a DNA microarray analysis was performed in which the transcriptional profiles of YK60-1 and the wild-type strains grown at 30 °C or 40 °C were compared (Satomura et al. 2013). The results showed that YK60-1 upregulates stress-responsive genes, such as HSP-encoding genes and trehalose synthesis genes, when compared to the wild-type strain. Among the 141 genes upregulated in YK60-1 at 30 °C, 24 and 18 genes are upregulated by Msn2p and Msn4p, respectively. In YK60-1, *MSN2* and *MSN4* themselves are upregulated, leading to the induction of genes encoding HSPs and trehalose synthetic enzymes.

Metabolome analysis via non-targeted approach was also performed to further investigate the mechanism of thermotolerance. Determination of the relative amounts of metabolites in the YK60-1 and wild-type strains grown at 30 °C or 40 °C revealed that trehalose is one of the metabolites increased in YK60-1. The trehalose concentration in YK60-1 grown at 40 °C is 5.1-fold higher than that in the wild-type strain grown at 40 °C, and the basal amount of trehalose in YK60-1 grown at 30 °C is 6.8-fold higher than that in the wild-type strain grown at 30 °C. Therefore, the increased amount of intracellular trehalose in YK60-1 and the induction of stress-responsive genes through upregulation of *MSN2* and *MSN4* could play an important role in the improved thermotolerance of YK60-1. Here, two mechanisms of improved thermotolerance in YK60-1 can be assumed: (i) improved thermotolerance is already provided in the absence of heat stress and (ii) improved thermotolerance is strongly induced in response to heat stress. Considering that YK60-1 is thermotolerant without mild heat shock and that thermotolerance is enhanced by mild heat shock, the thermotolerance of YK60-1 is improved by a combination of preexisting improved thermotolerance at 30 °C and heat stress-induced thermotolerance.

7 Genomic Analysis to Identify the Key Mutations for Improved Thermotolerance

In addition to the above-described comprehensive analyses, identification of the key mutations for improved thermotolerance is critical for understanding the underlying mechanism. Due to the stability of the improved thermotolerance phenotype of YK60-1 and the altered expression levels of numerous genes in YK60-1, we predicted that YK60-1 has a mutation in a gene affecting global transcriptional regulation. The recent development of next-generation sequencing technology has made it easier to identify such mutations through whole-genome analysis. To facilitate the identification of key mutations in YK60-1, 19 intermediate strains adapted to each elevated temperature in the stepwise adaptive evolution were preserved and analyzed by whole-genome sequencing (Fig. 1) (Satomura et al. 2016). Through this analysis, 49 genomic mutations, including SNPs, insertions, and deletions, were identified in the intermediate strains when compared to the genome of the wild-type strain; no inversions and other rearrangements were detected. Interestingly, four different point mutations in the *CDC25* gene were found in 16 intermediate strains. Based on phylogenetic tree analysis, there were at least five events in which these strains acquired mutations in the *CDC25* gene (T943P, G1459C, N1393T, and twice for W1416C;

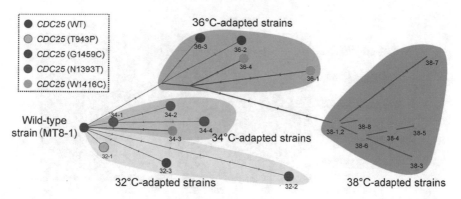

Fig. 4 Phylogenetic tree based on the mutational events in intermediate strains. The small scales indicate mutational events. The blue lines indicate cell lines harboring a *CDC25* (W1416C) mutation

Fig. 4). Because it is quite rare for different intermediate strains to have mutations in the same gene, we hypothesized that the *CDC25* mutations acquired in the stepwise adaptive evolution contribute to the improved thermotolerance of YK60-1.

To test this hypothesis, each of the four single point mutations was reintroduced into the parental wild-type strain (Satomura et al. 2016). These reconstructed *CDC25* mutants ($CDC25^{T943P}$, $CDC25^{G1459C}$, $CDC25^{N1393T}$, and $CDC25^{W1416C}$) showed improved growth at 38 °C compared to the wild-type strain, and this phenotype was stably maintained during cell growth (Fig. 5). In the reconstructed *CDC25* mutants, ethanol fermentation efficiency from glucose and galactose at 39 °C was maintained, whereas ethanol production by the wild-type strain was greatly inhibited (Fig. 6). Therefore, we concluded that these mutations in the *CDC25* gene were involved in the improved thermotolerance and efficient ethanol production from glucose and galactose at high temperature. *CDC25* encodes guanine nucleotide exchange factor (GEF), which indirectly regulates intracellular cAMP levels and the cAMP/PKA signaling pathway (Broek et al. 1987). In all the reconstructed *CDC25* mutants, intracellular cAMP levels were lower than that in the wild-type strain (Satomura et al. 2016). This indicates that the decreased activity of the mutant Cdc25p led to downregulation of the cAMP/PKA signaling pathway. Furthermore, lower cAMP levels activate stress-responsive transcription factors, such as Msn2p and Msn4p, by increasing their nuclear localization, leading to improved stress tolerance (Gorner et al. 1998). Transcriptome analysis of the reconstructed *CDC25* mutants showed that 50–60% of the genes upregulated more than 1.5-fold compared to the wild-type strain were regulated by Msn2p and Msn4p. However, excessive downregulation of the cAMP/PKA pathway leads to a decreased growth rate because the cell cycle is also regulated by this pathway (Mitsuzawa et al. 1989). A previous study reported that two *CDC25* mutations, *cdc25-21* and *cdc25-22*, showed depressed cAMP levels and improved thermotolerance, but also showed growth defects due to excessive downregulation of the cAMP/PKA pathway (Folch-Mallol et al. 2004). Therefore, the fine balance between thermotolerance and growth rate is probably maintained in

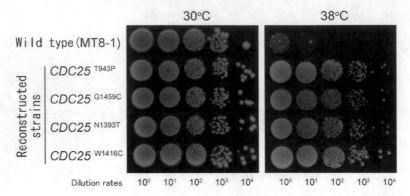

Fig. 5 Thermotolerance of the reconstructed *CDC25* mutants. Cell suspensions were spotted on a YPD plate at 30 °C or 38 °C

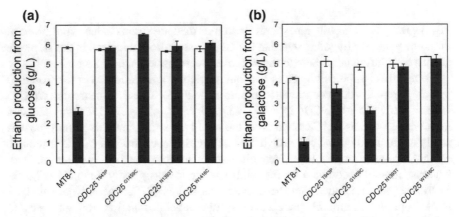

Fig. 6 Ethanol production from **a** glucose or **b** galactose at 30 °C (open bats) or 39 °C (filled bars)

YK60-1. Taken together, these results suggest that the *CDC25* mutations in YK60-1 alter global transcription regulation through an appropriate decrease in cAMP levels, and thereby improve thermotolerance (Fig. 7).

8 Conclusions and Perspectives

A novel yeast strain with improved thermotolerance, YK60-1, was successfully acquired by adaptive evolution. This strain grows faster than the wild-type strain, even though strains with improved stress tolerance previously generated by adaptive evolution often showed significantly lower growth rates. The fine balance between thermotolerance and growth rate was preserved by stepwise adaptation, in which the yeast cells were subjected to the selective pressures of heat stress and growth

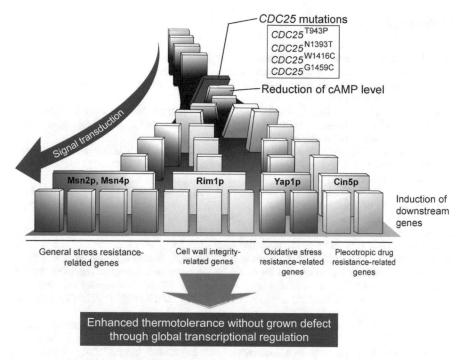

Fig. 7 Altered global transcription in the *CDC25* mutants acquired by adaptive evolution, leading to enhanced thermotolerance without growth defect

rate. Therefore, adaptive evolution under two selective pressures (stress and growth rate) may be one of the most efficient ways to improve stress tolerance through fine-tuning of signaling pathways by unpredictable mutations. Multi-omics of YK60-1, including genomics, transcriptomics, and metabolomics, identified key mutations in *CDC25*, and a thermotolerant yeast strain was successfully reconstructed by introducing *CDC25* mutations into the wild-type strain. Because Cdc25 regulates the expression of numerous genes through the cAMP/PKA pathway, such a global regulator is a promising target for modification to improve stress tolerance. Knowledge about the involvement of the identified *CDC25* mutations in thermotolerance will be beneficial for the rational design of improved thermotolerance in any yeast strain.

References

Abe H, Fujita Y, Takaoka Y, Kurita E, Yano S, Tanaka N, Nakayama K (2009) Ethanol-tolerant *Saccharomyces cerevisiae* strains isolated under selective conditions by over-expression of a proofreading-deficient DNA polymerase δ. J Biosci Bioeng 108(3):199–204. https://doi.org/10. 1016/j.jbiosc.2009.03.019

Almario MP, Reyes LH, Kao KC (2013) Evolutionary engineering of *Saccharomyces cerevisiae* for enhanced tolerance to hydrolysates of lignocellulosic biomass. Biotechnol Bioeng 110(10):2616–2623. https://doi.org/10.1002/bit.24938

Alpcr H, Moxley J, Nevoigt E, Fink GR, Stephanopoulos G (2006) Engineering yeast transcription machinery for improved ethanol tolerance and production. Science 314(5805):1565–1568. https:// doi.org/10.1126/science.1131969

Boy-Marcotte E, Perrot M, Bussereau F, Boucherie H, Jacquet M (1998) Msn2p and Msn4p control a large number of genes induced at the diauxic transition which are repressed by cyclic AMP in *Saccharomyces cerevisiae*. J Bacteriol 180(5):1044–1052

Bro C, Nielsen J (2004) Impact of 'ome' analyses on inverse metabolic engineering. Metab Eng 6(3):204–211. https://doi.org/10.1016/j.ymben.2003.11.005

Broek D, Toda T, Michaeli T, Levin L, Birchmeier C, Zoller M, Powers S, Wigler M (1987) The *S. cerevisiae CDC25* gene product regulates the *RAS*/adenylate cyclase pathway. Cell 48(5):789–799

Cakar ZP, Seker UO, Tamerler C, Sonderegger M, Sauer U (2005) Evolutionary engineering of multiple-stress resistant *Saccharomyces cerevisiae*. FEMS Yeast Res 5(6–7):569–578. https://do i.org/10.1016/j.femsyr.2004.10.010

Cakar ZP, Alkim C, Turanli B, Tokman N, Akman S, Sarikaya M, Tamerler C, Benbadis L, Francois JM (2009) Isolation of cobalt hyper-resistant mutants of *Saccharomyces cerevisiae* by *in vivo* evolutionary engineering approach. J Biotechnol 143(2):130–138. https://doi.org/10.1016/j.jbio tec.2009.06.024

Cakar ZP, Turanli-Yildiz B, Alkim C, Yilmaz U (2012) Evolutionary engineering of *Saccharomyces cerevisiae* for improved industrially important properties. FEMS Yeast Res 12(2):171–182. https://doi.org/10.1111/j.1567-1364.2011.00775.x

Chen J, Pederson DS (1993) A distal heat shock element promotes the rapid response to heat shock of the *HSP26* gene in the yeast *Saccharomyces cerevisiae*. J Biol Chem 268(10):7442–7448

Dragosits M, Mattanovich D (2013) Adaptive laboratory evolution—principles and applications for biotechnology. Microb Cell Fact 12:64. https://doi.org/10.1186/1475-2859-12-64

Eastmond DL, Nelson HC (2006) Genome-wide analysis reveals new roles for the activation domains of the *Saccharomyces cerevisiae* heat shock transcription factor (Hsf1) during the transient heat shock response. J Biol Chem 281(43):32909–32921. https://doi.org/10.1074/jbc.M60 2454200

Folch-Mallol JL, Martinez LM, Casas SJ, Yang R, Martinez-Anaya C, Lopez L, Hernandez A, Nieto-Sotelo J (2004) New roles for *CDC25* in growth control, galactose regulation and cellular differentiation in *Saccharomyces cerevisiae*. Microbiology 150(9):2865–2879. https://doi.org/1 0.1099/mic.0.27144-0

Giardina C, Lis JT (1995) Dynamic protein-DNA architecture of a yeast heat shock promoter. Mol Cell Biol 15(5):2737–2744

Gibney PA, Lu C, Caudy AA, Hess DC, Botstein D (2013) Yeast metabolic and signaling genes are required for heat-shock survival and have little overlap with the heat-induced genes. Proc Natl Acad Sci USA 110(46):E4393–4402. https://doi.org/10.1073/pnas.1318100110

Gorner W, Durchschlag E, Martinez-Pastor MT, Estruch F, Ammerer G, Hamilton B, Ruis H, Schuller C (1998) Nuclear localization of the C_2H_2 zinc finger protein Msn2p is regulated by stress and protein kinase A activity. Genes Dev 12(4):586–597

Gorsich SW, Dien BS, Nichols NN, Slininger PJ, Liu ZL, Skory CD (2006) Tolerance to furfural-induced stress is associated with pentose phosphate pathway genes *ZWF1*, *GND1*, *RPE1*, and *TKL1* in *Saccharomyces cerevisiae*. Appl Microbiol Biotechnol 71(3):339–349. https://doi.org/ 10.1007/s00253-005-0142-3

Hahn JS, Hu Z, Thiele DJ, Iyer VR (2004) Genome-wide analysis of the biology of stress responses through heat shock transcription factor. Mol Cell Biol 24(12):5249–5256. https://doi.org/10.112 8/MCB.24.12.5249-5256.2004

Kida K, Kume K, Morimura S, Sonoda Y (1992) Repeated-batch fermentation process using a thermotolerant flocculating yeast constructed by protoplast fusion. J Ferment Bioeng 74(3):169–173. https://doi.org/10.1016/0922-338X(92)90078-9

Kuroda K, Ueda M (2013) Arming technology in yeast-novel strategy for whole-cell biocatalyst and protein engineering. Biomolecules 3(3):632–650. https://doi.org/10.3390/biom3030632

Kuroda K, Ueda M (2017) Engineering of global regulators and cell surface properties toward enhancing stress tolerance in *Saccharomyces cerevisiae*. J Biosci Bioeng 124(6):599–605. https://doi.org/10.1016/j.jbiosc.2017.06.010

Kuyper M, Toirkens MJ, Diderich JA, Winkler AA, van Dijken JP, Pronk JT (2005) Evolutionary engineering of mixed-sugar utilization by a xylose-fermenting *Saccharomyces cerevisiae* strain. FEMS Yeast Res 5(10):925–934. https://doi.org/10.1016/j.femsyr.2005.04.004

Martinez-Pastor MT, Marchler G, Schuller C, Marchler-Bauer A, Ruis H, Estruch F (1996) The *Saccharomyces cerevisiae* zinc finger proteins Msn2p and Msn4p are required for transcriptional induction through the stress response element (STRE). EMBO J 15(9):2227–2235

Matsui K, Kuroda K, Ueda M (2009) Creation of a novel peptide endowing yeasts with acid tolerance using yeast cell-surface engineering. Appl Microbiol Biotechnol 82(1):105–113. https://doi.org/10.1007/s00253-008-1761-2

Matsumoto K, Uno I, Oshima Y, Ishikawa T (1982) Isolation and characterization of yeast mutants deficient in adenylate cyclase and cAMP-dependent protein kinase. Proc Natl Acad Sci USA 79(7):2355–2359

Mitsuzawa H, Uno I, Oshima T, Ishikawa T (1989) Isolation and characterization of temperature-sensitive mutations in the *RAS2* and *CYR1* genes of *Saccharomyces cerevisiae*. Genetics 123(4):739–748

Pacheco A, Pereira C, Almeida MJ, Sousa MJ (2009) Small heat-shock protein Hsp12 contributes to yeast tolerance to freezing stress. Microbiology 155(6):2021–2028. https://doi.org/10.1099/mic.0.025981-0

Sakurai H, Takemori Y (2007) Interaction between heat shock transcription factors (HSFs) and divergent binding sequences: binding specificities of yeast HSFs and human HSF1. J Biol Chem 282(18):13334–13341. https://doi.org/10.1074/jbc.M611801200

Satomura A, Katsuyama Y, Miura N, Kuroda K, Tomio A, Bamba T, Fukusaki E, Ueda M (2013) Acquisition of thermotolerant yeast *Saccharomyces cerevisiae* by breeding via stepwise adaptation. Biotechnol Prog 29(5):1116–1123. https://doi.org/10.1002/btpr.1754

Satomura A, Kuroda K, Ueda M (2014) Environmental stress tolerance engineering by modification of cell surface and transcription factor in *Saccharomyces cerevisiae*. Curr Environ Eng 1(3):149–156. https://doi.org/10.2174/221271780103150522154913

Satomura A, Miura N, Kuroda K, Ueda M (2016) Reconstruction of thermotolerant yeast by one-point mutation identified through whole-genome analyses of adaptively-evolved strains. Sci Rep 6:23157. https://doi.org/10.1038/srep23157

Schmitt AP, McEntee K (1996) Msn2p, a zinc finger DNA-binding protein, is the transcriptional activator of the multistress response in *Saccharomyces cerevisiae*. Proc Natl Acad Sci USA 93(12):5777–5782

Shi DJ, Wang CL, Wang KM (2009) Genome shuffling to improve thermotolerance, ethanol tolerance and ethanol productivity of *Saccharomyces cerevisiae*. J Ind Microbiol Biotechnol 36(1):139–147. https://doi.org/10.1007/s10295-008-0481-z

Thevelein JM, de Winde JH (1999) Novel sensing mechanisms and targets for the cAMP-protein kinase A pathway in the yeast *Saccharomyces cerevisiae*. Mol Microbiol 33(5):904–918

Ueda M, Tanaka A (2000) Genetic immobilization of proteins on the yeast cell surface. Biotechnol Adv 18(2):121–140

Wallace-Salinas V, Gorwa-Grauslund MF (2013) Adaptive evolution of an industrial strain of *Saccharomyces cerevisiae* for combined tolerance to inhibitors and temperature. Biotechnol Biofuels 6(1):151. https://doi.org/10.1186/1754-6834-6-151

Winkler JD, Kao KC (2014) Recent advances in the evolutionary engineering of industrial biocatalysts. Genomics 104(6):406-411. https://doi.org/10.1016/j.ygeno.2014.09.006

Wisselink HW, Toirkens MJ, Wu Q, Pronk JT, van Maris AJ (2009) Novel evolutionary engineering approach for accelerated utilization of glucose, xylose, and arabinose mixtures by engineered *Saccharomyces cerevisiae* strains. Appl Environ Microbiol 75(4):907–914. https://doi.org/10.11 28/AEM.02268-08

Yazawa H, Iwahashi H, Uemura H (2007) Disruption of *URA7* and *GAL6* improves the ethanol tolerance and fermentation capacity of *Saccharomyces cerevisiae*. Yeast 24(7):551–560. https://doi.org/10.1002/yea.1492

Zou W, Ueda M, Yamanaka H, Tanaka A (2001) Construction of a combinatorial protein library displayed on yeast cell surface using DNA random priming method. J Biosci Bioeng 92(4):393–396

Zou W, Ueda M, Tanaka A (2002) Screening of a molecule endowing *Saccharomyces cerevisiae* with *n*-nonane-tolerance from a combinatorial random protein library. Appl Microbiol Biotechnol 58(6):806–812. https://doi.org/10.1007/s00253-002-0961-4

The Domestication Syndrome in *Phaseolus* Crop Plants: A Review of Two Key Domestication Traits

María Isabel Chacón-Sánchez

Abstract Plant domestication can be seen as a long-term experiment that involves a complex interplay among demographic processes and evolutionary forces. Long-standing questions about plant domestication are the number of times a species was domesticated, the extent of the domestication bottleneck, and the genetic basis of adaptive processes. Crops such as *Phaseolus* beans offer an excellent opportunity to answer these questions, especially the ones related to the genetic basis of the adaptive domestication syndrome. In the genus *Phaseolus*, five species have been domesticated: the common bean (*P. vulgaris*), the runner bean (*P. coccineus*), the tepary bean (*P. acutifolius*), the year bean (*P. dumosus*) and the Lima bean (*P. Lunatus*). These five species were domesticated in seven independent events that resulted in phenotypic convergence for several traits of the domestication syndrome. Two of these traits, namely reduced pod shattering and increased seed weight, are of special interest due to their role in the initial adaptation of these species during domestication. The objective of the present study was to review current evidence about (1) the effects of domestication on phenotypic variation of these two domestication traits in *Phaseolus* beans to understand whether adaptation led to clear-cut differences among wild and domestic forms or to a phenotypic continuum, and (2) the genetic basis of these two traits in *Phaseolus* beans to understand whether phenotypic convergence has been driven by parallel or convergent evolution. Research on these subjects in *Phaseolus* beans has been very scarce, and areas in need of urgent development are highlighted in this review.

1 Introduction

Domestication may be considered a dynamic and continuous process that began when humans started to exploit, manage, and/or deliberately cultivate wild populations (Pickersgill 2007). With time, cultivated populations diverged from their

M. I. Chacón-Sánchez (✉)
Facultad de Ciencias Agrarias, Departamento de Agronomía, Universidad Nacional
de Colombia - Bogotá, Carrera 30 no. 45-03 - Edificio 500, Bogotá, D.C., 111321, Colombia
e-mail: michacons@unal.edu.co

© Springer International Publishing AG, part of Springer Nature 2018   37
P. Pontarotti (ed.), *Origin and Evolution of Biodiversity*,
https://doi.org/10.1007/978-3-319-95954-2_3

wild ancestors and became better adapted to agro-ecological environments and more dependent on humans for their own survival (Meyer and Purugganan 2013). Many of the genetic modifications underlying the adaptation process have led to a set of phenotypic changes of all kind (morphological, physiological, etc.) between domesticated species and their wild ancestors, which are collectively known as the domestication syndrome. Sometimes, these differences are not clear-cut, but rather a phenotypic continuum is observed among crop and wild progenitors, which makes it difficult to distinguish between traits that were crucial for the domestication episode (e.g., loss of seed dispersal mechanisms and reduced seed dormancy) and traits that evolved after domestication during landrace diversification (e.g., changes in seed color) (Abbo et al. 2014).

The genus *Phaseolus* contains five domesticated bean species, namely *Phaseolus vulgaris* L., *Phaseolus coccineus* L., *Phaseolus acutifolius* A. Gray, *Phaseolus dumosus* Macfady, and *Phaseolus lunatus* L. These five domesticated *Phaseolus* species offer the opportunity to investigate traits of the domestication syndrome and their genetic bases. For these five crop species, similar morphological changes have been observed for several traits of the domestication syndrome. For legume species, reduced pod shattering, increased seed germination, and increase in seed size are among the most important traits that allowed adaptation of domesticated populations (Harlan 1992); other traits are changes in growth habit and photoperiod sensitivity. In spite of the importance of these traits in legume domestication, not only from an evolutionary perspective but also for practical applications in crop production, very little is known about their genetic control, as noted by Dong and Wang (2015).

The objective of the present study is to review the evidence available for two domestication traits, namely pod shattering and seed size, in *Phaseolus* crop species, in order to gain insights into the magnitude of phenotypic changes associated with domestication and their genetic control. I will first review current evidence about domestication areas for these five *Phaseolus* species to establish how many independent domestication events were involved in the origin of *Phaseolus* crop species. Then, information available on phenotypic evaluations of these two key domestication traits in wild and domestic forms in *Phaseolus* beans will be reviewed. I finalize with a review of the information available on the genetic control of these two key domestication traits to establish whether or not there is evidence in the literature that support common genetic basis for these traits. This review is expected to stimulate further research into these matters.

2 Domestication Areas of *Phaseolus* Beans

Pinpointing crop domestication areas requires an interdisciplinary approach based on sources of evidence from fields such as archaeology, botany, and genetics. Plant remains that have been retrieved in archaeological sites, mainly in the form of carbonized remains, imprints of part plants on pottery, or plant remains preserved by desiccation in arid areas, provide information about where the earliest cultivation activities for a crop occurred, the spread of a crop to other areas, and how early culti-gens looked like (Zohary and Hopf 2000). However, archaeological evidence does not necessarily offer information on the wild progenitors from which domesticated plants derived, and comparisons of domesticated plants with their wild progenitors are key to understand the kind of changes (morphological, physiological, and genetic) that occurred during domestication.

Identification of domestication areas requires understanding of the genetic struc-ture of the wild progenitor, along its range of distribution, and also of its domesticated counterpart. Areas of domestication are usually pinpointed by identifying the wild populations that are most closely related to the cultivated varieties. The reasoning behind this, as noted by Zohary and Hopf (2000), is that domestication is a very recent event (10,000 years ago) compared to evolutionary times of wild relatives. Therefore, most of genetic variations present in cultivated varieties have been mainly inherited from their wild progenitors. This means that establishing the genetic relationships among wild relatives and domesticated populations may reveal the wild stocks from which domesticated varieties arose, thus the domestication areas. This approach assumes that extant wild relatives have not appreciably changed in their genetic con-stitution and geographic distribution since the time of domestication. Advances in molecular marker technology have allowed, in many cases, delimitation of domesti-cation areas when traditional approaches have not been successful. Below, I provide a summary of current knowledge on areas of domestication for the five *Phaseolus* crop species that is mainly based on molecular markers.

P. vulgaris, the common bean, occurs in its wild form from Northern Mexico to northern Argentina, where three gene pools have been recognized: the Mesoameri-can, the Andean, and the Ecuadorian–Northern Peruvian one (Debouck et al. 1993; Gepts and Debouck 1991; Kwak and Gepts 2009). Common bean was domesticated at least twice: once from wild Mesoamerican populations and once from wild pop-ulations in the Andes of South America (Chacón et al. 2005; Gepts and Bliss 1986; Gepts and Debouck 1991; Kwak and Gepts 2009; Kwak et al. 2009; Nanni et al. 2011). For Mesoamerica, Kwak et al. (2009) proposed the Lerma-Santiago Basin in the state of Jalisco, Mexico, as the putative domestication area of Mesoamerican landraces based on data on 26 microsatellite loci (Fig. 1a). The evidence was based on the fact that all Mesoamerican varieties clustered in a single group, thus indicating a single domestication, and also that the wild populations most closely related to the domesticated cluster where those distributed on Rio Lerma-Rio Grande de Santiago Basin, in western-central Mexico. In contrast, Bitocchi et al. (2012) on the basis of sequencing of five DNA fragments concluded that common bean was domesticated in Oaxaca Valley in Mexico (Fig. 1a). This conclusion was reached on the basis of a close genetic relationship observed between two wild accessions from Oaxaca to a cluster formed by most of Mesoamerican domesticated accessions analyzed. For the Andean gene pool, Beebe et al. (2001) and Bitocchi et al. (2012), on the basis of AFLP and DNA sequence markers, respectively, proposed Bolivia–north-ern Argentina as a possible area of domestication for Andean landraces (Fig. 1a). These two studies found a close genetic relationship between wild accessions from this geographical region and Andean landraces. In contrast, Chacón et al. (2005) on

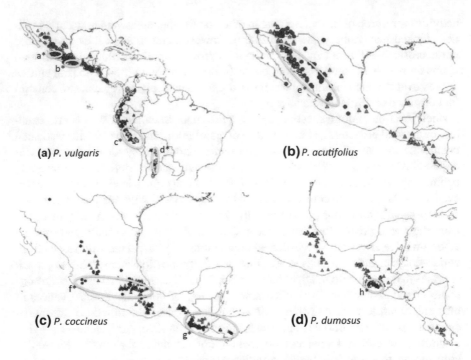

Fig. 1 Geographic distribution of wild relatives (red or light blue circles), landraces (pink or dark blue triangles), and putative domestication areas (areas highlighted in yellow) of four *Phaseolus* species. Putative domestication areas: for *P. vulgaris*: a* Lerma-Santiago Basin in the state of Jalisco, Mexico (Kwak et al. 2009), b* Oaxaca Valley in Mexico (Bitocchi et al. 2012), c* central-southern Peru (Chacón et al. 2005), d* Bolivia–northern Argentina (Beebe et al. 2001 and Bitocchi et al. 2012); for *P. acutifolius*: e* area spanning from north to south the states of Sonora, Sinaloa, and Jalisco in Mexico (Blair et al. 2012); for *P. coccineus*: f* Trans-Mexican Volcanic Belt (Guerra-García et al. 2017), g* area of Guatemala–Honduras (Rodriguez et al. 2013; Spataro et al. 2011); for *P. dumosus*: h* western Guatemala. See text for details

the basis of chloroplast DNA polymorphisms suggested central-southern Peru as the domestication area for Andean landraces (Fig. 1a), on the basis that the chloroplast DNA haplotype that was predominant among Andean landraces was found in wild accessions from this region. These two domestication events followed by diversification resulted in a large diversity of Mesoamerican and Andean eco-geographic races (Beebe et al. 2000, 2001; Singh et al. 1991).

P. acutifolius, the tepary bean, is distributed as a wild plant in Southwestern United States and Northern Mexico, and according to previous studies, this species was probably domesticated only once in Mexico, in the states of Jalisco and Sinaloa on the basis of isozyme markers, in a region between Sonora and Sinaloa as suggested by phaseolin markers (although not a conclusive result) (Schinkel and Gepts 1988) or in a region between Jalisco, Sinaloa, and Sonora on the basis of the closer relationship between cultivars and wild accessions from this geographical region as shown by microsatellite (STR, short tandem repeats) molecular markers (Blair et al. 2012)

(Fig. 1b). A recent survey of SNP markers also showed that tepary beans were likely domesticated only once, although the authors could not precise the place of domestication (Gujaria-Verma et al. 2016).

P. coccineus, the runner bean, occurs in its wild form from Northern Mexico to probably not further south of Lake Nicaragua as claimed by Freytag and Debouck (2002) because this lake is a natural barrier for many floristic elements. Two subspecies have been recognized based on morphological characters: subspecies *coccineus* and subspecies *striatus*, the former including wild and domestic forms (Freytag and Debouck 2002). Previous studies have suggested at least two domestication events in Mesoamerica: one event in the area Guatemala–Honduras and the second event in Mexico (Rodriguez et al. 2013; Spataro et al. 2011) (Fig. 1c). Based on six chloroplast SSR markers, Rodriguez et al. (2013) found that domesticated accessions grouped in two clusters that also contained wild accessions. Although the authors suggested this pattern might indicate multiple domestications, they authors also suggested that a more focused study should address this question. On the basis of 12 SSR loci, Spataro et al. (2011) found that most of landraces resembled wild genotypes from the area Guatemala–Honduras. However, a recent genomic analysis of ten wild, three feral, and eleven domesticated populations of runner bean (242 individuals in total), with more than 40,000 SNP markers, suggested that this species was domesticated only once in Mexico, maybe in the area of the Trans-Mexican Volcanic Belt (TMVB) (Guerra-García et al. 2017) (Fig. 1c). This conclusion was based on the observation that domesticated individuals formed a monophyletic group and the results of an approximate Bayesian computation approach that supported the sce-

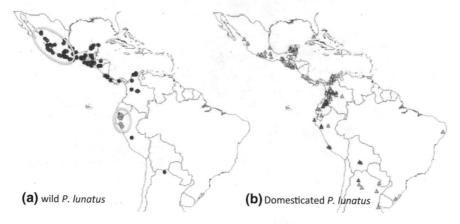

(a) wild *P. lunatus*   **(b)** Domesticated *P. lunatus*

Fig. 2 Geographic distribution of **a** wild and **b** domesticated accessions of *P. lunatus* and putative domestication areas (enclosed in yellow). Wild accessions are shown as circles with different colors according to their gene pool: blue: Mesoamerican I, red: Mesoamerican II, bright green: Andean I, bright pink: Andean II. Domesticated accessions are shown in triangles of different color according to their gene pool: light blue: Mesoamerican, dark green: Andean. The putative domestication area for Mesoamerican landraces is central-western Mexico and for Andean landraces is the Andes of Ecuador–Northern Peru

nario where wild populations from TMVB were closely related to the domesticated populations.

P. dumosus, the year bean, is found in its wild state in western Guatemala and was domesticated in that place in line with evidence from isozyme patterns and linguistics (Schmit and Debouck 1991) (Fig. 1d). There are no other studies based on DNA molecular markers that provide additional evidence about the domestication of this species.

Wild populations of *P. lunatus*, the Lima bean, are geographically widespread from Northern Mexico to northern Argentina (Fig. 2a). Along its distribution range, wild Lima bean is structured into four non-overlapping gene pools: the Mesoamerican pool MI occurs at the northern end of the distribution range mainly in Northern and central-western Mexico, Mesoamerican pool MII is the most widely distributed and is found from southern Mexico to northern Argentina, the Andean gene pool AI is restricted to the western slope of the Andes of Ecuador–Northern Peru, and the

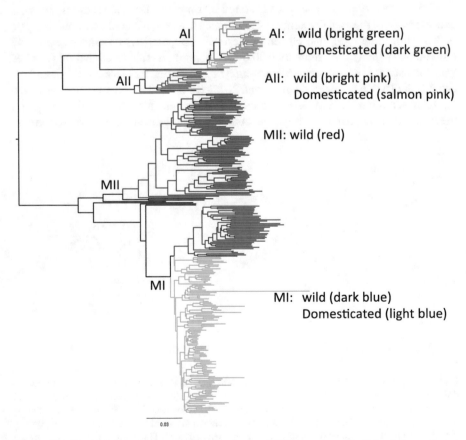

Fig. 3 Neighbor-joining topology, built on the basis of 317 accessions of wild and domesticated Lima bean and 19,387 SNP markers, showing the genetic relationships among accessions (data not published)

Andean gene pool AII is found in the central departments of Cundinamarca and Boyacá in Colombia (Fig. 2a) (Chacón-Sánchez and Martínez-Castillo 2017; Serrano-Serrano et al. 2010). The genetic relationships among these four gene pools are shown in Fig. 3. The genetic relationships along with the almost non-overlapping distribution of these four gene pools (Fig. 2a) suggest a strong phylogeographic structure. Lima bean was domesticated at least twice and independently (Fig. 2b) (Andueza-Noh et al. 2015; Andueza-Noh et al. 2013; Chacón-Sánchez and Martínez-Castillo 2017; Motta-Aldana et al. 2010; Serrano-Serrano et al. 2012). One domestication occurred from gene pool AI and gave rise to large-seeded Andean landraces known as "Big Lima." Another domestication occurred in central-western Mexico from gene pool MI and gave rise to small-seeded Mesoamerican landraces known as "Sieva" and "Potato."

3   Phenotypic Evaluations of Two Key Domestication Traits in *Phaseolus* Beans

3.1   Pod Shattering

The success of wild plants depends on their capacity to disperse their offspring and thus assure survival. Paradoxically, for domesticated plants, seed dispersal is an undesired trait that was either reduced or lost during their adaptation to novel agroecosystems not only during the domestication of *Phaseolus* beans but also during domestication of seed crops in general (Abbo et al. 2014; Harlan 1992). According to Harlan (1992), this trait is "crucial in establishing the disruptive selection that effectively maintains separation of the two kinds of populations" (wild and domesticated). This is because with shattering, seeds will escape harvest by humans and therefore will not contribute to the next cultivated generation; therefore, reduced fruit shattering is a trait that may show up in seed crops under unconscious or unintentional selection associated with harvesting (Harlan 1992).

In common bean, pod shattering is a trait that was not completely fixed during domestication. Wild populations are highly dehiscent and at maturity pod valves separate and twist dispersing the seeds a few meters away from the mother plant (Gepts and Debouck 1991). Some domesticated varieties of dry beans (e.g., some landraces of Nueva Granada race) still show some degree of pod shattering, especially when they are grown in dry environments (Acosta-Gallegos et al. 2007). Modern varieties known as snap beans, which have been bred to produce pods that are eaten as vegetables, are completely indehiscent (Gepts and Debouck 1991). In common bean, the twisting movement of pods has been attributed to the presence of fibers in pod walls and thus a reduced content of fibers would have been a selective advantage during domestication because retardation of seed dispersal facilitates harvest (Gepts and Debouck 1991).

Koinange et al. (1996) reported that pod shattering in a wild common bean geno-type (accession G12873) was due to the presence of pod fibers in both the suture and pod walls, while pods of a domesticated non-shattering genotype (cultivar MIDAS, a snap variety) completely lacked these fibers (Koinange et al. 1996). Recently, Murgia et al. (2017) studied pod shattering in common bean in a population of 267 intro-gression lines derived from a cross between two contrasting parents, MG38 (a RIL with 55% wild background and high shattering) and the recurrent parent MIDAS. The authors analyzed pod shattering in the field, measured the chemical composition of the pods (total fiber content, lignin, cellulose, and hemicellulose), and carried out anatomical and histological analyses of pod valves. The authors found, as expected, that MG38 had 65% shattering pods while MIDAS was completely indehiscent, that MG38 had in the pod valves a higher carbon content (43.8% dry weight) than MIDAS (41% dry weight), that MG38 also had a higher fiber content in the pods (62%) than MIDAS (42%), and that MG38 always had higher contents of lignin, hemicellulose, and cellulose than MIDAS. Histological analyses revealed that in 20-day-old pods, shattering and non-shattering phenotypes showed striking differences in lignin depo-sition in the ventral sheath of the pod valves (lignification is more pronounced in shattering types), a trait that was highly correlated with carbon content (higher in shattering phenotypes and lower in non-shattering types). In the ventral sheath of the pod valves, MG38 (the shattering phenotype) showed a higher proportion of cells with thick secondary cell-wall formation compared to MIDAS (the non-shattering type) and this thickness tended to be reduced toward the dehiscence zone (a zone of easy fracture). In addition, MG38 showed a high degree of lignification in the inner cells of the pod walls (sclerenchymatic cells) compared to the absence of lig-nification in MIDAS. Murgia et al. (2017) discussed the histological differences between non-shattering and shattering types in common bean and soybean. In soy-bean, non-shattering types showed a higher degree of lignification in the dehiscence zone compared to the wild type (Dong et al. 2014), thus making the pod resistant to shatter, and also non-shattering types showed a lower degree of lignification in inner sclerenchyma cells of the pod wall, thus reducing valve twisting (Funatsuki et al. 2014). In contrast, there were no clear differences in the dehiscence zone in common bean; rather the differences between shattering and non-shattering pheno-types are in the degree of lignification in the ventral sheath of the pod valves and the presence/absence of lignification of the inner layer of sclerenchymatic cells of pod walls. These findings show clear histological differences in pod tissues associated with pod shattering between common bean and soybean, suggesting that different molecular mechanisms might underlie this trait in these two species. To establish whether similar or different mechanisms are responsible for pod shattering within the genus *Phaseolus*, a comprehensive method as the one proposed by Murgia et al. (2017) to phenotype pod shattering in *Phaseolus* beans should be implemented in a comparative framework. This method is very promising to disentangle the genetic architecture of this trait.

In spite of being a key domestication trait for *Phaseolus* beans, there is a great lack of studies comparing wild and domesticated populations for pod shattering (even for the common bean). For this reason, we have made a first attempt to analyze the variation of this trait in a collection of 19 wild and 62 domesticated accessions of Lima bean (data from Paola Hurtado et al. not published). In brief, a total of 50 plants per accession were planted in one location in Colombia chosen to match the ecological conditions of their place of origin. Pods were harvested daily at physiological maturity and were placed in a desiccator during 48 h at 60 °C and constant airflow. A sample of 10 pods per accession was taken for measuring the minimum force (in Newtons (N)) required to fracture the pod in an AMETEK LS1 texture analyzer. In these accessions, about 50% needed a pod shattering force between 5 and 10 N, less than 5% required a force below 5 N, and less than 5% required a force beyond 15 N.

For investigating whether the biological status (wild or domesticated) was significant to explain variation in pod shattering force, a linear model was built and a Breusch Pagan test ($p=0.1147$) and an ANOVA test ($p=0.1525$) were applied; the results suggest that the biological status is not significant to explain variation in this trait and rather a phenotypic continuum is observed (data from Paola Hurtado et al., not published) (see Fig. 4a). This result is somehow unexpected given that pod shattering is one of the traits under disruptive selection during domestication. In contrast, gene pool of the landraces resulted to be significant to explain variation in this trait (Breusch Pagan test, $p=0.03$, ANOVA test, $p=1.7e-06$) (data from Paola Hurtado et al., not published). Mesoamerican (MI) landraces required less pod shattering force (average of 8.22 N) than Andean landraces (average of 13.91 N)

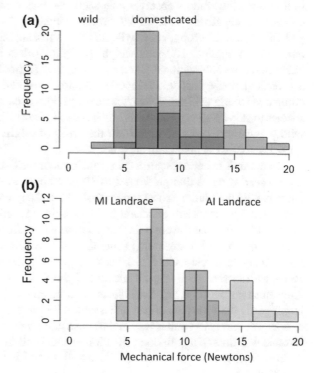

Fig. 4 Changes in pod shattering force (measured in Newtons) in the domestication of Lima bean based on data from Paola Hurtado et al. (not published). **a** Mechanical force applied to wild (blue bars) and domesticated (pink bars) accessions to break the pods. **b** Mechanical force applied to Mesoamerican (gray bars) and Andean (green bars) landraces to break the pods

(Fig. 4b). These results suggest that pod lignification patterns might be different for Andean and Mesoamerican landraces, a question of high interest that should be thoroughly investigated. Future studies in Lima bean should include more robust methods to evaluate this trait in order to investigate phenotypic convergence in both domestication events.

4 Seed Size

Increase in seed size during domestication may be an adaptation related to seedling competition (seedlings of the same species) in the cultivation field (Harlan 1992), because not only seeds that germinate first but also those that produce more vigorous seedlings (namely larger seeds) are more likely to contribute with progeny to the next generation. In cereals, increase in seed size is due to an increase in the endosperm (not in the embryo) with a tendency to elevate carbohydrate content and to lower protein content (Harlan 1992). In common bean, Singh et al. (2014) investigated whether the increase in seed size during common bean domestication was accompanied by an increase in starch or protein content. For this, a RIL population (168 individuals) derived from a cross between a wild parent and a domesticated parent from the Andean gene pool was analyzed using near-infrared reflectance (NIR). As in cereals, in this study the authors observed that seed size was positively correlated to starch content and negatively correlated to protein content.

In all *Phaseolus* crop species, wild relatives show smaller seeds than domesticated ones. In this section, 100-seed weight (in grams) data taken from the *Phaseolus* database at the International Center for Tropical Agriculture (CIAT) (http://geneb ank.ciat.cgiar.org/genebank/beancollection.do, consulted on April 7, 2018) will be compared among wild and domesticated accessions in each one of the five *Phaseolus* species to gain insights on how this trait has been affected during domestication. Only wild populations and landraces from the areas of origin of each crop species (see Figs. 1 and 2) were taken into account.

In common bean, Mesoamerican wild populations show seed sizes (measured as 100-seed weight) that go from 2 to 19 g, with an average of around 6 g, while Andean wild populations show ranges from 3.2 to 26 g, with average of around 12 g. Domestication increased considerably the variation in seed size in both domestication events (Fig. 5a). Seed weight in Mesoamerican landraces goes from 12 to 72 g, with average of 29 g, an average increase of 4–6 times in comparison with wild Mesoamerican ancestors (Fig. 5b). Seed weight in Andean landraces goes from 16 to 108 g, with an average of 48 g, an increase of about 4–5 times in comparison with wild Andean ancestors (Fig. 5c). So, it seems that in spite that Andean ancestors exhibit larger seeds than Mesoamerican ancestors, wild populations apparently responded to domestication in a similar way in the sense that a comparable increase in seed size (around 4–5 times) in both domestication events is observed.

In runner bean (*P. coccineus*), wild populations (that occur only in Mesoamerica) have a wide range in seed size that go from 0.6 to 37.9 g, with an average of

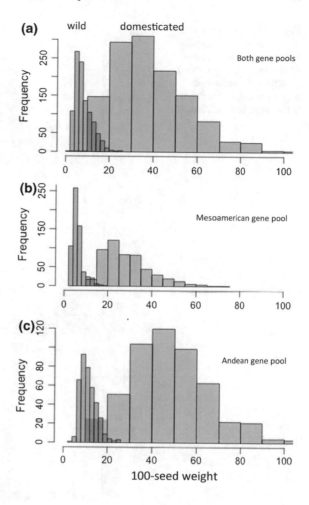

Fig. 5 Changes in seed weight in the domestication of common bean. Blue bars: wild populations. Pink bars: domesticated populations. **a** Wild and domesticated populations of common bean from both gene pools (Mesoamerican and Andean). **b** Wild and domesticated populations from Mesoamerican gene pool. **c** Wild and domesticated populations from Andean gene pool

around 14 g. In this species, domestication also increased variation in seed size and domesticated populations exhibit an average increase of about 5–6 times (Fig. 6a) in relation to their wild ancestors, with sizes that go from 20.1 to 199.5 g, with an average of around 89 g (see Fig. 6a). It is interesting to note that runner beans show the largest variation in seed size among all five domesticated species.

Tepary bean (*P. acutifolius*) shows the smallest seeds of all five species, in wild and domesticated types. Wild populations that occur mainly in central-northern Mexico and Southwestern United States contain seed sizes between 1.6 and 8.2 g, with an average of 3.34 g (Fig. 6b). Domesticated tepary beans show seed sizes with a non-overlapping distribution with wild tepary beans, with a range that goes from 9.2 to 27.2 g and an average of 15.28 g, which represents an increase of about 3–6 times. It is worth noticing that among the five species, tepary bean is the only one that shows a clear-cut difference in seed size among wild and domesticated forms, which may

Fig. 6 Changes in seed weight in the domestication of three *Phaseolus* species. Blue bars: wild populations. Pink bars: domesticated populations. **a** *P. coccineus*. **b** *P. acutifolius*. **c** *P. dumosus*

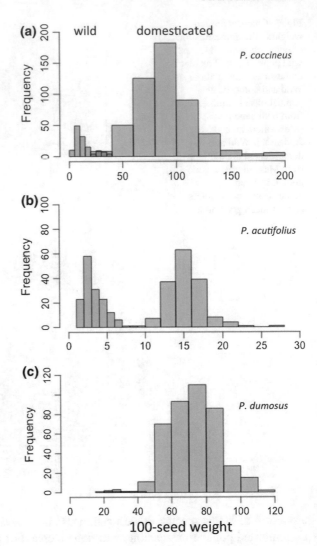

suggest that in this species seed size was a crucial domestication trait as defined by Abbo et al. (2014).

For year bean (*P. dumosus*), according to seed size data available at CIAT (only nine accessions for the wild), wild populations show seed sizes from 19 to 42.8 g (even larger than in *P. coccineus*), with an average of 29.42 g. Domestication increased seed size about 2–3 times only, and domesticated seeds go from 26 to 116.2 g, with an average of 73.32 g (Fig. 6c).

In Lima bean, wild populations show seed sizes that go from 4 to around 33 g with an average of 11.2 g (Fig. 7a). In this species, two major gene pools exist: the Mesoamerican and the Andean. The seed weight of Mesoamerican gene pool ranges

Fig. 7 Changes in seed weight in the domestication of Lima bean. Blue bars: wild populations. Pink bars: domesticated populations. **a** Wild and domesticated populations of Lima bean from both gene pools (Mesoamerican and Andean). **b** Wild and domesticated populations from Mesoamerican gene pool. **c** Wild and domesticated populations from Andean gene pool

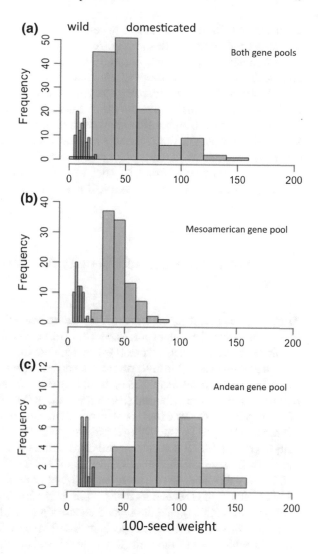

from 4 to about 21, with an average of around 9.4 g. In the Andean wild gene pool, seeds are heavier and their weights go from 11 to 24 g with an average of 15.4 g. Lima bean landraces have seeds whose weight go from 21 to around 144 g, with average of 56 g. As mentioned before, this species was domesticated in Mesoamerica and the Andes. Mesoamerican landraces (gene pool MI) show seed sizes that go from 21 to 107 g, with average of around 44 g. The increase in seed size for Mesoamerican landraces in comparison with their wild relatives has been of around 5 times (Fig. 7b). On the other hand, Andean landraces show seeds that weight from 32 to 144 g, with an average of around 84.2 g, almost doubling the size of Mesoamerican landraces. In comparison with the wild Andean ancestor, Andean landraces have increased their size about 5–6 times in average (Fig. 7c).

In summary, it appears that domestication in these five species did not affect seed size in the same way; in some species like *P. dumosus,* increase in seed size was only 2–3 times, while in the other species the average increase was up to 6 times. In all species examined. there are clear differences in seed size between wild and domestic forms, although different degrees of overlap were observed. Very little is known about what nutritional changes have accompanied the increase in seed size in domesticated populations (except for common bean), a subject that should be further investigated. Another important question to solve is whether the same genes underlie changes in seed size among these five domesticated species, a subject that will be revised below.

5   Molecular Genetics and Genetic Architecture of Pod Shattering

Arabidopsis thaliana has been the model plant to study the cascade of genes involved in pod shattering. This plant possesses a dry fruit called silique that shatters at maturity through the dehiscence zone (DZ) to disperse the seeds (Estornell et al. 2013). DZ is constituted by a thin layer of cells that differentiates at the valve/replum margin of the fruit; the separation of these cells allows the detachment of the valves from the replum, and seeds are dispersed (Ferrándiz et al. 2000). Two genes that are at the top of the regulatory cascade that control fruit shattering in *Arabidopsis* are the closely related MADS-box genes *SHATERPROOF* (*SHP1*) and *SHATERPROOF2* (*SHP2*) (Liljegren et al. 2000) that control cell differentiation in DZ. Acting downstream of the *SHP1/2* genes are the genes known as *INDEHISCENT* (*IND*) and *ALCATRAZ* (*ALC*)—the former codes for an atypical bHLH protein that specifies the differentiation of cells in the DZ, and its mutation in *Arabidopsis* results in indehiscent fruits (Liljegren et al. 2004), and the second encodes a myc/bHLH protein, necessary for the formation of a non-lignified specialized cell layer within the DZ and its mutation prevents dehiscence of the fruit (Rajani and Sundaresan 2001). On the other hand, *FRUITFULL* (*FUL*), a MADS-box gene, is required for the expansion and differentiation of fruit valves, is a negative regulator of *SHP1/2* and restricts the expression of *SHP1/2* and *IND* to the DZ (Ferrándiz et al. 2000). The constitutive expression of FUL produces indehiscent fruits because of the conversion of DZ cells into valve cells (Ferrándiz et al. 2000). The *NST1* gene (*NAC SECONDARY WALL THICKENING PROMOTING FACTOR 1*) acts downstream of *SHP1/2* and is expressed in the lignified layer of cells in the DZ, and its mutation results in indehiscent fruits because the cells of the valve margins lose lignification (Mitsuda and Ohme-Takagi 2008). Finally, for silique dehiscence to take place, the action of enzymes known as endo-polygalacturonases (PGs) that degrade pectin to promote cell separation is indispensable. The genes *ARABIDOPSIS DEHISCENCE ZONE POLYGALAC-*

TORUNASE1 (*ADPG1*) and *ADGP2* encode for PGs that are expressed specifically in DZ (Ogawa et al. 2009). It has been demonstrated that *IND* is necessary for normal expression of *ADPG1* in DZ (Ogawa et al. 2009).

In legume crops (chickpea, lentils, soybean, common bean, Lima bean, etc.), indehiscent pods appeared in all domestic forms apparently as a result of selection associated with harvesting (Harlan 1992). The question to investigate here is whether or not the same genes and molecular mechanisms explain the observed phenotypic convergence. In the last few years, some evidence has appeared on this subject but is still very poor. In soybean, for example, Dong et al. (2014) showed that loss of seed dispersal in domesticated soybean is caused by the excessive lignification of fiber cap cells (FCC) in the abscission layer that prevents pod shattering. The thickening of FCC secondary walls is controlled by a gene known as *SHATTERING1-5* (*SHAT1-5*), a homolog of *AtNST1/2*, which is expressed at much higher levels in domesticated soybean compared to wild soybean due to the loss in the former of a repressive *cis*-regulatory element at the 5' promoter region. While in soybean Dong et al. (2014) concentrated their efforts in the study of the genes responsible for differences in binding strength of the pod among wild and domestic forms, Funatsuki et al. (2014) studied the genes responsible for the generation of dehiscing forces in the pod. These authors identified a major QTL that encodes a protein called *Pdh1*. The gene *PDH1* is expressed in the inner sclerenchyma of pod walls at the beginning of lignin deposition, and this gene is associated with an increase in the torsion of dried pod walls that promotes pod dehiscence. In domesticated soybean, a functional protein of this gene is not produced due to a premature stop codon (Funatsuki et al. 2014).

The first study that addressed the genetic architecture of several morphological and physiological traits of the domestication syndrome in *Phaseolus* beans is the one that Koinange et al. (1996) carried out in common bean. Among the traits the authors studied were seed dispersal (that was measured as presence or absence of pod suture fibers and pod wall fibers), seed dormancy, growth habit, gigantism (pod length and 100-seed weight), earliness, photoperiod sensitivity, harvest index, and seed pigmentation. The presence of pod suture fibers and pod wall fibers was determined visually by breaking the pod beak and pod wall, respectively. The authors built a population of 65 F8 recombinant inbred lines (RILs) derived from the cross between a wild (G12873, Mesoamerican origin) and a domesticated accession (cultivar Midas, a wax snap bean of Andean origin), chosen to show the broadest range of domestication syndrome traits. The authors also built a linkage map with 83 DNA and biochemical markers to identify quantitative trait loci (QTL) governing domestication-related traits. For seed dispersal, measured as lack of suture fibers, the authors detected a single locus (*St*) on linkage group (LG) 2, and for the lack of pod wall fibers, they also detected a single locus in the same linkage group that was either tightly linked to *St* or identical to *St* (pleiotropy) (Table 1). These results suggest that the genetic control of pod shattering in common bean might be simple; however, this result might be due to the way this trait was measured.

Recent attempts to identify the causal gene underlying *St* in common bean have been reported. Gioia et al. (2013) adopted a candidate-gene approach to identify the gene in *St* responsible for pod dehiscence in common bean, selecting *IND* as a

Table 1 QTLs for pod shattering and seed weight detected in crosses between wild × domesticated, domesticated × domesticated, or in a set of wild and domesticated accessions

Domestication attribute	Trait	Linkage group	QTL or gene or linked marker	R^2 (%) per locus	R^2 (%) per trait	Reference
Pod shattering (wild × domesticated cross)	Pod suture fibers	2	St	–	–	Koinange et al. (1996)
	Pod wall fibers	2	St?	–	–	
Gigantism (wild × domesticated cross)	100-seed weight	1	D1492-3	18	57	Koinange et al. (1996)
		7	Phs	27		
		7	Uri-2	16		
		11	D0252	15		
		1	QTL1	–	42.3	Guzmán-Maldonado et al. (2003)
		2	QTL2	–		
		3	QTL3	–		
		4	QTL4	–		
		4	QTL5	–		
		2	sw2.2 Popayan	16	–	Blair et al. (2006)
		2	sw2.2 Darien	17		
		7	Phs Popayan	7		
		7	Phs Darien	6		
		8	sw8.1 Popayan	5		
		8	sw8.1 Darien	8		
		8	sw8.2 Popayan	8		
		8	sw8.2 Darien	6		
		11	sw11.1 Popayan	8		
		11	sw11.1 Darien	9		

(continued)

Table 1 (continued)

Domestication attribute	Trait	Linkage group	QTL or gene or linked marker	R^2 (%) per locus	R^2 (%) per trait	Reference
Gigantism (domesticated × domesticated cross)	100-seed weight	4a	BC500.400	18	44	Park et al. (2000)
		5	Q05.850	10		
		6	Y07.1200	9		
		3	G03.1150	9		
		8	D12.700	5		
		7	J09.950–1.0	3		
		4b	BC406.450	5		
		8	BC457.400	4		
Gigantism GWAS	100-seed weight	1	3 genes	–	–	Schmutz et al. (2014)
		2	1 gene	–		
		3	10 genes	–		
		4	2 genes	–		
		6	2 genes	–		
		7	42 genes	–		
		8	9 genes	–		
		9	3 genes	–		
		10	2 genes	–		
		11	1 gene	–		

R^2: percent of phenotypic variance explained by QTL or trait
–: lack of information

potential candidate due to its key role in *Arabidopsis*. For this, the authors identified by sequence comparison the homologue in common bean (*PvIND*) of the *AtIND* gene. The authors sequenced this gene in a collection of 77 wild and 80 domesticated accessions of common bean and three related species (*P. acutifolius*, *P. coccineus*, and *P. lunatus*). In addition, 105 accessions of common bean were scored for the presence of fibers in pod sutures and pod walls with the same method reported by Koinange et al. (1996). For mapping this candidate gene, the authors developed two RIL populations (BAT93 × Jalo EEP 558 and Midas × G12873). This candidate gene was mapped to chromosome 2, and although it was located close to the *St* locus, these two loci did not show cosegregation. In addition, the authors did not find any SNP within the sequence of *PvIND* associated with pod shattering. These results showed that *PvIND* does not play a role in pod shattering in common bean. As noted by Dong and Wang (2015), the role of *IND* in valve margin cell lignification might be specific to the Brassicaceae clade and other *AtIND* homologues in common bean should be investigated. Another study that implemented a candidate-gene approach was the one reported by Nanni et al. (2011) who sequenced in a set of wild and domesticated accessions of common bean, the homologue of *AtSHP1* in common bean, called *PvSHP1*. The authors found that this gene mapped to chromosome six and was not related to pod shattering.

As it can be seen, in spite of being a key trait for domestication of *Phaseolus* beans, almost nothing is known about its genetic control; therefore, to unravel the genetic basis of pod shattering in *Phaseolus* beans, more research is needed as noted by Dong and Wang (2015), especially in fiber cell differentiation and regulation of secondary cell-wall deposition given that fibers are mainly composed of sclerenchyma cells with well-developed secondary cell walls. To carry out studies in *Phaseolus* beans comparable to those reported in other legume crops, it will be very important to understand the differences in pod anatomy and functional mechanisms governing pod shattering among species and accordingly develop phenotyping methods to score this trait (the methods reported by Murgia et al. (2017) are very promising), taking into account specific tissues from which to take samples (in the case of expression analyses) and the time of sampling (pod developmental stage).

6 Genetic Architecture of Seed Weight

Very few studies in *Phaseolus* beans have addressed the identification of genes involved in changes in seed weight during domestication. In the study of Koinange et al. (1996) mentioned above, four QTLs were detected for seed weight that explained 57% of phenotypic variation: one QTL in LG 1, two QTLs in LG 7, and one QTL in LG 11 (Table 1). In LG 7, the nearest marker associated with one of the detected QTLs for seed weight was the Phaseolin (*Phs*, the major seed storage protein) locus.

Another study that analyzed QTLs associated with seed weight in a cross between wild and domesticated genotypes of common bean was the one of Guzmán-Maldonado et al. (2003). In that study, 120 $F_{2:3}$ lines derived from a cross between the cultivar Bayo Baranda (race Durango, Mesoamerican origin, 100-seed weight of 43.9 g) and a wild genotype (G22387, Mesoamerican origin, 100-seed weight of 4.7 g) were analyzed for 100-seed weight and other seed characters. This cross was therefore informative for detection of QTLs for the Mesoamerican domestication event. Five QTLs were significantly associated with seed mass that together explained about 42.3% of variation: one in LG1, one in LG2, one in LG3, and two in LG4 (Table 1).

An additional study that analyzed QTLs associated with seed weight in a cross between wild and domesticated genotypes was reported by Blair et al. (2006) in a $BC_2F_{3:5}$ population derived from a cross between a wild genotype (G24404, Andean origin, the donor parent, average 100-seed weight of 16 g) and an Andean cultivar (ICA Cerinza, the recurrent parent, average 100-seed weight of 53 g). This cross was therefore informative about QTLs associated with seed size in the Andean domestication event. A total of five QTLs associated with seed weight were consistently detected in the two locations where populations were grown (Darién and Popayán in Colombia). These QTLs were located in LG 2 (one QTL), LG 7 (one QTL), LG 8 (two QTLs), and LG 11 (one QTL) (Table 1). The QTL in LG7 was linked to the *Phs* locus, as was also reported by Koinange et al. (1996).

Other studies in common bean have investigated QTLs contributing to seed weight but from crosses between domesticated genotypes, not between wild and domesticated genotypes; therefore, QTLs detected in these studies may differ from those acting on the domestication syndrome. For example, Park et al. (2000) analyzed 63 RILs derived from a cross between two domesticated genotypes of Andean origin ('PC-50' and XAN-159) for seed characters including 100-seed weight (PC-50: 100-seed weight 35.2–45.1 g, XAN-159: 100-seed weight 17.8–24.7 g), seed length, and seed height. For seed weight, the authors found five QTLs in linkage groups 3, 4, 6, 7, and 8 that together explained about 44% of variation in this trait (Table 1). In this study, one of the seed weight QTLs detected in LG7 was linked to the *Phs* locus, a similar finding to that reported by Koinange et al. (1996).

Schmutz et al. (2014) published the first reference genome of common bean and in the context of this genome carried out a genome-wide association analysis (GWAS) to detect genes associated with the increase in seed size during domestication in Mesoamerica. They found 15 genes associated with improvement of Mesoamerican cultivars, including multiple members of the cytokinin synthesis whose orthologs in *Arabidopsis* are known to be involved in seed size. These 15 genes were found in chromosomes 1 (one gene), 3 (five genes), 4 (one gene), 6 (one gene), 7 (two genes), 8 (one gene), 10 (one gene), and 11 (three genes). The authors also carried out a GWAS to identify genes associated with domestication of common bean in Mesoamerica and were able to locate 75 candidate genes that included the 15 improvement genes described above. These 75 candidate genes were located in chromosomes 1 (three genes), 2 (one gene), 3 (ten genes), 4 (two genes), 6 (two genes), 7 (42 genes, 33 of which are in a sweep window 9.662–10.662 Mb), 8 (nine genes), 9 (three genes), 10 (two genes), and 11 (one gene).

It is interesting to note that while the earliest studies identified rather few genes for seed weight, the study of Schmutz et al. (2014) that implemented a genome approach was able to detect much more genes in almost all chromosomes (except in chromosome 5), which suggests a polygenic control for this trait. Due to its polygenic nature, in order to advance in the detection of genes that control seed size among the five domesticated species in the *Phaseolus* genus, a genomic perspective should be implemented in future studies.

7 Conclusions

We can conclude that the five domesticated *Phaseolus* species offer the opportunity to evaluate the genetic basis of the domestication syndrome in at least seven independent domestication events. Wild ancestors and areas of domestication for all these species have been identified which enable correct comparisons in future studies. In all these species, phenotypic convergence is observed in pod shattering and seed weight, as adaptations during domestication. The question that arises here is whether this phenotypic convergence may be explained by parallel or convergent evolution. The phenotypic evidence revised here for seed weight suggests that these species have responded in a similar way during domestication in the sense that domesticated forms show larger seeds than their wild ancestors, but there are evident differences in the magnitude of seed size increase and the degree of phenotypic overlap among wild and domestic forms. In this last aspect, only in *P. acutifolius* a clear pattern of disruptive selection with no overlapping phenotypic distributions among wild and domestic forms was observed; in the other four species, a rather phenotypic continuum is present. For pod shattering, evidence is very scarce. Evaluations of this trait among wild and domestic forms have been carried out in common bean and Lima bean only. This trait is expected to be under disruptive selection; however, the studies show that this trait has not been fixed among domesticated forms, and some landraces still show a high degree of pod shattering, sometimes overlapping wild forms. Very little is known about the genetic control of these two domestication traits in *Phaseolus* beans, and the few studies have been carried out only in common bean. In common bean, two loci (or only one with pleiotropic effects) have been detected in linkage group 2, but so far the genes underlying these loci have not been identified. A major limitation is the way this trait has been phenotyped, but recent advances have been reported. For seed weight, recent studies using a genomic perspective show a highly polygenic nature, with QTLs present in all chromosomes (except in chromosome 5). Future studies should involve more sophisticated approaches for the phenotyping of these traits, especially pod shattering, and genomic tools to detect loci involved in their control; only in this way we could answer the question of how the observed phenotypic convergence arose, either by parallel or convergent evolution.

Funding The present study was partially funded by Colciencias, Colombia, under contract number 009-2015 and project code 1101-658-42502.

Acknowledgements Thanks are due to the Genetic Resources Unit of CIAT, especially to Dr. Daniel Debouck and Dr. Peter Wenzl, for providing the Lima bean seed material used in this study and for all their support in the phenotypic characterization of pod shattering in Lima bean. Thanks are also due to Paola Hurtado for sharing unpublished pod shattering data in Lima bean.

References

Abbo S, Pinhasi van-Oss R, Gopher A, Saranga Y, Ofner I, Peleg Z (2014) Plant domestication versus crop evolution: a conceptual framework for cereals and grain legumes. Trends Plant Sci 19:351–360

Acosta-Gallegos JA, Kelly J, Gepts P (2007) Prebreeding in common bean and use of genetic diversity from wild germplasm. Crop Sci 47:S44–S59

Andueza-Noh RH, Martínez-Castillo J, Chacón-Sánchez MI (2015) Domestication of small-seeded Lima bean (*Phaseolus lunatus* L.) landraces in Mesoamerica: evidence from microsatellite markers. Genetica 143:657–669

Andueza-Noh RH, Serrano-Serrano ML, Sánchez MIC, del Pino IS, Camacho-Pérez L, Coello-Coello J, Cortes JM, Debouck DG, Martínez-Castillo J (2013) Multiple domestications of the Mesoamerican gene pool of Lima bean (*Phaseolus lunatus* L.): evidence from chloroplast DNA sequences. Genet Resour Crop Evol 60:1069–1086

Beebe S, Rengifo J, Gaitan E, Duque MC, Tohme J (2001) Diversity and origin of Andean landraces of common bean. Crop Sci 41:854–862

Beebe S, Skroch PW, Tohme J, Duque MC, Pedraza F, Nienhuis J (2000) Structure of genetic diversity among common bean landraces of Middle American origin based on correspondence analysis of RAPD. Crop Sci 40:264–273

Bitocchi E, Bellucci E, Giardini A, Rau D, Rodriguez M, Biagetti E, Santilocchi R, Spagnoletti Zeuli P, Gioia T, Logozzo G, Attene G, Nanni L, Papa R (2012) Molecular analysis of the parallel domestication of the common bean (*Phaseolus vulgaris*) in Mesoamerica and the Andes. New Phytol 197:300–313

Blair MW, Iriarte G, Beebe S (2006) QTL analysis of yield traits in an advanced backcross population derived from a cultivated Andean × wild common bean (*Phaseolus vulgaris* L.) cross. Theor Appl Genet 112:1149–1163

Blair MW, Pantoja W, Carmenza Muñoz L (2012) First use of microsatellite markers in a large collection of cultivated and wild accessions of tepary bean (*Phaseolus acutifolius* A. Gray). Theor Appl Genet 125:1137–1147

Chacón S. MI, Pickersgill B, Debouck DG (2005) Domestication patterns in common bean (*Phaseolus vulgaris* L.) and the origin of the Mesoamerican and Andean cultivated races. Theor Appl Genet 110:432–444

Chacón-Sánchez MI, Martínez-Castillo J (2017) Testing domestication scenarios of Lima bean (*Phaseolus lunatus* L.) in Mesoamerica: insights from genome-wide genetic markers. Front Plant Sci 8:1551

Debouck DG, Toro O, Paredes OM, Johnson WC, Gepts P (1993) Genetic diversity and ecological distribution of *Phaseolus vulgaris* (Fabaceae) in northwestern South America. Econ Bot 47:408–423

Dong Y, Wang Y-Z (2015) Seed shattering: from models to crops. Front Plant Sci 6:476

Dong Y, Yang X, Liu J, Wang B-H, Liu B-L, Wang Y-Z (2014) Pod shattering resistance associated with domestication is mediated by a NAC gene in soybean. Nat Commun 5:3352

Estornell LH, Agustí J, Merelo P, Talón M, Tadeo FR (2013) Elucidating mechanisms underlying organ abscission. Plant Sci 199–200:48–60

Ferrándiz C, Liljegren SJ, Yanofsky MF (2000) Negative regulation of the SHATTERPROOF genes by FRUITFULL during arabidopsis fruit development. Science 289:436–438

Freytag GF, Debouck DG (2002) Taxonomy, distribution, and ecology of the genus *Phaseolus* (Leguminosae–Papilionoideae) in North America, Mexico, and Central America. Botanical Research Institute, Fort Worth, TX

Funatsuki H, Suzuki M, Hirose A, Inaba H, Yamada T, Hajika M, Komatsu K, Katayama T, Sayama T, Ishimoto M, Fujino K (2014) Molecular basis of a shattering resistance boosting global dissemination of soybean. Proc Natl Acad Sci 111:17797

Gepts P, Bliss FA (1986) Phaseolin variability among wild and cultivated common bean (*Phaseolus vulgaris*) from Colombia. Econ Bot 40:469–478

Gepts P, Debouck D (1991) Origin, domestication, and evolution of the common bean (*Phaseolus vulgaris*, L.). In: van Schoonhoven A, Voysest O (eds) Common beans: research for crop improvement. United Kingdom, Commonwealth Agricultural Bureaux International, Wallingford, pp 7–53

Gioia T, Logozzo G, Kami J, Spagnoletti Zeuli P, Gepts P (2013) Identification and characterization of a homologue to the Arabidopsis INDEHISCENT gene in common bean. J Hered 104:273–286

Guerra-García A, Suárez-Atilano M, Mastretta-Yanes A, Delgado-Salinas A, Piñero D (2017) Domestication genomics of the open-pollinated scarlet runner bean (*Phaseolus coccineus L.*). Front Plant Sci 8:1891

Gujaria-Verma N, Ramsay L, Sharpe AG, Sanderson L-A, Debouck DG, Tar'an B, Bett KE (2016) Gene-based SNP discovery in tepary bean (*Phaseolus acutifolius*) and common bean (*P. vulgaris*) for diversity analysis and comparative mapping. BMC Genomics 17:239

Guzmán-Maldonado SH, Martínez O, Acosta-Gallegos JA, Guevara-Lara F, Paredes-López O (2003) Putative quantitative trait loci for physical and chemical components of common bean. Crop Sci 43:1029–1035

Harlan JR (1992) Crops & man, 2nd edn. American Society of Agronomy, Inc., Crop Science Society of America Inc., Madison, Wisconsin, USA

Koinange EMK, Singh SP, Gepts P (1996) Genetic control of the domestication syndrome in common bean. Crop Sci 36:1037–1045

Kwak M, Gepts P (2009) Structure of genetic diversity in the two major gene pools of common bean (*Phaseolus vulgaris* L., Fabaceae). Theor Appl Genet 118:979–992

Kwak M, Kami JA, Gepts P (2009) The putative Mesoamerican domestication center of *Phaseolus vulgaris* is located in the Lerma-Santiago Basin of Mexico. Crop Sci 49:554–563

Liljegren SJ, Ditta GS, Eshed Y, Savidge B, Bowman JL, Yanofsky MF (2000) SHATTERPROOF MADS-box genes control seed dispersal in Arabidopsis. Nature 404:766–770

Liljegren SJ, Roeder AHK, Kempin SA, Gremski K, Østergaard L, Guimil S, Reyes DK, Yanofsky MF (2004) Control of fruit patterning in *Arabidopsis* by INDEHISCENT. Cell 116:843–853

Meyer RS, Purugganan MD (2013) Evolution of crop species: genetics of domestication and diversification. Nat Rev Genet 14:840–852

Mitsuda N, Ohme-Takagi M (2008) NAC transcription factors NST1 and NST3 regulate pod shattering in a partially redundant manner by promoting secondary wall formation after the establishment of tissue identity. Plant J 56:768–778

Motta-Aldana JR, Serrano-Serrano ML, Hernández-Torres J, Castillo-Villamizar G, Debouck DG (2010) Multiple origins of Lima bean landraces in the Americas: evidence from chloroplast and nuclear DNA polymorphisms. Crop Sci 50:1773–1787

Murgia ML, Attene G, Rodriguez M, Bitocchi E, Bellucci E, Fois D, Nanni L, Gioia T, Albani DM, Papa R, Rau D (2017) A comprehensive phenotypic investigation of the "pod-shattering syndrome" in common bean. Front Plant Scie 8:251

Nanni L, Bitocchi E, Bellucci E, Rossi M, Rau D, Attene G, Gepts P, Papa R (2011) Nucleotide diversity of a genomic sequence similar to SHATTERPROOF (PvSHP1) in domesticated and wild common bean (*Phaseolus vulgaris* L.). Theor Appl Genet 123:1341–1357

Ogawa M, Kay P, Wilson S, Swain SM (2009) ARABIDOPSIS DEHISCENCE ZONE POLY-GALACTURONASE1 (ADPG1), ADPG2, and QUARTET2 Are Polygalacturonases required for cell separation during reproductive development in *Arabidopsis*. Plant Cell 21:216

Park SO, Coyne DP, Jung G, Skroch PW, Arnaud-Santana E, Steadman JR, Ariyarathne HM, Nienhuis J (2000) Mapping of QTL for seed size and shape traits in common bean. J Am Soc Hortic Sci 125:466–475

Pickersgill B (2007) Domestication of plants in the Americas: insights from Mendelian and molecular genetics. Ann Bot 100:925–940

Rajani S, Sundaresan V (2001) The *Arabidopsis* myc/bHLH gene ALCATRAZ enables cell separation in fruit dehiscence. Curr Biol 11:1914–1922

Rodriguez M, Rau D, Angioi SA, Bellucci E, Bitocchi E, Nanni L, Knüpffer H, Negri V, Papa R, Attene G (2013) European *Phaseolus coccineus* L. landraces: population structure and adaptation, as revealed by cpSSRs and phenotypic analyses. PLoS ONE 8:e57337

Schinkel C, Gepts P (1988) Phaseolin Diversity in the Tepary Bean, *Phaseolus acutifolius* A. Gray. Plant Breeding 101:292–301

Schmit V, Debouck DG (1991) Observations on the origin of *Phaseolus polyanthus* Greenman. Econ Bot 45:345–364

Schmutz J, McClean PE, Mamidi S, Wu GA, Cannon SB, Grimwood J, Jenkins J, Shu S, Song Q, Chavarro C, Torres-Torres M, Geffroy V, Moghaddam SM, Gao D, Abernathy B, Barry K, Blair M, Brick MA, Chovatia M, Gepts P, Goodstein DM, Gonzales M, Hellsten U, Hyten DL, Jia G, Kelly JD, Kudrna D, Lee R, Richard MMS, Miklas PN, Osorno JM, Rodrigues J, Thareau V, Urrea CA, Wang M, Yu Y, Zhang M, Wing RA, Cregan PB, Rokhsar DS, Jackson SA (2014) A reference genome for common bean and genome-wide analysis of dual domestications. Nat Genet 46:707–713

Serrano-Serrano ML, Hernández-Torres J, Castillo-Villamizar G, Debouck DG, Chacón Sánchez MI (2010) Gene pools in wild Lima bean (*Phaseolus lunatus* L.) from the Americas: evidences for an Andean origin and past migrations. Mol Phylogenet Evol 54:76–87

Serrano-Serrano ML, Andueza-Noh RH, Martínez-Castillo J, Debouck DG, Chacón SMI (2012) Evolution and domestication of Lima bean in Mexico: evidence from ribosomal DNA. Crop Sci 52:1698–1712

Singh J, Gustin J, Baier J, Settles AM, Vallejos E (2014) Mapping QTLs for seed weight, starch and protein content in common bean (*Phaseolus vulgaris* L.). In: International plant and animal genome conference XXII San Diego, USA

Singh SP, Gepts P, Debouck DG (1991) Races of common bean (*Phaseolus vulgaris*, Fabaceae). Econ Bot 45:379–396

Spataro G, Tiranti B, Arcaleni P, Bellucci E, Attene G, Papa R, Spagnoletti Zeuli P, Negri V (2011) Genetic diversity and structure of a worldwide collection of *Phaseolus coccineus* L. Theor Appl Genet 122:1281–1291

Zohary D, Hopf M (2000) Domestication of plants in the Old World, 3rd edn. Oxford University Press, New York, 316 pp

Tracing the Evolutionary Origin of the Gut–Brain Axis

Thomas C. G. Bosch

Abstract Colonization of body epithelial surfaces with a rather specific microbial community is a fundamental feature of all animals. Recent studies suggest that in mammals the enteric microbiota has a bidirectional communication with the nervous system. The complexity of these interactions is referred to as "gut–brain axis". Highlighting the significance of these interactions are studies in mice which show that the intestinal microbiota can directly affect complex behaviour. The origin and ancestral function of these interactions are not well understood. Here, I review findings that neurons in the early emerging metazoan Hydra secrete neuropeptides which shape the microbiome on the body surface. I also discuss recent observations which indicate that symbiotic bacteria modulate spontaneous body contractions in Hydra. Germ-free polyps show strongly reduced and less regular spontaneous contraction frequencies. The effects on contraction frequency were partially restored by reconstituting the natural microbiota. These findings strongly suggest that the influence of bacteria on neuronal activity is the outcome of an evolutionary ancient interaction between bacteria and metazoans, opening a window into investigating the basic mechanisms of, for example, neurological disorders in vertebrates.

1 Introduction—The Holobiont Imperative

There are convincing arguments that bacteria have been present since early Precambrian time, about 3.5 billion years ago (Fig. 1). This long history has resulted in ecological interactions among microbes that are broadly diverse and flexible, features enabled by their rapid generation times and large population sizes, in addition to their proclivity for horizontal gene transfer. Since animals diverged from their

T. C. G. Bosch (✉)
Zoological Institute, Christian-Albrechts-University Kiel, Am Botanischen Garten 1-9,
24118 Kiel, Germany
e-mail: tbosch@zoologie.uni-kiel.de

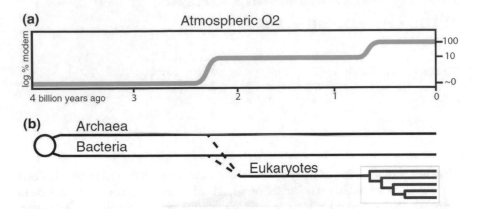

Fig. 1 Evolution of plants and animals in Earth's history has occurred as a patina upon a microbe-dominated landscape. **a** Upper atmospheric oxygen concentration, as a per cent of current levels, plotted against geological time. **b** Phylogenetic history of life on Earth, scaled to match the oxygen timeline. Note that the origin of the eukaryotes and the subsequent diversification of animals both correspond to periods of increasing atmospheric oxygen. From Bosch and Miller (2016)

protistan ancestor some 3 billion years after bacterial life originated and as much as 1 billion years after the first appearance of eukaryotic cells, relationships of animals with bacteria were likely already operating when animals first appeared near the end of the Proterozoic aeon. Animal evolution, therefore, is intimately linked to the presence of microbes such as bacteria, archaea, and viruses (Bosch and Miller 2016).

Due to the emergence and rapid technological advances in culture-independent techniques to identify and characterize microbes, particularly genomic approaches, we learned in the last two decades that all animals, from Hydra to human, are stably associated with bacteria, but also including viruses, archaea, fungi, and protists. Eukaryotic evolution has probably never seen a period without the presence of microbes. Appreciation grows for the fact that animals live in a bacterial world (McFall-Ngai et al. 2013), and no longer can the animal be viewed as separate from the microbes it requires to subsist, reproduce, and evolve over time. Consequently, symbiosis appears as a major component of eukaryotic fitness and evolution. Based on these new findings, all multicellular organisms must be considered an association of the macroscopic host in synergistic interdependence with bacteria, archaea, fungi, and numerous other microbial and eukaryotic species. We refer to these associations that can be analysed, measured, and sequenced, as "holobionts" or "metaorganisms" (Fig. 2).

Box 1: Terminology Metaorganism (Taken from Deines et al. 2017)

Holobiont: Is an eukaryotic host with all its associated microbial partners. This multispecies assemblage includes viruses, phages, eubacteria, archaea, fungi, and protozoa.

Hologenome: Genetic information encoded in the eukaryotic host and all of its associated partners. This collective genome forms the theoretical genetic repertoire of a holobiont.

Metaorganism: Includes the function of a holobiont in a given environment. The function of a holobiont depends on (I) presence and composition of the associated partners, framing the genetic potential of the holobiont, the hologenome; (II) the activity, abundance, and the transcriptional active part of the genome of every single partner of the holobiont; (III) this subsequently results in interactions between host–microbes and microbe–microbe which finally must be retained at homeostasis in order to maintain a stable holobiont.

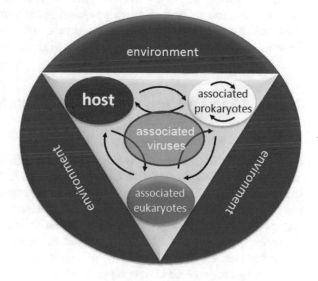

Fig. 2 Multicellular organisms are holobionts or metaorganisms, a complex community of many species which have been, and are being, evolved

Interactions between the members of the holobiont, i.e. bacteria, viruses, eukaryotic symbionts, and host cells, have probably been critical to enabling the key transitions in animal evolution (Bosch and Miller 2016; Theis et al. 2016). However, the reciprocal is also true—animals have dramatically transformed the physical environment that is available for bacterial colonization. Along the way, some animals have also formed close relationships with other eukaryotes such as algal symbionts, but these macro-symbiotic bonds have been forged in the context of pre-existing host–bacterial interactions. The increasing realization that animals cannot be considered in isolation but only as a partnership of animals and symbionts has lead to three important realizations. First, it is becoming increasingly clear that to understand the evolution and biology of a given species, we cannot study the species in isolation. Second, the health of animals, including humans, appears to be fundamental multiorganismal. Any disturbance within the complex community has drastic consequences for the well-being of the members (see, e.g., Rosenberg et al. 2007). And third, the holobiont may be an important unit of evolutionary selection, a selection of "teams" containing many genomes and species (Theis et al. 2016; Catania et al. 2017).

The metaorganism concept considers the dynamic communities of bacteria on epithelial surfaces as an integral part of the functionality of the respective organism itself. The holobiont, therefore, appears as dynamic system being characterized by functional redundancy and fast adaptations to altered environmental conditions. It is this modularity and interoperability of the components of the holobiont which allows rapid adaptation to changing environmental conditions by altering the associated microbiota. This view is conceptualized by Rosenberg's "hologenome theory of evolution" (Zilber-Rosenberg and Rosenberg 2008). Depending on the variety of different niches provided by the host, which can change with developmental stage, diet or other environmental factors, a more or less diverse microbial community can be established within a given host species. The dynamic relationship between symbiotic microorganisms and environmental conditions results in the selection of the most advantageous holobiont (Catania et al. 2017). The strategies implied in these interactions are, at their core, strategies of resilience.

2 Developmental Symbiosis and the Evolution of New Organs

More than just the product of co-evolution, the holobiont is a performance of co-development (Theis et al. 2016; Catania et al. 2017). Functional studies clearly demonstrate that microorganisms play a major role in the animal host's development, physiological functions, and adaptation. For instance, many marine invertebrates require environmental bacterial clues for the settlement and further development of the larvae (Shikuma et al. 2016). In the squid Euprymna scolopes, the development of the light organ, which is involved in camouflage and anti-predator defence, is triggered by the presence of lipopolysaccharide and peptidoglycan fragments of the symbiont Vibrio fisheri (Koropatnick et al. 2014). The light organ fails to mature in squids raised without V. fischeri implicating a specific association between V. fischeri and the squid (McFall-Ngai 2014, 2015). In insects, the gut microbiota is crucial for immune and epithelial cell homeostasis of the gut, as well as food supplier (Engel and Moran 2013). Similarly in vertebrates, the gut bacteria provide the host with metabolic products such as short-chain fatty acids, which promote intestinal gluconeogenesis and in turn improve glucose metabolism (De Vadder et al. 2014). In humans and mice, the gut microbiota affect fat uptake, leading to weight loss or gain depending on the bacterial composition harboured by the host (Ley et al. 2006; Ridaura et al. 2013; Sonnenburg and Bäckhed 2016). On the other hand, in mice the presence of specific bacteria is necessary for tuning up the host immune system, promoting lymphoid organogenesis, activating the T regulatory cells and balancing pro- and anti-inflammatory cytokines, preventing inflammatory diseases (Mazmanian et al. 2005; Round and Mazmanian 2009). In addition, in zebrafish, the gut microbiota is required for the development of the digestive tract and the differentiation of gut epithelia (Bates et al. 2006). Similar results have been reported in mice, where leaky gut and abnormal mucosa are observed in the absence of gut bacteria (Smith et al. 2007). In summary, these studies show that associated microbiota, in particular bacteria, are involved in the host's normal metabolism, development and immune system maturation. Striking similarities in response to the absence of microbes between fish (zebrafish) and mammalian (mouse) model systems suggest that animals possess a conserved programme of interactions with the symbiotic microbes with which they have coevolved. Brain development in mouse also appears to be modulated by the gut microbiota, and evidence exists for a role of the microbiome in mouse behaviour (see below) (Sampson et al. 2016; Sharon et al. 2016). Taken together, in numerous animals, symbiotic interactions are essential to development; and it seems possible that all animals form some of their organs through symbiosis. Just as we shape the microbiome, it seems to shape us. While these considerations suggest that microbes have contributed to host development and evolution, an equally fascinating question is: How does the invention of a novel organ such as the nervous system affect the interactions within a given holobiont?

3 Thoughts About the Evolution of the Nervous System

The nervous system plays three major roles in animals: control of behaviour, physiology, and development (Jékely et al. 2015a, b). This implies sensing inputs from the environment as well as coordinating internal activities of the organism. The ability to rapidly respond to environmental cues is a feature well developed in members of the phyla Cnidaria and Ctenophora (Bosch et al. 2014), pre-bilaterian animals that diverged close to the base of the metazoan radiation (Fig. 3). Comparisons across the animal kingdom imply that the neurogenic machinery was largely in place in the common bilaterian ancestor and that some of its foundations were already present in the cnidarian/bilaterian ancestor.

The nervous systems of cnidarians are structurally simple (Fig. 4). Textbooks usually describe it as diffuse nerve net, and it is assumed to reflect the prototypes of bilaterian nervous systems. Although it is certainly true that cnidarian nervous systems are often largely diffuse, in many cases dense masses of nerve cells or ganglia are also present. Despite its structural simplicity, the nervous system in cnidarians offers a rich chemical complexity (for review, see Bosch et al. 2017). The ectodermal and endodermal net is constructed by two types of neurons: peripheral sensory cells and sensory-motor interneurons. Muscle contraction is achieved via

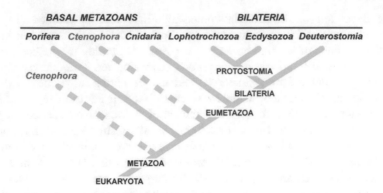

Fig. 3 Relationships between the five metazoan clades. The phylogenetic position of Ctenophores is still controversial

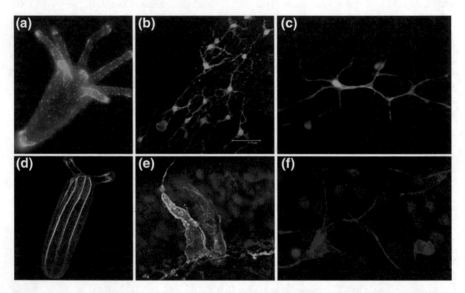

Fig. 4 Cnidarian nerve net. Example of transgenic neurons in cnidarians. **a–c** *Hydra* nerve cells labelled by GFP. **d–f** *Nematostella* ectodermal sensory cells labelled by a NvSoxB(2)::EGFP transgene, the exterior is to the top. In red is the signal from the stable NvSoxB(2)::mOrange line, into which the soxB2::EGFP plasmid was injected, so the EGFP is mosaic. *Picture credit* Alexander Klimovich

chemical synapses using large dense-cored vesicles containing a large variety of neuropeptides. Neurotransmitters include catecholamines, serotonin, acetylcholine, glutamate, and γ-aminobutyric acid. In addition to peptide-gated ion channels, two putative acetylcholine receptors (mAchRs) have been identified as well as numerous genes coding for nicotinic acetylcholine receptor (nAchR) subunits.

4   Of Neurological Disorders and Animals as Complex Systems: Why Bacteria Matter

As pointed out first by Bach (2002), epidemiologic data provide strong evidence of a steady rise in the incidence of allergic and chronic diseases in developed countries over the last 50 years. The incidence of complex neurological disorders, such as autism, apparently began to increase in the 1950s and continues to do so today (Doenyas 2018). When the disorder was first described in 1943, it was uncommon. Today, about one in eighty-eight children has autism or autism spectrum disorder (ASD). Although overdiagnosis certainly contributes to the rise in cases, it is not enough to explain the enormous increase. Multiple theories try to explain the increase in autism cases, including toxins in water, food, and air; exposures to chemicals and pesticides during pregnancy. But no one knows for sure. Correlation does not equal causation. Evidence, however, is mounting that intestinal microbes exacerbate or perhaps even cause some of autism's symptoms (Hsiao et al. 2013; Strati et al. 2017; Doenyas 2018). Recent observations in animal models show that autism-like syndromes are not developing in germ-free animals and that syndromes can be cured by addition of certain bacterial compounds (Strati et al. 2017; Doenyas 2018). Again, it is not clear whether these microbial differences drive the development of the condition or are instead a consequence of it. However, and strongly supporting the former idea, when researchers at the California Institute of Technology incited autism-like symptoms in mice using an established paradigm that involved infecting their mothers with a virus-like molecule during pregnancy, they found that after birth, the mice had altered gut bacteria compared with healthy mice (Hsiao et al. 2013). By treating the sick rodents with Bacteroides fragilis, the researchers were able to attenuate some, but not all, of their behavioural symptoms. The treated mice had less anxious and stereotyped behaviours and became more vocally communicative. This elegant study teaches us at least two important lessons. First, when searching for general concepts, simple animal models such as germ-free mice may help to study mechanisms and identify key players or mediators even when they only partly reflect the human situation. And second, the composition of the gut microbes and their metabolic activity seems to be an important factor to keep in mind when attempting to understand why the incidence of autism is increasing so dramatically in the last few years.

5   The Gut–Brain Axis

Recent studies provide evidence that the natural gut microbiota is required for the normal development of the enteric nervous system, including enteric neurogenesis and gliogenesis (Collins et al. 2014; Kabouridis et al. 2015). This interaction between the gut microbiota and the nervous system is known as the "microbiota–gut–brain axis" (Grenham et al. 2011). Interestingly, gut bacteria seem to regulate host serotonin levels (Yano et al. 2015; Reigstad et al. 2015; Hata et al. 2017) and also tryptophan signalling (Krishnan et al. 2018). This suggests that microbiota can act as an endocrine organ. The absence of gut microbiota or any disturbance in the natural microbiota composition can cause gut motility problems (Abrams and Bishop 1967; Husebye et al. 2001; Quigley 2011). In particular, chronic diseases such as irritable bowel syndrome that are associated with both gut dysmotility and dysbiosis (Kostic et al. 2014) have raised the interest to understand the role of commensal microbiota in the host's health and disease. This increasing interest has also revealed that the influence of commensal gut bacteria on animal physiology extends beyond the gut to other systems. For example, one of the most surprising findings was the recognition that in vertebrates, the gut microbiota is involved in the normal development of the central nervous system.

The presence of gut microbiota appears necessary for the establishment of the blood–brain barrier, adult hippocampal neurogenesis, and proper microglia function (Braniste et al. 2014; Erny et al. 2015; Möhle et al. 2016). Moreover, cognitive, emotional, and behavioural processes may also at least in part depend on the gut microbiota (Cryan and Dinan 2012; Forsythe and Kunze 2013; Sharon et al. 2016). For instance, compared with animals harbouring undisturbed gut microbiota, germ-free mice show anxiolytic behaviour, exaggerated stress response, impaired memory, learning and sociability, and altered brain function. These effects are associated with changes in brain chemistry and differential gene expression such as abnormal neurotransmitter signalling and higher levels of stress hormones as corticosterone and lower levels of brain-derived neurotrophic factor (BDNF), glutamate, dopamine, and serotonin receptors (Diaz Heijtz et al. 2011; Gareau et al. 2011; Neufeld et al. 2011; Sudo et al. 2004). Similar results are obtained when the gut microbiota is disrupted by antibiotic treatment (Möhle et al. 2016; O'Mahony et al. 2014). In addition, sociability in rodents, as shown by reduced time spent with conspecifics and increased engagement in repetitive behaviours like self-grooming, is also affected in the absence of gut microbiota (Desbonnet et al. 2014; Hsiao et al. 2013). Notably, restoration of native gut microbiota or administration of probiotics ameliorates and might even reverse the behavioural abnormalities observed in these animals (Bravo et al. 2011; Hsiao et al. 2013; Liu et al. 2015; Sudo et al. 2004).

The microbiota–gut–brain axis concept is gaining acceptance as it opens new ways to interpret not only gastrointestinal pathologies but also neurological disorders. In fact, emerging evidence suggests that diseases such as Alzheimer's, Parkinson's, and multiple sclerosis are associated with gut microbiota dysbiosis (Cekanaviciute et al. 2017; Harach et al. 2017; Jangi et al. 2016; Sampson et al. 2016). Therefore, it is of clinical importance to understand how the commensal microbiota affects the host brain function. Although endocrine, immune, nervous, and metabolic signalling as well as epigenetic regulation are involved in establishing this communication (Cryan and Dinan 2012; Stilling et al. 2014), the basic molecular and cellular mechanisms have not been discovered yet. In part, this might be due to the fact that mainly vertebrate models, such as rodents, have been employed so far to study the microbiota–gut–brain axis. The complexity of the candidate pathways involved and the still limited knowledge about vertebrate microbiota function and composition (Kowarsky et al. 2017) has proven to be an obstacle to achieve a complete understanding of this communication. For this reason, disentangling the roots of the microbiota–gut–brain axis requires the use of simple organism that allows manipulations of both the host and the microbiota (Chen et al. 2013).

6 The *Hydra* Holobiont—A Model Organism to Study the Microbiota–Gut–Brain Axis

The freshwater polyp Hydra belongs to the phylum Cnidaria, the sister group of Bilateria (Fig. 5a), which is considered one the first groups where the nervous system first appeared (Jékely et al. 2015a, b; Pisani et al. 2015). Hydra's body plan is very simple: two epithelial layers, the ecto- and the endoderm, form the body column, which is a tube with an apical head and a basal foot. The body column lumen is generally considered a gastric cavity because it has digestive functions and therefore may represent an evolutionary ancient intestine (Schröder and Bosch 2016). *Hydra* is a classical model organism in developmental biology which was introduced by Abraham Trembley as early as 1744. Because of its simple body plan, *Hydra* served for many years as model in developmental biology to approach basic mechanisms underlying de novo pattern formation, regeneration, and cell differentiation. Novel computational tools and genomic resources have brought a molecular perspective on the *Hydra* holobiont. The genome sequence of *Hydra magnipapillata* (Chapman et al. 2010) revealed an unexpectedly high genetic complexity and a conserved genome structure between *Hydra* and other animals including humans. In spite of the fact, however, that *Hydra* belongs to one of the phylogenetically oldest eumetazoan lineages, this organism certainly is not a "living fossil"; its genome contains a rather unique combination of ancestral, novel, and "borrowed" (e.g. via horizontal gene transfer) genes, similar to the genomes of other animals. In the sequencing process, we accidentally also sequenced the genome of a bacterial species in the *Curvibacter* genus that is stably associated with *Hydra*. For analytical purposes, an important technical breakthrough was the development of a transgenic procedure allowing efficient generation of transgenic *Hydra* lines by embryo microinjection. This not only allows functional analysis of genes controlling development and immune reactions but also in vivo tracing of cell behaviour (Wittlieb et al. 2006).

Although the Hydra nerve net possesses no centralized structures, nerve cells are concentrated in the head and foot region (Fig. 5b–d). The nervous system is made up of about 6.000 nerve cells belonging to two morphological types—sensory and ganglion neurons (Bode et al. 1973; Koizumi 2002). The neurons constantly develop from the multipotent interstitial stem cells and are assembled into two apparently autonomous nerve nets intercalated into the ectodermal and endodermal epithelial layers (Bode 1992; Koizumi et al. 2004). In spite of this morphological simplicity, *Hydra* demonstrates a surprisingly rich behavioural repertoire: spontaneous periodic contractions and extension of the body and tentacles, contractions in response to mechanical stimuli and light, complex feeding behaviour, and somersaulting loco-

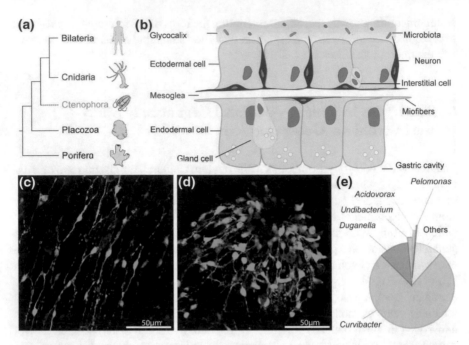

Fig. 5 Hydra as a model to study host–symbiont interactions. **a** Simplified view of relationships between the five metazoan clades. Hydra belongs to the phylum Cnidaria, the sister group of Bilateria. **b** The body wall of Hydra is composed of three cell lineages: the ectodermal and the endodermal epithelia that are separated by an extracellular matrix (mesoglea), and the lineage of interstitial cells that differentiate into neurons, gland cells, cnidocytes, and gametes. The outer surface of the ectoderm is covered by a glycocalyx, the habitat of symbiotic bacteria. The endoderm lining the gastric cavity is free of glycocalyx and stable microbiota. **c, d** Hydra has a simple nerve net. Here, ganglion neurons (**c**) in the body column and sensory neurons in the hypostome (**d**) are revealed by expression of RFP (green pseudo-colour) under the actin promoter. **e** 90% of H. vulgaris AEP microbiota is composed by five bacterial strains, with less abundant strains making up the remainder. *Picture credit* Andrea Murillo-Rincon

motion (Wagner 1905; Loomis 1955; Passano and McCullough 1964, 1965). All these behaviours are under control of the nervous system, as evidenced by experiments on neuron-ablated polyps (Campbell 1976; Marcum and Campbell 1978). In addition to this conventional role in coordinating motor activities, the nervous system in *Hydra* appears to be orchestrating other organism's functions, such as development, tissue homeostasis, and immune function. However, the mechanisms behind these "non-conventional functions" beyond simple sensory and motor coordination remain largely unclear. The combination of diverse in vitro techniques, such as in situ hybridization (Hansen et al. 2000), immunofluorescence (Koizumi et al. 2004) and electrophysiological measurements (Gonzales et al. 2017; Vitale et al. 2017), and recently developed in vivo imaging (Dupre and Yuste 2017) and behavioural recordings (Murillo-Rincon et al. 2017) promises deep insights into the molecular architecture and function of the nervous system in *Hydra*. This has renewed the interest in cnidarian neuroscience and the use of cnidarians as model to study the origin and function of the nervous system (Bosch et al. 2017).

As any other holobiont, Hydra harbours species-specific microbiota (Fraune and Bosch 2007; Franzenburg et al. 2013a, b), mainly found in the glycocalyx, a mucus layer covering the outside surface of the ectoderm (Fig. 5b) (Fraune et al. 2015; Schröder and Bosch 2016). Compared to vertebrates, Hydra's microbiota is composed by relatively few species most of which can be cultured (Fig. 5e) and thus amenable for experimental manipulation. In fact, it is possible, for example, to create germ-free Hydras, devoid of symbiotic bacteria, and to conventionalize them animals by recolonizing them with native microbiota (Fraune and Bosch 2007; Fraune et al. 2015). The presence of symbiotic bacteria is critical for tissue homeostasis and health of the polyp. In the absence of symbiotic bacteria, Hydra polyps are more susceptible to fungal infection with lethal consequences (Fraune et al. 2015). The abundance of Hydra's symbiotic bacteria can be altered by both epithelial derived antimicrobial peptides and neuropeptides with antimicrobial activity secreted by Hydra (Augustin et al. 2017). In summary, a simple body plan, the presence of an anatomically simple but still complex enough nerve net that coordinates several behaviours and physiological functions, together with the presence of stable associated microbiota, supports Hydra as an attractive model to study the causal effect of symbiotic bacteria and the host neuronal activity.

7 Hydra Neurons Shape the Microbiota

Early observations revealed that eliminating nerve cells dramatically changes the composition of the microbial community on *Hydra* in vivo (Fig. 6) (Fraune et al. 2009a). Animals lacking neurons demonstrated a tenfold reduced abundance of β-Proteobacteria accompanied by a tenfold increased abundance of Bacteroidetes. Interestingly, the overall density of bacteria on *Hydra* remained unaffected. Although these findings provided evidence for the significance of nerve cells in host–microbiome interactions, it remained unclear, whether the observed changes were due to the loss of some neuron-derived antibacterial factors, or due to altered epithelial immune response unleashed from the nerve cell control. A recent study by Augustin et al. provided evidence supporting the first scenario (Augustin et al. 2017) and showed some sensory and ganglion neurons express a cationic neuropeptide called NDA-1, secrete it into the mucus layer, and regulate the spatial distribution of the main colonizer, the Gram-negative bacterium *Curvibacter*, along *Hydra* body. The density of *Curvibacter* colonization is relatively low in the foot and tentacles of *Hydra*, where NDA-1 is strongly expressed, compared to the body column. Additionally, NDA-1 is highly potent against Gram-positive bacteria. Strikingly, other neuropeptides, such as *Hydra*-specific Hym-357 and Hym-370 and a member of the highly conserved RFamide family, all previously characterized as neuromodulators eliciting motor activity, turned out to be also potent against Gram-positive bacteria (Augustin et al. 2017). Taken together, these findings indicate that distinct nerve cells

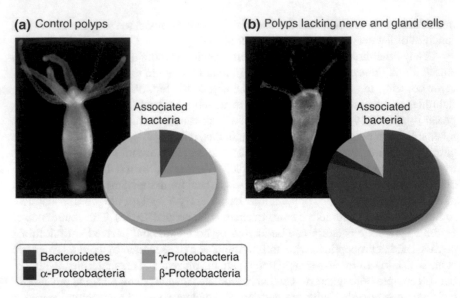

(a) Control polyps

(b) Polyps lacking nerve and gland cells

Associated bacteria

Associated bacteria

Bacteroidetes   γ-Proteobacteria
α-Proteobacteria   β-Proteobacteria

Fig. 6 Loss of nerve cells affects the associated bacterial community. **a** Control polyp with its associated bacterial community. **b** Polyp lacking nerves and gland cells with its associated bacterial community. From Fraune et al. (2009b)

contribute to the composition and spatial structure of *Hydra's* microbial community by expressing a variety of neuropeptides with distinct antimicrobial activities.

8 Microbes Affect the Nervous System and Behaviour in Hydra

The communication between the microbiota and the nerve cells in *Hydra* appears to be bidirectional. Recently, by focusing on Hydra's spontaneous body contractions, we showed that symbiotic bacteria modulate the nervous system activity and behaviour in *Hydra* (Murillo-Rincon et al. 2017). As in other hydrozoan polyps, Hydra's spontaneous contractions are shrinkages of the body column that occur periodically in the absence of any exogenous stimuli (Stokes and Rushforth 1979). A full body column contraction is also called a contraction burst because it reduces the polyp to a tight ball in a series of stepwise contractions (Fig. 7). Although the cell population(s) responsible for generating the spontaneous contractions have not been characterized yet, the evidence suggests that they are neurons, as the spontaneous contractions are abolished in nerve-free Hydras (Campbell 1976) and drastically reduced when gap junction communication among neurons is inhibited (Takaku et al. 2014). Moreover, contraction bursts are always preceded by periodic endogenous electrical activity generated in the head and the foot region, suggesting the existence of underlying

Fig. 7 Sequence of pictures demonstrating the spontaneous contraction of the body column, Hydra's most common behaviour. *Picture credit* Andrea Murillo-Rincon

pacemaker activity (Kass-Simon 1972; Passano and McCullough 1963, 1964, 1965). Early behavioural and electrophysiological studies on Hydra showed that pacemaker activity is located in the hypostome and foot region and suggested that the pacemakers were neurons (Kass-Simon 1972; Passano and McCullough 1963, 1964, 1965). Recently, this idea was confirmed by functional alteration of neurons in the foot (Takaku et al. 2014) as well as by calcium imaging of neuronal activity (Dupre and Yuste 2017). Using genetically engineered Hydra, the later study identified separate non-overlapping neuronal networks, which are associated with specific behaviours, including the spontaneous contractions (Dupre and Yuste 2017).

Interestingly, germ-free animals display strongly reduced and less regular spontaneous body contraction frequencies. Importantly, the effects can be partially restored by reconstituting the native microbiota in its natural species composition and proportions. These findings strongly suggest that not the presence of microbiota per se, but its precise species structure modulates the behaviour of *Hydra*. Further, we showed that a soluble molecule produced by bacteria may be involved in the contraction frequency modulation. Since disturbances in the microbiota primarily affect the regularity of contractions, we suggested that bacterial products may target the pacemaker neurons, responsible for timing of contractions in *Hydra* (Murillo-Rincon et al. 2017). Since the absence of microbiota affects the contraction frequency and contraction regularity but not the amplitude of the contractions, we propose that bacteria influence the pacemaker activity that is responsible for timing of the contractions.

Hydra spontaneous contractile activity resembles one kind of peristaltic movement of the vertebrate gastrointestinal tract, the fasting intestinal motor activity, also called the migrating motor complex (MMC). It rises from endogenous electrical pacemaker activity and is abolished by food ingestion (Deloose et al. 2012). Interestingly, there is a strong correlation between the MMC and the microbial assemblage in the gut of humans and rodents. However, it is still unclear whether dysmotility is cause or consequence of an altered microbiota composition (Hadizadeh et al. 2017; Vandeputte et al. 2016; Vantrappen et al. 1977).

In our recent study, we were able to evaluate the relative effect of less abundant bacteria as well as the role of potential bacterial interactions in the spontaneous contraction pattern. Our results provide evidence for the causal role of bacteria in modulating spontaneous contractile behaviour, which is the basis for gut motility and the according to these results, the outcome of an ancient interaction between bacteria and early emerging animals.

9   Where to from Here? Understanding the Gut–Brain Axis Requires Brainless Models

The time has come to replace the purely reductionist "eyes-down" molecular perspective with a new and genuinely holistic, eyes-up, view of the living world, one whose primary focus is on evolution, emergence, and biology's innate complexity. (Carl R. Woese 2004)

A growing body of evidence in animals supports the concept that the gut microbiota influences cognitive processes, emotional states, and behaviour (Cryan and Dinan 2012; Forsythe and Kunze 2013; Sharon et al. 2016). Changes in the gut microbiota or intestinal exposure to specific bacteria can modulate the peripheral nervous system in animals, as evidenced, for instance, by gut motility disturbance in germ-free mice (Abrams and Bishop 1967; Thompson and Trexler 1971; Collins et al. 2012). More strikingly, the intestinal microbes extend their effects onto the central nervous systems (CNSs), resulting in altered brain functioning and suggesting the existence of a microbiota gut–brain axis (Bienenstock et al. 2015). There is strong evidence from animal studies that gut bacteria influence brain chemistry and development and that the enteric nervous system, including the sensory vagus nerve, appears to be able to differentiate between non-pathogenic and potentially pathogenic bacteria and may play a critical role in mediating the effects of gut microorganisms on behaviour (Collins et al. 2012, 2014; Cryan and Dinan 2012; Dinan et al. 2013; Kabouridis et al. 2015; Erny et al. 2015). Because the nervous system has constant bidirectional communication with the immune system, the effects of bacteria on the nervous system cannot be disassociated from effects on the immune system. This type of crosstalk occurs regularly and can have profound neurological and immunological effects. However, the exact molecules responsible for host–microbe communication remain largely unknown.

Simple animal models such as *Hydra* therefore may help to model and understand the basic principles governing the microbiome-host crosstalk mediated by the neurons—even when they only partly reflect the human situation. The unique experimental accessibility of *Hydra* provides a way to identify molecular mechanisms and to prove causalities. Combining this mechanistic understanding with a body of knowledge on development, stem cell biology, and neurophysiology promises a truly holistic understanding of the *Hydra* holobiont. This may be of particular importance in the light of *Hydra's* non-senescence and unlimited lifespan (Mortzfeld and Bosch 2017). Furthermore, the understanding of fundamental mechanisms and a comparative evolutionary approach will clarify the role of neuron–microbe interactions in animal's health and disease. In particular, there is increasing evidence that that multiple neurologic diseases in human, such as Parkinson's disease, multiple sclerosis, and gut motility disorders (Kerschensteiner et al. 2009; Cekanaviciute et al. 2017; Jangi et al. 2016; Sampson et al. 2016), are correlated with gut microbiota disturbances. The management of such disorders will benefit from a deep understanding of the cellular and molecular pathways of neuron–microbe dialogue.

Our recent studies have shown that neuronal activity in the early emerging metazoan Hydra is modulated by its symbiotic bacteria. The ability of microbiota to modulate the neuronal activity of the host appears, therefore, to be an evolutionary ancient feature. Neuron–bacteria interactions apparently have a deep evolutionary origin, dating back to the emergence of the nervous system itself. This thinking provides a new scope onto the ancestral role of the nervous systems and suggests that the nervous system evolved as much to coordinate movements as to sense and control bacteria.

Acknowledgements I am particularly thankful to René Augustin, Sebastian Fraune, Alexander Klimovich, and Andrea Murillo-Rincon for their pioneering contributions towards uncovering microbe–neuron interactions in Hydra. I thank Andrea Murillo-Rincon and Alexander Klimovich for providing pictures and figures for this review. And I gratefully appreciate the support from the Canadian Institute for Advanced Research (CIFAR). This work was supported by the Deutsche Forschungsgemeinschaft (DFG) (CRC1182 "Origin and Function of Metaorganisms", DFG grant BO 848/15-3, and grants from the DFG Cluster of Excellence programme "Inflammation at Interfaces").

References

Abrams GD, Bishop JE (1967) Effect of the normal microbial flora on gastrointestinal motility. Exp Biol Med 126:301–304

Augustin R, Schröder K, Murillo Rincón AP, Fraune S, Anton-Erxleben F, Herbst E-M, Wittlieb J, Schwentner M, Grötzinger J, Wassenaar TM et al (2017) A secreted antibacterial neuropeptide shapes the microbiome of Hydra. Nat Commun 8:698

Bach JF (2002) The effect of infections on susceptibility to autoimmune and and allergic diseases. N Engl J Med 347(2002):911–920

Bates JM, Mittge E, Kuhlman J, Baden KN, Cheesman SE, Guillemin K (2006) Distinct signals from the microbiota promote different aspects of zebrafish gut differentiation. Dev Biol 297:374–386

Bienenstock J, Kunze W, Forsythe P (2015) Microbiota and the gut–brain axis. Nutr Rev 73:28–31

Bode H, Berking S, David CN, Gierer A et al (1973) Quantitative analysis of cell types during growth and morphogenesis in Hydra. Wilhelm Roux'Archiv für Entwicklungsmechanik der Org 171:269–285

Bode HR (1992) Continuous conversion of neuron phenotype in hydra. Trends Genet 8:279–284

Bosch TCG, Miller DJ (2016) The holobiont imperative. Springer, Vienna

Bosch TCG, Adamska M, Augustin R, Domazet-Loso T, Foret S, Fraune S, Funayama N, Grasis J, Hamada M, Hatta M, Hobmayer B, Kawai K, Klimovich A, Manuel M, Shinzato C, Technau U,

Yum S, Miller DJ (2014) How do environmental factors influence life cycles and development? An experimental framework for early-diverging metazoans. BioEssays 36(12):1185–1194

Bosch TCG, Klimovich A, Domazet-Lošo T, Gründer S, Holstein TW, Jékely G, Miller DJ, Murillo-Rincon AP, Rentzsch F, Richards GS et al (2017) Back to the basics: cnidarians start to fire. Trends Neurosci 40:92–105

Braniste V, Al-Asmakh M, Kowal C, Anuar F, Abbaspour A, Toth M, Korecka A, Bakocevic N, Ng LG, Kundu P et al (2014) The gut microbiota influences blood-brain barrier permeability in mice. Sci Transl Med 6:263ra158–263ra158

Bravo JA, Forsythe P, Chew MV, Escaravage E, Savignac HM, Dinan TG, Bienenstock J, Cryan JF (2011) Ingestion of Lactobacillus strain regulates emotional behavior and central GABA receptor expression in a mouse via the vagus nerve. Proc Natl Acad Sci 108:16050–16055

Campbell RD (1976) Elimination by Hydra interstitial and nerve cells by means of colchicine. J Cell Sci 21:1–13

Catania F, Krohs U, Chittò M, Ferro D, Ferro K, Lepennetier G, Görtz HD, Schreiber RS, Kurtz J, Gadau J (2017) The hologenome concept: we need to incorporate function. Theory Biosci 136(3–4):89–98

Cekanaviciute E, Yoo BB, Runia TF, Debelius JW, Singh S, Nelson CA, Kanner R, Bencosme Y, Lee YK, Hauser SL et al (2017) Gut bacteria from multiple sclerosis patients modulate human T cells and exacerbate symptoms in mouse models. Proc Natl Acad Sci 201711235

Chapman JA, Kirkness EF, Simakov O et al (2010) The dynamic genome of Hydra. Nature (7288):592–596

Chen X, D'Souza R, Hong S-T (2013) The role of gut microbiota in the gut-brain axis: current challenges and perspectives. Protein Cell 4:403–414

Collins J, Borojevic R, Verdu EF, Huizinga JD, Ratcliffe EM (2014) Intestinal microbiota influence the early postnatal development of the enteric nervous system. Neurogastroenterol Motil 26:98–107

Collins SM, Surette M, Bercik P (2012) The interplay between the intestinal microbiota and the brain. Nat Rev Microbiol 10:735

Cryan JF, Dinan TG (2012) Mind-altering microorganisms: the impact of the gut microbiota on brain and behaviour. Nat Rev Neurosci 13:701–712

De Vadder F, Kovatcheva-Datchary P, Goncalves D, Vinera J, Zitoun C, Duchampt A, Bäckhed F, Mithieux G (2014) Microbiota-generated metabolites promote metabolic benefits via gut-brain neural circuits. Cell 156:84–96

Deines P, Lachnit T, Bosch TCG (2017) Competing forces maintain the Hydra metaorganism. Immunol Rev 279(1):123–136

Deloose E, Janssen P, Depoortere I, Tack J (2012) The migrating motor complex: control mechanisms and its role in health and disease. Nat Rev Gastroenterol Hepatol 9:271–285

Desbonnet L, Clarke G, Shanahan F, Dinan TG, Cryan JF (2014) Microbiota is essential for social development in the mouse. Mol Psychiatry 19:146–148

Diaz Heijtz R, Wang S, Anuar F, Qian Y, Björkholm B, Samuelsson A, Hibberd ML, Forssberg H, Pettersson S (2011) Normal gut microbiota modulates brain development and behavior. Proc Natl Acad Sci USA 108:3047–3052

Dinan TG, Stanton C, Cryan JF (2013) Psychobiotics: a novel class of psychotropic. Biol Psychiatry 74:720–726

Doenyas C (2018) Gut microbiota, inflammation, and probiotics on neural development in autism spectrum disorder. Neuroscience 374:271–286

Dupre C, Yuste R (2017) Non-overlapping neural networks in Hydra vulgaris. Curr Biol 27:1085–1097

Engel P, Moran NA (2013) The gut microbiota of insects—diversity in structure and function. FEMS Microbiol Rev 37:699–735

Erny D, Hrabě de Angelis AL, Jaitin D, Wieghofer P, Staszewski O, David E, Keren-Shaul H, Mahlakoiv T, Jakobshagen K, Buch T et al (2015) Host microbiota constantly control maturation and function of microglia in the CNS. Nat Neurosci 18:965–977

Forsythe P, Kunze WA (2013) Voices from within: gut microbes and the CNS. Cell Mol Life Sci 70:55–69

Franzenburg S, Walter J, Künzel S, Wang J, Baines JF, Bosch TCG, Fraune S (2013a) Distinct antimicrobial peptide expression determines host species-specific bacterial associations. Proc Natl Acad Sci USA 110:E3730–E3738

Franzenburg S, Fraune S, Altrock PM, Künzel S, Baines JF, Traulsen A, Bosch TCG (2013b) Bacterial colonization of Hydra hatchlings follows a robust temporal pattern. ISME J 7(4):781–790

Fraune S, Bosch TCG (2007) Long-term maintenance of species-specific bacterial microbiota in the basal metazoan Hydra. Proc Natl Acad Sci USA 104:13146–13151

Fraune S, Abe Y, Bosch TCG (2009a) Disturbing epithelial homeostasis in the metazoan Hydra leads to drastic changes in associated microbiota. Environ Microbiol 11:2361–2369

Fraune S, Augustin R, Bosch TCG (2009b) Exploring host-microbe interactions in hydra. Microbe 4(10):457–462

Fraune S, Anton-Erxleben F, Augustin R, Franzenburg S, Knop M, Schröder K, Willoweit-Ohl D, Bosch TCG (2015) Bacteria-bacteria interactions within the microbiota of the ancestral metazoan Hydra contribute to fungal resistance. ISME J 9:1543–1556

Gareau MG, Wine E, Rodrigues DM, Cho JH, Whary MT, Philpott DJ, MacQueen G, Sherman PM (2011) Bacterial infection causes stress-induced memory dysfunction in mice. Gut 60:307–317

Gonzales DL, Badhiwala KN, Vercosa DG, Avants BW et al (2017) Scalable electrophysiology in intact small animals with nanoscale suspended electrode arrays. Nat Nanotechnol 12:684

Grenham S, Clarke G, Cryan JF, Dinan TG (2011) Brain-gut-microbe communication in health and disease. Front Physiol 2:94

Hadizadeh F, Walter S, Belheouane M, Bonfiglio F, Heinsen F-A, Andreasson A, Agreus L, Engstrand L, Baines JF, Rafter J et al (2017) Stool frequency is associated with gut microbiota composition. Gut 66:559–560

Hansen GN, Williamson M, Grimmelikhuijzen CJP (2000) Two-color double-labeling in situ hybridization of whole-mount Hydra using RNA probes for five different Hydra neuropeptide preprohormones: evidence for colocalization. Cell Tissue Res 301:245–253

Harach T, Marungruang N, Duthilleul N, Cheatham V, Mc Coy KD, Frisoni G, Neher JJ, Fåk F, Jucker M, Lasser T et al (2017) Reduction of Abeta amyloid pathology in APPPS1 transgenic mice in the absence of gut microbiota. Sci Rep 7:41802

Hata T, Asano Y, Yoshihara K, Kimura-Todani T, Miyata N, Zhang X-T et al (2017) Regulation of gut luminal serotonin by commensal microbiota in mice. PLoS ONE 12(7):e0180745

Hsiao EY, McBride SW, Hsien S, Sharon G, Hyde ER, McCue T, Codelli JA, Chow J, Reisman SE, Petrosino JF, Patterson PH, Mazmanian SK (2013) Microbiota modulate behavioral and physiological abnormalities associated with neurodevelopmental disorders. Cell 155(7):1451–1463

Husebye E, Hellström PM, Sundler F, Chen J, Midtvedt T, Hellstrom PM, Sundler F, Chen J, Midtvedt T, Hellström PM et al (2001) Influence of microbial species on small intestinal myoelectric activity and transit in germ-free rats. Am J Physiol Gastrointest Liver Physiol 280:G368–G380

Jékely G, Paps J, Nielsen C (2015a) The phylogenetic position of ctenophores and the origin(s) of nervous systems. Evodevo 6:1

Jékely G, Keijzer F, Godfrey-Smith P, Jékely G, Keijzer F, Godfrey-Smith P (2015b) An option space for early neural evolution. Philos Trans R Soc B-Biol Sci 370:1–12

Jangi S, Gandhi R, Cox LM, Li N, von Glehn F, Yan R, Patel B, Mazzola MA, Liu S, Glanz BL et al (2016) Alterations of the human gut microbiome in multiple sclerosis. Nat Commun 7:12015

Kabouridis PS, Lasrado R, McCallum S, Chng SH, Snippert HJ, Clevers H, Pettersson S, Pachnis V (2015) Microbiota controls the homeostasis of glial cells in the gut lamina propria. Neuron 85:289–295

Kass-Simon G (1972) Longitudinal conduction of contraction burst pulses from hypostomal excitation loci in Hydra attenuata. J Comp Physiol 80:29–49

Kerschensteiner M, Meinl E, Hohlfeld R (2009) Neuro-immune crosstalk in CNS diseases. In: Molecular basis of multiple sclerosis. Springer, pp 197–216

Koizumi O (2002) Developmental neurobiology of hydra, a model animal of cnidarians. Can J Zool 80:1678–1689

Koizumi O, Sato N, Goto C (2004) Chemical anatomy of hydra nervous system using antibodies against hydra neuropeptides: a review. Hydrobiologia 530:41–47

Koropatnick T, Goodson Michael S, Heath-Heckman Elizabeth AC, McFall-Ngai M (2014) Identifying the cellular mechanisms of symbiont-induced epithelial morphogenesis in the squid-vibrio association. Biol Bull 226(1):56–68

Kostic AD, Xavier RJ, Gevers D (2014) The microbiome in inflammatory bowel disease: current status and the future ahead. Gastroenterology 146:1489–1499

Kowarsky M, Camunas-Soler J, Kertesz M, De Vlaminck I, Koh W, Pan W, Martin L, Neff NF, Okamoto J, Wong RJ et al (2017) Numerous uncharacterized and highly divergent microbes which colonize humans are revealed by circulating cell-free DNA. Proc Natl Acad Sci USA 201707009

Krishnan S, Ding Y, Saedi N, Choi M, Sridharan GV, Sherr DH, Yarmush ML, Alaniz RC, Jayaraman A, Lee K (2018) Gut microbiota-derived tryptophan metabolites modulate inflammatory response in hepatocytes and macrophages. Cell Rep 23(4):1099–1111

Ley RE, Turnbaugh PJ, Klein S, Gordon JI (2006) Microbial ecology: human gut microbes associated with obesity. Nature 444:1022–1023

Liu J, Sun J, Wang F, Yu X, Ling Z, Li H, Zhang H, Jin J, Chen W, Pang M et al (2015) Neuroprotective effects of Clostridium butyricum against vascular dementia in mice via metabolic butyrate. Biomed Res Int 2015:412946

Loomis WF (1955) Glutathione control of the specific feeding reactions of hydra. Ann NY Acad Sci 62:211–227

Marcum BA, Campbell RD (1978) Development of Hydra lacking nerve and interstitial cells. J Cell Sci 29:17 LP-33

Mazmanian SK, Liu CH, Tzianabos AO, Kasper DL (2005) An immunomodulatory molecule of symbiotic bacteria directs maturation of the host immune system. Cell 122:107–118

McFall-Ngai M (2014) Divining the essence of symbiosis: insights from the squid-vibrio model. PLoS Biol 12(2):e1001783

McFall-Ngai M (2015) Giving microbes their due—animal life in a microbially dominant world. J Exp Biol 218:1968–1973

McFall-Ngai M, Hadfield MG, Bosch TCG, Carey HV, Domazet-Lošo T, Douglas AE, Dubilier N, Eberl G, Fukami T, Gilbert SF et al (2013) Animals in a bacterial world, a new imperative for the life sciences. Proc Natl Acad Sci USA 110:3229–3236

Möhle L, Mattei D, Heimesaat MM, Bereswill S, Fischer A, Alutis M, French T, Hambardzumyan D, Matzinger P, Dunay IR et al (2016) Ly6Chi monocytes provide a link between antibiotic-induced changes in gut microbiota and adult hippocampal neurogenesis. Cell Rep 15(1945–1956):58

Mortzfeld BM, Bosch TCG (2017) Eco-aging: stem cells and microbes are controlled by aging antagonist FoxO. Curr Opin Microbiol 38:181–187

Murillo-Rincon AP, Klimovich A, Pemöller E, Taubenheim J et al (2017) Spontaneous body contractions are modulated by the microbiome of Hydra. Sci Rep 7:15937

Neufeld KM, Kang N, Bienenstock J, Foster JA (2011) Reduced anxiety-like behavior and central neurochemical change in germ-free mice. Neurogastroenterol Motil 23(255–64):e119

O'Mahony SM, Felice VD, Nally K, Savignac HM, Claesson MJ, Scully P, Woznicki J, Hyland NP, Shanahan F, Quigley EM et al (2014) Disturbance of the gut microbiota in early-life selectively affects visceral pain in adulthood without impacting cognitive or anxiety-related behaviors in male rats. Neuroscience 277:885–901

Passano LM, McCullough CB (1963) Pacemaker hierarchies controlling the behaviour of Hydras. Nature 199:1174–1175

Passano LM, McCullough CB (1964) Co-and behaviour in Hydra: I. Pacemaker system of the periodic contractions. J Exp Biol 41:643–664

Passano LM, McCullough CB (1965) Co-ordinating systems and behaviour in Hydra II. The rhythmic potential system. J Exp Bwl 43:205–231

Pisani D, Pett W, Dohrmann M, Feuda R, Rota-Stabelli O, Philippe H, Lartillot N, Wörheide G (2015) Genomic data do not support comb jellies as the sister group to all other animals. Proc Natl Acad Sci USA 112:15402–15407

Quigley EMM (2011) Microflora modulation of motility. J Neurogastroenterol Motil 17:140–147

Reigstad CS, Salmonson CE, Rainey JF, Szurszewski JH, Linden DR, Sonnenburg JL, Farrugia G, Kashyap PC (2015) Gut microbes promote colonic serotonin production through an effect of short-chain fatty acids on enterochromaffin cells. FASEB J 29(4):1395–1403

Ridaura VK, Faith JJ, Rey FE, Cheng J, Duncan AE, Kau AL, Griffin NW, Lombard V, Henrissat B, Bain JR et al (2013) Gut microbiota from twins discordant for obesity modulate metabolism in mice. Science (80):341, 1241214–1241214

Rosenberg E, Koren O, Reshef L, Efrony R, Zilber-Rosenburg I (2007) The role of microorganisms in coral health, disease and evolution. Nat Rev Microbiol 5:355–362

Round JL, Mazmanian SK (2009) The gut microbiota shapes intestinal immune responses during health and disease. Nat Rev Immunol 9:313–323

Sampson TR, Debelius JW, Thron T, Janssen S, Shastri GG, Ilhan ZE, Challis C, Schretter CE, Rocha S, Gradinaru V et al (2016) Gut microbiota regulate motor deficits and neuroinflammation in a model of Parkinson's disease. Cell 167(1469–1480):e12

Schröder K, Bosch TCG (2016) The origin of mucosal immunity: lessons from the holobiont Hydra. mBio 7:e01184–16

Sharon G, Sampson TR, Geschwind DH, Mazmanian SK (2016) The Central Nervous System and the Gut Microbiome. Cell 167:915–932

Shikuma NJ, Antoshechkin I, Medeiros JM, Pilhofer M, Newman DK (2016) Stepwise metamorphosis of the tubeworm *Hydroides elegans* is mediated by a bacterial inducer and MAPK signaling. Proc Natl Acad Sci 113:10097–10102

Smith K, McCoy KD, Macpherson AJ (2007) Use of axenic animals in studying the adaptation of mammals to their commensal intestinal microbiota. Semin Immunol 19:59–69

Sonnenburg JL, Bäckhed F (2016) Diet-microbiota interactions as moderators of human metabolism. Nature 535(7610):56–64

Stilling RM, Dinan TG, Cryan JF (2014) Microbial genes, brain & behaviour—epigenetic regulation of the gut-brain axis. Genes Brain Behav 13:69–86

Stokes DR, Rushforth NB (1979) Contraction pulse system in hydroids. Comp Biochem Physiol 64(A):207–212

Strati F, Cavalieri D, Albanese D, De Felice C, Donati C, Hayek J, Jousson O, Leoncini S, Renzi D, Calabrò A, De Filippo C (2017) New evidences on the altered gut microbiota in autism spectrum disorders. Microbiome 5:24

Sudo N, Chida Y, Aiba Y, Sonoda J, Oyama N, Yu X-NX, Kubo C, Koga Y (2004) Postnatal microbial colonization programs the hypothalamic-pituitary-adrenal system for stress response in mice. J Physiol 558:263–275

Takaku Y, Hwang JS, Wolf A, Böttger A, Shimizu H, David CN, Gojobori T (2014) Innexin gap junctions in nerve cells coordinate spontaneous contractile behavior in Hydra polyps. Sci Rep 4:3573

Theis KR, Dheilly NM, Klassen J, Brucker RM, Bosch TCG, Cryan JF, Gilbert SF, Goodnight CJ, Sapp J, Vandenkoornhuyse P, Zilber-Rosenberg I, Rosenberg E, Bordenstein SR (2016) Getting the hologenome concept right: an eco-evolutionary framework for hosts and their microbiomes. mSystems 1(2):e00028-16

Thompson GR, Trexler PC (1971) Gastrointestinal structure and function in germ-free or gnotobiotic animals. Gut 12:230–235

Trembley A (1744) Mémoires, Pour Servir à l'Histoire d'un Genre de Polypes d'Eau Douce, à Bras en Frome de Cornes. Verbeek, Leiden (Netherlands)

Vandeputte D, Falony G, Vieira-Silva S, Tito RY, Joossens M, Raes J (2016) Stool consistency is strongly associated with gut microbiota richness and composition, enterotypes and bacterial growth rates. Gut 65:57–62

Vantrappen G, Janssens J, Hellemans J, Ghoos Y (1977) The interdigestive motor complex of normal subjects and patients with bacterial overgrowth of the small intestine. 59

Vitale F, Vercosa D, Rodriguez AV, Pamulapati SS et al (2017) Fluidic microactuation of flexible electrodes for neural recording. Nano Lett

Wagner G (1905) Memoirs: on some movements and reactions of Hydra. J Cell Sci 2:585–622

Wittlieb J, Khalturin K, Lohmann JU, Anton-Erxleben F, Bosch TCG (2006) Transgenic Hydra allow in vivo tracking of individual stem cells during morphogenesis. Proc Natl Acad Sci USA 103(16):6208–6211

Woese CR (2014) A new biology for a new century. Microbiol Mol Biol Rev 68(2):173–186

Yano JM, Yu K, Donaldson GP, Shastri GG, Ann P, Ma L, Nagler CR, Ismagilov RF, Mazmanian SK, Hsiao EY (2015) Indigenous bacteria from the gut microbiota regulate host serotonin biosynthesis. Cell 161(2):264–276

Zilber-Rosenberg I, Rosenberg E (2008) Role of microorganims in the evolution of animals and plants: the hologenome theory of evolution. FEMS Microbiol Rev 32:723–735

Mini-bioreactors as Tools for Adaptive Laboratory Evolution for Antibiotic Drug Resistance and Evolutionary Tuning of Bacterial Optogenetic Circuits

Ya-Tang Yang

Abstract Adaptive laboratory evolution for microbial cells has become an indispensable tool for metabolic engineering and system and synthetic biology. Commercial bioreactors, however, are cumbersome to use and difficult to meet special needs for adaptive laboratory evolution, and there has been the recent renaissance of mini-bioreactor of working volume in the milliliter range. Leveraging on availability of low-cost electronic components such as light-emitting diode and microcontroller board, we have developed a low-cost, flexible, and robust bioreactor platform. We detail two examples of our design, namely the morbidostat and optogenetic bioreactors for antibiotic drug resistance study and optically controlled evolution, respectively. These mini-bioreactors will be useful tools for adaptive laboratory evolution for microbial cells.

1 Introduction

In adaptive laboratory evolution (ALE) experiments, bacteria are cultured for prolonged periods of time under a chosen evolutionary pressure, by either serially diluting the culture or in a chemostat mode (Lenski 2017; Dragosits and Mattanovich 2013). Microbial cells offer advantages for adaptive laboratory studies because microbial cells generally grow very fast with typical specific growth rate in the range of 0.05–1 h^{-1}. They can be cultivated for several hundred generations within weeks or months. Coupling with next-generation sequencing (NGS) technique, phenotype–genotype relation can also be easily obtained. Subsequently, the identified mutations over the course of evolution can further be re-synthesized and inserted into the ancestral strain to measure and verify the phenotype response. The power of ALE is perhaps best illustrated by the recent work reported by Milo's group to engineer a

Y.-T. Yang (✉)
Department of Electrical Engineering, National Tsing Hua University, Hsinchu 30013,
Taiwan, Republic of China
e-mail: ytyang@ee.nthu.edu.tw

© Springer International Publishing AG, part of Springer Nature 2018
P. Pontarotti (ed.), *Origin and Evolution of Biodiversity*,
https://doi.org/10.1007/978-3-319-95954-2_5

carbon fixation cycle (Antonovsky et al. 2016). They demonstrate that a combination of metabolic rewiring and adaptive laboratory evolution has led to the biosynthesis of sugars by a fully functional Calvin–Benson–Bassham (CBB) cycle in *E. coli*. In short, ALE has become an indispensable tool for metabolic engineering and system and synthetic biology. Currently, a vast amount of data is available for ALE under different nutrient and environmental stress and is reviewed elsewhere (Dragosits and Mattanovich 2013). Bioreactors are devices to grow microbial cells with controlled growth conditions and therefore often serve as workhorses for ALE. In general, bioreactors are operated in so-called batch or chemostat mode (Novick and Szilard 1950). In the batch mode, the cells are given a fixed amount of nutrients and allowed to grow until nutrients are depleted. Batch culture has the advantage of easy implementation, but it is difficult to maintain constant growth condition. In the chemostat mode, fresh medium is constantly supplied and microbial cells are removed at constant rate to maintain constant population. The advantage of chemostatic operation is tight control over growth conditions such as nutrient supply, pH, and oxygenation. However, commercial bioreactors at liter or subliter size are often cumbersome and difficult to modify to meet special needs for ALE. Recent renaissance of mini-bioreactor of working volume in the milliliter range fills in such a gap (Toprak et al. 2012, 2013; Takahashi et al. 2014; Liu et al. 2016; Wang and Yang 2017). Such development leverages on the increasing availability and low cost of electronic components such as light-emitting diodes and photodetectors and 3D printing technology for custom parts (Baden et al. 2015). For example, Takashi et al. have demonstrated an open-source turbidostat design and the whole setup including the 3D printer is ~$2000 (Takahashi et al. 2014). The "morbidostat," developed by Kishony's group, is a custom culturing solution that uses a feedback to progressively increase antibiotic drug concentration to maintain constant microbial population and at the same time keep the microbes constantly challenged (Toprak et al. 2012, 2013). Using trimethoprim as a test case antibiotic drug for the morbidostat, stepwise increases of antibiotic drug resistance have been observed and corresponding mutations in the drug's target have been identified with whole genome sequencing (WGS). The identified mutations have been synthesized and retrospectively combined to the ancestral strain to reconstruct the antibiotic resistance fitness landscape (Palmer et al. 2015). Such results have also led to the recent development of the microbial evolution and growth arena (MEGA) plate, in which bacteria spread and evolved on a large antibiotic landscape (120 × 60 cm) that allowed visual observation of mutation and selection in a migrating bacterial front (Baym et al. 2016). The optogenetic bioreactor, developed in my group in National Tsing Hua University, offers in vivo, real-time gene expression measurement with control of input light stimulus for intervention of optogenetic circuits (Wang and Yang 2017).

In this chapter, we detail two examples of mini-bioreactor design—morbidostat and optogenetic bioreactor. The morbidostat is based on chemostat with the addition of feedback scheme and design for antibiotic drug resistance. The optogenetic bioreactor operates in batch mode and uses light for intervention of gene expression. In terms of engineering design, one striking feature of these bioreactor designs is the low footprint, low power consumption due to extensive usage of low power electronic components such as LEDs, fans, and micropumps. This allows additional functionalities to be added with ease and high degree of integration to be achieved.

2 Experiment Setup for Morbidostat

We design the morbidostat with a culture vial of working volume of ~10 ml with feedback scheme implemented (Fig. 1). The snapshots of the assembly of components and experimental setup is shown in Fig. 2. The tube holder has two openings drilled for LED light sources and photodetectors to measure optical density. An LED light source with a wavelength of 940 nm and a photodetector (ST-2L2B; Kodenshi) are mounted on each tube holder and connected to the circuits. The photodetector is connected to a bias resistor of 100 kΩ and 5 V power supply. (Note that no amplifier is needed, but we implement a median filter scheme, in which 100 data points for voltage are taken and the median of these data is calculated and used in the feedback algorithm.) For the fluidic mixing, a tube holder for a culture vial sits on top of an electronic fan attached with a magnet and a magnetic stirring bar is placed inside the culture vial. (Note that the distance between the magnet and the magnetic stirring bar is very critical to success of the magnetic stirring and it is very important to do a test to determine the optimal distance before one designs the custom acrylic housing.) The entire setup is placed in a shaker incubator with temperature set to 30 °C (Fig. 2h).

Overall, the entire operation including the plumbing, fluidic mixing, and data acquisition is orchestrated by a microcontroller board (Arduino, Inc.). Of particular importance is the feedback algorithm based on measured optical density and implemented in the micro controller board (Fig. 1). The morbidostat operates as a chemostat with two modes, namely drug injection mode and drug dilution mode. The continuous magnetic stirring of bacterial culture is briefly interrupted every 12 min for the measurement of optical density. The algorithm in the microcontroller

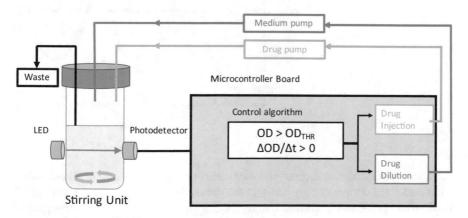

Fig. 1 Schematics of morbidostat setup with feedback scheme

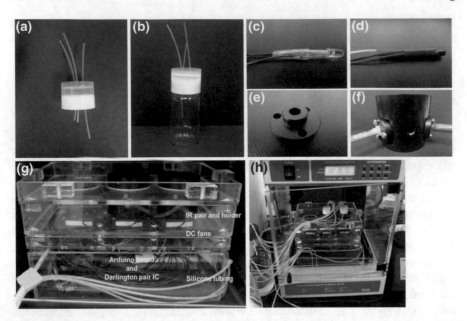

Fig. 2 Tube holder design and experimental setup. **a** Cap of the culture vial with Tygon tubing attached. **b** Culture vial with the cap. **c** LED soldered with resistor and wires. **d** Photodetector soldered with resistor and wrapped in insulating tape. **e** Cap of the culture vial holder for mounting LED and photodetector. **f** Main body of the culture vial holder. **g** Tube holder with microcontroller board in the acrylic housing. h. the whole setup placed in shaker incubator

Fig. 3 Growth curve during morbidostat operation. The red and purpose arrow indicates the onset and ending of drug injection, respectively

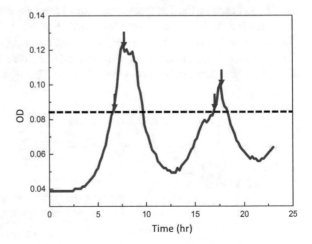

board then compares the measured optical density with the preset threshold value. If the measured optical density exceeds the preset threshold, the board switches off medium pump and switch on micropump for drug injection. A representative growth curve during the morbidostat operation is shown in Fig. 3. Each day, the culture is expected to grow above the preset optical density threshold and triggers the drug injection. After drug injection is turned off, the culture will grow because mutants take over the microbial cells as the drug is diluted. Note that the plumbing and drug injection scheme used here is based on continuous chemostatic operation and is a simplified version of the periodic dilution scheme by Kishony's group (Toprak et al. 2012, 2013).

3 Results and Discussions of Measurements in Morbidostat

The general procedure for morbidostat involves daily cultivation of bacterial cells, and daily samples are frozen at −80 °C to serve as the "fossil" record for evolution. After the entire course of evolution, antibiotic susbitibility measurement on the preserved samples is done. We measure the antibiotic drug resistance, quantified by IC_{50}, defined as the inhibitor concentration at which growth rate is half the maximal growth rate. OD reads from the microplates are used to calculate the IC_{50} values of the evolving strains. For every well in every plate, the growth rate during the mid-exponential phase was measured by fitting the growth curves. The IC_{50} values were calculated by interpolating the drug concentrations corresponding to growth rates of 50% (Fig. 4).

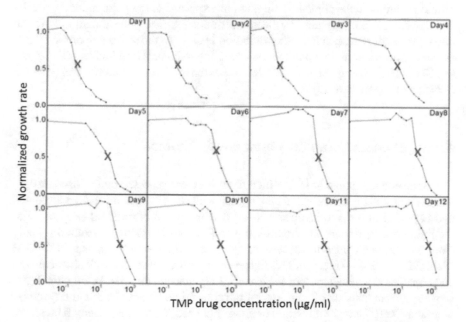

Fig. 4 Drug dose response of daily samples. The growth rate is normalized to the maximal growth rate at Day 0. The blue cross-marks the point at which the growth rate is half of maximal growth rate and the corresponding drug concentration is IC_{50}

Fig. 5 Plot of IC$_{50}$ of daily samples over the course of 12 days

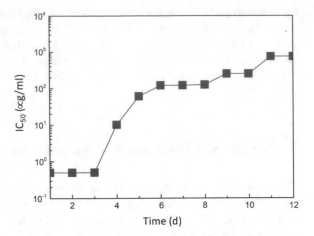

To validate the morbidostat, the *E. coli* MG 1655 is exposed to trimethoprim (TMP), a commonly used antibiotic drug. TMP induces very distinctive stepwise increase in drug resistance within few days, and mutations are well clustered around dihydrofolate reductase (DHFR) gene. The acquired drug resistance increased approximated by 1500-fold to ~1000 μg/ml over 12 days (Fig. 5). This result is consistent with clinical isolates (Heikkila et al. 1990) and proves its effectiveness to mimic evolution in the clinical setting. In our experiment, the point mutations in the last-day sample are measured by Sanger sequencing. One point mutation is found in the promoter region, and another mutation is close to the active site of the DHFR enzyme. These identified mutations are consistent with previous findings (Flensburg and Skold 1987; Ohmae et al. 2001). A more extensive list of mutations can be found elsewhere (Toprak et al. 2012).

4   Experiment Setup for Optogenetic Bioreactor

The optogenetic bioreactor is modified from previous reported morbidostat design, and the whole setup consists of a tube holder with custom housing and a culture vial and is placed in a shake incubator covered with optical blackboard (Fig. 6a). For the light-sensing system, we chose CcaS-CcaR, a two-component system originally from the chromatic adaption system of cyanobacteria *Synechocystis* sp. PCC 6803 (Schmidl et al. 2014) (Fig. 6b). This light-sensing system has been well characterized and considered a model system for bacterial optogenetic circuits. Briefly, the light-sensing system consists of membrane-bound histidine kinase CcaS and a response regulator CcaR. Upon illumination of green light at 520 nm in wavelength increases the autophosphorylation rate of CcaS. Phosphotransfer from CcaS to CcaR results in the transcription from the promoter region. Two plasmids (pSR43.6 and pSR58.6;

(a)   **(b)**

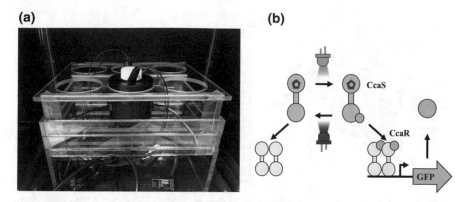

Fig. 6 Optogenetic bioreactor and the CcaS-CcaR optogenetic bacterial circuit. **a** The whole assembly placed in the shaker incubator with optical blackboard. **b** The CcaS-CcaR light-sensing system used in this work

Addgene Inc.) originally developed in Tabor's group to harbor the CcaS-CcaR light-sensing system were used to transform *E. coli* BW29655 strain.

For the optogenetic bioreactor, the major engineering challenge is to integrate light sources for turning on and off gene expression and optical density and fluorescence measurement. Careful mechanical design for the tube holder is executed to ensure such holder houses all LED light sources, photodetectors, and fluorescence filter set (Fig. 7a). All of these operations are again orchestrated by an Arduino microcontroller board (Arduino, Inc.) as shown in Fig. 7b. The continuous illumination of bacterial cultures to activate gene expression is briefly interrupted every 12 min for measurements of optical density and fluorescence with the magnetic stirring unit switched off. (Such operation therefore has ~12 min temporal resolution.) The tube holder has four openings drilled for LED light sources and photodetectors to measure optical density and fluorescence, respectively. Two LED sources with wavelengths of 520 and 625 nm in strip format (3M) are mounted on the tube holder for the light intervention of gene expression. For optical density measurements, two of the openings are positioned at an angle of 135° to maximize the detection of scattered light (Fig. 7a). An LED light source with a wavelength of 940 nm and a photodetector (ST-2L2B; Kodenshi) are mounted on each tube holder and connected to the circuits. For fluorescence measurements, two of the openings are positioned at an angle of 90° and designed to hold the excitation filter (FGB25; Thorlabs Inc.) and emission filter (FGL530; Thorlabs Inc.) for GFP, and an LED light source with wavelength of 470 nm and a photodetector (ST1KLA; Kodenshi) are also mounted (Fig. 7a).

A circuit is built to measure the voltage across the photodetectors with bias resistors. (The photodetectors for optical density and fluorescence are connected to bias resistors of 100 kΩ and 20 MΩ, respectively. Both uses 5 V power supply from microcontroller board.) All of the LEDs and an electronic fan for magnetic stirring are powered by a Darlington pair chip (ULN2803APG; Toshiba). Light intensity for intervention of optogenetic circuit is controlled via a pulse width modulation (PWM)

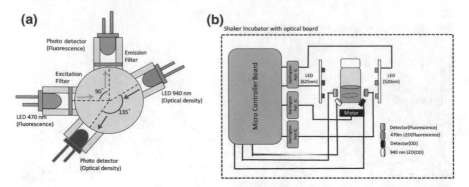

Fig. 7 Optical design and detailed schematics of the optogenetic bioreactor. **a** Top view of the holder design for optical density and fluorescence detection. **b** Schematics of the electronics for light-emitting diodes, photodetector, and magnetic stirring

scheme by microcontroller board. The PWM in the board has an 8-bit resolution (256 levels) in duty cycle setting, and linear relation between the duty cycle and average intensity is found. Calibration of light intensity is done with a silicon photodiode power sensor (S120C, Thorlabs).

5 Evolutionary Tuning in Optogenetic Bioreactor

We employed the bioreactor for evolutionary tuning of the optogenetic circuit by carrying out sequential serial dilution passages (Fig. 8a). The bacterial cultures were grown each day at 37 °C in shaker incubator. Cells grow by 100-fold each day, corresponding to $\log_2 100 = 6.6$ generations so for 9 days, the culture evolves for ~59 generations. After each day of growth, the cells are diluted 1:100 in a tube with fresh medium. Daily samples are frozen (−80 °C) with the addition of glycerol and stored for subsequent measurement. In total, five replicate evolution experiments are carried out for 9 days with light illumination. During the course of evolution, we expect that the gene expression level will monotonically decrease over time because the optogenetic circuit expresses wasteful protein with no benefit for growth or equivalently fitness according to cost-benefit theory (Dekel and Alon 2005). The fitness and gene expression (quantified as GFP fluorescence divided by optical density at the end of each day) are displayed in Fig. 8b and 8c, respectively. Both the fitness and gene expression data were normalized by the corresponding values on Day 1. At the end of the experiment, the fitness had increased by $21 \pm 0.33\%$ and the gene expression had decreased on average by $49 \pm 9.3\%$ in the five replicate runs. The interdependence between fitness and gene expression level is displayed in Fig. 8d. (As a control, triplicate evolution experiments are also carried out under the dark condition for 9 days and the result is compared to the effect due to light illumination. On Day 10, both

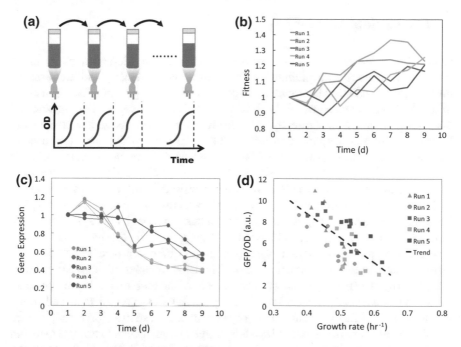

Fig. 8 Evolutionary tuning of gene expression in optogenetic bioreactor. **a** Serial dilution transfer with the optogenetic circuit activated with light illumination. **b** Measured fitness over the entire course of the evolution experiment. **c** Gene expression over time. **d** Interdependence of gene expression and growth rate

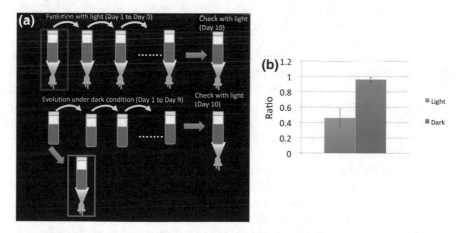

Fig. 9 Control evolution experiment with dark condition. **a** Schematics of the experiment. **b** Gene expression of samples with light illumination and dark condition at Day 10

samples are illuminated again to measure the gene expression. The gene expression for evolution under dark condition shows only ~4% degradation.) (Fig. 9).

6 Conclusion and Outlook

In conclusion, we have demonstrated a low-cost, versatile bioreactor platform for ALE by meticulously integrating fluidic mixing, optical density and fluorescence measurement, and other functionalities such as feedback algorithm. The robustness of our design is demonstrated through two sophisticated examples, morbidostat and optogenetic bioreactors. Currently, the mini-bioreactor reported in this chapter can be constructed with material cost under $300 with most of the cost from custom machined parts and microcontroller board. The $300 components cost will drop even more with access to a 3D printer. Such low costs make scaling up to large array of bioreactors feasible for parallel evolution experiments and its widespread usage in the research communities. Currently, three laboratories have at least one complete instrument (see acknowledgments), some with custom modifications. Because each bioreactor can be configured, addressed, and programmed separately, one can set up each of them to run evolution experiment under different growth conditions with different nutrient or environmental stress. Extension to anaerobic operation is also straightforward by placing the bioreactor either in an anaerobic chamber or enclosure with oxygen adsorbing materials with usage of wireless control (Liu and Yang, unpublished). Adding the light source for photosynthesis, such a bioreactor platform can also be used for microalgae and cyanobacteria. Still in our bioreactor, oxygenation is still the limiting factor for growth and oxygenation is related to the magnetic stirring. Measurement has been done to verify the relation between rotation speed of magnetic stirring and the specific growth rate (Dou and Yang, unpublished). Recently, enhancement of bacterial growth has been reported with the help of immiscible oxygenated oils and may be implemented with our current design to achieve full oxygenation (Sklodowska and Jakiela 2017).

The morbidostat reported here is demonstrated through a test case of trimethoprim. The morbidostat is particularly useful when the microbe acquires multiple mutations during the evolution. On the mechanistic level, the description of the morbidostat is recapitulated in differential equations (Chen et al. 2017). The global dynamics of a microbial species undergoing sequential evolution is studied, and uniform persistence of all species which wins a sequential evolution scenario is proven. Such rigorous mathematical analysis paves the basis for in silico simulation for the morbidostat. The microbial mutation and competition in the actual morbiodostat is much more complicated than the simple sequential evolution scenario so for a more realistic simulation, one need to input the evolutionary pathway, phenotype-genotype correlation, and take into account of mutation rate changes. We expect data from the morbidostat, and MEGA plate study will be very useful to build such simulation. This approach can also be generalized to the case of multiple antibiotic drugs.

Although optogenetic bioreactors reported here are used for evolution experiment, they are a general tool for bacterial optogenetics. With changes of light sources, it can accommodate the vast variety of light-sensing system with a wavelength ranging from UV to visible and near-infrared (Levskaya et al. 2005; Ohlendorf et al. 2012; Ryu and Gomelsky 2014; Ramakrishnan and Tabor 2016). To demonstrate the scaling up, we have also built an array of six mini-photobioreactor and used them to characterize the interdependence of optogenetic gene expression and microbial growth under various nutrients and oxygenation (Zheng and Yang unpublished). Compared to the current state of art tools such as light tube array (LTA) reported by Tabor's group in the field of bacterial optogenetics (Olson et al. 2014; Gerhardt et al. 2016), such a device represents a major technological advance due to its capability of simultaneous recording grow curves and gene expression. For the evolution experiment reported here for ~10 days with 12 min temporal resolution, LTA will require ~1000 manual runs in the flow cytometer in order to obtain the fluorescence data and entail enormous efforts for data processing. The optogenetic bioreactor presented here also facilitates any optogenetic measurement related to metabolism and/or antibiotic drug resistance (Davidson et al. 2013; Ohlendorf et al. 2012) and multiplexed optogenetic control (Fernandez-Rodriguez et al. 2017).

Acknowledgements The author thanks Dr. Teuta Pilizota's laboratory at the University of Edinburgh, Han-Jia Lin's laboratory at the National Ocean University, and Prof. Charles Carter's laboratory at the University of Carolina at Chapel Hill for their becoming early adopters as well as for their input on our design and assembly instructions. The author would like to acknowledge funding support from the Ministry of Science and Technology under grant numbers MOST 105-2221-E-007-MY3.

References

Antonovsky N et al (2016) Sugar synthesis from CO_2 in *Escherichia coli*. Cell 166:115–125

Baden T, Chagas A, Gage G, Marzullo T, Prieto Godino L, Euler T (2015) Open labware-3D printing your own lab equipment. PLoS Biol 13:e1002175

Baym M, Lieberman TD, Kelsic ED, Chait R, Gross R, Yelin I, Kishony R (2016) Spatiotemporal microbial evolution on antibiotic landscapes. Science 353:1147–1151

Chen Z, Hsu SB, Yang YT (2017) The morbidostat: a bioreactor that promotes selection for drug resistance in bacteria. SIAM J Appl Math 77(2):470–490

Davidson EA, Basu AS, Bayer TS (2013) Programming microbes using pulse width modulation of optical signals. J Mol Biol 425:4161–4166

Dekel E, Alon U (2005) Optimality and evolutionary tuning of the expression level of a protein. Nature 436:588–592

Dragosits M, Mattanovich D (2013) Adaptive laboratory evolution—principles and applications for biotechnology. Microbial Cell Fact 12:64

Fernandez-Rodriguez J, Moser F, Song M, Voigt CA (2017) Engineering RGB color vision into *Escherichia coli*. Nat Chem Biol 13:706–710

Flensburg J, Skold O (1987) Massive overproduction of dihydrofolate reductase in bacteria as a response to the use of trimethoprim. Eur J Biochem 162:473–476

Gerhardt KP et al (2016) An open-hardware platform for optogenetics and photobiology. Sci Rep. https://doi.org/10.1038/srep35363

Heikkila E, Sundstrom L, Huovinen P (1990) Trimethoprim resistance in *Escherichia coli* isolates from a geriatric unit. Antimicrob Agents Chemother 34:2013–2015

Lenski RE (2017) Experimental evolution and the dynamics of adaptation and genome evolution in microbial populations. ISME J 11:2181–2194

Levskaya A, Chevalier AA, Tabor JJ, Simpson ZB, Lavery LA, Levy M, Davidson EA, Scouras A, Ellington AD, Marcotte EM, Voigt CA (2005) Synthetic biology: engineering Escherichia coli to see light. Nature 438:441–442

Liu PC, Lee YT, Wang CY, Yang YT (2016) Design and use of a low cost, automated morbidostat for adaptive evolution of bacteria under antibiotic drug selection. J Vis Exp 115. https://doi.org/10.3791/54426

Novick A, Szilard L (1950) Description of the chemostat. Science 112:715–716

Ohmae E, Sasaki Y, Gekko K (2001) Effects of five-tryptophan mutations on structure, stability and function of Escherichia coli dihydrofolate reductase. J Biochem 130:439–447

Ohlendorf R, Vidavski RR, Eldar A, Moffat K, Möglich A (2012) From dusk till dawn: one-plasmid systems for light- regulated gene expression. J Mol Biol 416:534–542

Olson EJ, Hartsough LA, Landry BP, Shroff R, Tabor JJ (2014) Characterizing bacterial gene circuit dynamics with optically programmed gene expression signals. Nat Methods 11:449–455

Palmer AC, Toprak E, Baym MHS, Kim S, Veres A, Bershtein SR, Kishony SR (2015) Delayed commitment to evolutionary fate in antibiotic resistance fitness landscapes. Nat Commun 6:7385. https://doi.org/10.1038/ncomms8385

Ryu MH, Gomelsky M (2014) Near-infrared light responsive synthetic c-di-GMP module for optogenetic applications. ACS Synth Biol 3:802–810

Ramakrishnan P, Tabor JJ (2016) Repurposing Synechocystis PCC6803 UirS–UirR as a UV-Violet/Green photoreversible transcriptional regulatory tool in E. coli. ACS Synth Biol 5:733–740

Schmidl SR, Sheth RU, Wu A, Tabor JJ (2014) Refactoring and optimization of light-switchable Escherichia coli two-component systems. ACS Synth Biol 3:820–831

Sklodowska K, Jakiela S (2017) Enhancement of bacterial growth with the help of immiscible oxygenated oils. RSC Adv 7:40990

Takahashi CN, Miller AW, Ekness F, Dunham MJ, Klavins E (2014) A low cost, customizable turbidostat for use in synthetic circuit characterization. ACS Synth Biol 4:32–38

Toprak E, Veres A, Michel JB, Chait R, Hartl DL, Kishony R (2012) Evolutionary paths to antibiotic resistance under dynamically sustained drug selection. Nat Genet 44:101–106

Toprak E, Veres A, Yildiz S, Pedraza JM, Chait R, Paulsson J, Kishony R (2013) Building a morbidostat: an automated continuous-culture device for studying bacterial drug resistance under dynamically sustained drug inhibition. Nat Protoc 8:555–567

Wang HK, Yang YT (2017) Mini photobioreactors for in vivo real-time characterization and evolutionary tuning of bacterial optogenetic circuit. ACS Synth Biol. https://doi.org/10.1021/acssynbio.7b00091

Part II
Self/Nonself Evolution

Deciphering the Evolution of Vertebrate Immune Cell Types with Single-Cell RNA-Seq

Santiago J. Carmona and David Gfeller

Abstract Single-cell RNA-seq is revolutionizing our understanding of the cell type heterogeneity and evolution in many fields of biology, ranging from neuroscience to cancer to immunology. In immunology, one of the main promises of this approach is the ability to define cell types as clusters in the whole transcriptome space (i.e., without relying on specific surface markers), thereby providing an unbiased classification of immune cell types. So far, this technology has been mainly applied in mouse and human. However, technically it could be used for immune cell type identification in any species without requiring the development and validation of species-specific antibodies for cell sorting. Here, we review recent developments using single-cell RNA-seq to characterize immune cell populations in non-mammalian vertebrates, with a focus on zebrafish (*Danio rerio*). We advocate that single-cell RNA-seq technology is likely to provide key insights into our understanding of the evolution of the adaptive immune system.

1 Introduction—Immune Cell Types in Vertebrates

The adaptive immune system, as defined in mammals, is centered on lymphocytes-bearing antigen receptors generated by recombination-activating gene (RAG)-mediated rearrangement. This highly complex system arose early in the evolution of jawed vertebrates approximately 500 million years ago (Cooper and Alder 2006; Flajnik and Kasahara 2010). The adaptive immune system is able to generate an enormous diversity of B- and T-cell receptors that can recognize and eliminate newly encountered pathogens and develop memory to protect against previously encountered pathogens. The adaptive immune system of jawed vertebrates has a functional

S. J. Carmona · D. Gfeller (✉)
Department of Oncology, Ludwig Institute for Cancer Research, University of Lausanne, Lausanne, Switzerland
e-mail: david.gfeller@unil.ch

S. J. Carmona · D. Gfeller
Swiss Institute of Bioinformatics (SIB), Lausanne, Switzerland

© Springer International Publishing AG, part of Springer Nature 2018
P. Pontarotti (ed.), *Origin and Evolution of Biodiversity*,
https://doi.org/10.1007/978-3-319-95954-2_6

counterpart in jawless vertebrates (lampreys and hagfish), involving somatic rearrangement of variable regions in leucine-rich repeat proteins (Pancer et al. 2004). Mammalian species also have an innate immune system that does not possess the ability to somatically rearrange receptors but relies on largely expanded families of genes such as Toll-like receptors (TLRs) or killer cell immunoglobulin-like receptors (KIRs). The innate immune system is a key for mounting a rapid response against pathogens and for modulating the adaptive immune response. The innate and adaptive immune systems further play central roles in wound healing, tissue remodeling, and homeostasis.

Different types of immune responses are mediated by different types of immune cells. In mammals, immunologists have been able to study the immune system at a very high resolution, delineating more than ten main cell types, including CD4/CD8 T cells, B cells, dendritic cells (DC), macrophages, granulocytes, natural killers (NK), or other innate lymphoid cells (ILC). These can be further split into tens of distinct cellular subtypes or states [e.g., Th1, Th2, Th17 CD4 T cells, M1-/M2-polarized macrophages, several DCs subtypes (Manh et al. 2013; Villani et al. 2017)]. Immunologists take advantage of monoclonal antibodies for more than 300 cell surface markers that enable fine dissection of immune cell types (Clark et al. 2016). Once a large number of such markers are available, new populations can be discovered by exploring T-cell subsets expressing different combinations of them [e.g., Lin$^-$ CD123+CD127 low innate lymphoid cells (Mora-Velandia et al. 2017)]. This approach has proved very powerful in mouse and human and resulted in a detailed phenotypic and functional characterization of immune cell populations in these two species. Gene expression profiling of sorted cell populations revealed both distinct and conserved molecular signatures of immune cells in mouse and human (Crozat et al. 2010; Robbins et al. 2008).

Historically, lymphocytes were first discriminated from myeloid cells among white blood cells in 1879 (Vivier et al. 2016). Almost a century later, B and T cells were distinguished in birds and mammals (Warner et al. 1962) and later NK cells were identified (Kiessling et al. 1975) as non-T, non-B lymphocytes with the capacity to spontaneously kill tumor cells without the need for prior immunization. In 1997, lymphoid tissue inducer (LTi) cells were identified (Mebius et al. 1997). More recently, new types of non-T, non-B lymphocytes were identified and have been classified as 'innate lymphoid cells' including subsets ILC1, ILC2, ILC3, NK cells, and LTi (Spits et al. 2013; Vivier et al. 2016).

Since many decades, people realized that studying the immune system of evolutionary distant vertebrates would be a powerful approach to define selection pressures that have shaped mechanisms, molecules, and specialized cells or structures during the evolution of the adaptive immune system (Cooper and Alder 2006; Flajnik and Kasahara 2010; Pancer et al. 2004). Approaches based on anatomy and histology were historically the first to investigate immune cells in different species. Birds were often used for such studies (Cooper et al. 1965), and it is noteworthy to recall that B cells were first discovered in birds as cells developing in bursa of Fabricius, hence the "B" of B cells (Cooper et al. 1966). Jawed fish were observed to possess lymphocyte populations that are analogous to T cells, B cells, and non-specific cytotoxic cells, macrophages, and polymorphonuclear leukocytes (Iwama and Nakanishi 1996). Comparative studies also revealed that the same organs perform different roles in the development of immune cells in different species. For instance, in teleost fish, kidneys play the role of bone marrow in mammalian species for the development of T and B cells (Zapata and Amemiya 2000).

With the availability of genome sequencing technologies, most of our current understanding of immune cell types in non-mammalian species and their evolution comes from comparative genomic studies. These studies have provided very important insights into the evolution of the vertebrates' immune system, including tracing the origin of RAG (Morales Poole et al. 2017) or the discovery of the evolutionary distant cytokines, chemokines, and their receptors (Jacobson et al. 2017). Many genes used as markers of mammalian immune cell populations are conserved in all jawed vertebrates, suggesting that the corresponding cell types are also conserved (Cooper and Alder 2006; Flajnik and Kasahara 2010; Vivier et al. 2016). Unfortunately, comparative genomics studies are limited to orthology-based relationships and cannot provide phenotypic or functional information about immune cell populations.

Gene expression profiling and molecular fingerprints have been successfully used to investigate the evolution of organs and tissues (Arendt 2005, 2008; Brawand et al. 2011). In particular, gene expression analyses of lymphatic tissues using cDNA microarrays and RNA-seq have greatly contributed to our understanding of immune responses across multiple vertebrates. These include studying immune responses of teleost fish to pathogens (Sudhagar et al. 2018) or nutrition (Martin and Król 2017), or the discovery of new cytokines (Jacobson et al. 2017). Moreover, cell sorting strategies combined with gene expression profiling have been used to identify and characterize distinct populations of mononuclear phagocytes in distant mammalian species and chicken (Contreras et al. 2010; Vu Manh et al. 2014, 2015).

Unfortunately, in 'bulk' gene expression profiling of multicellular organisms, the mixing of numerous cell types results in the dilution and masking the heterogeneity of immune cell types. Furthermore, a deep understanding of the evolution of the immune system will ultimately require the study of individual immune cell types, their ontogeny, differentiation, function, and interactions with other cell types (Arendt et al. 2016).

To overcome these limitations, different tools have been developed such as antibodies or transgenic lines. In the field of comparative immunology of jawed vertebrates, the teleost zebrafish (*Danio rerio*) has become a reference species for most immunological studies in non-mammalian vertebrates. This organism presents many important advantages such as the possibility to conduct in vivo imaging and the availability of powerful genetic tools (Renshaw and Trede 2012). The genome of zebrafish is also among the best-annotated genomes of non-mammalian vertebrates (Howe et al. 2013).

Classical studies of hematopoiesis in zebrafish have defined major blood lineages. In the kidney marrow, the site of adult hematopoiesis, erythroid, lymphoid, and myeloid cells is present and can be separated based on light-scattering characteristics (Traver et al. 2003). However, subsets of cells within each of these lineages are poorly defined, primarily because of the lack of markers to distinguish them, therefore limiting our understanding of immune cell types.

Immune cell-specific monoclonal antibodies can be in principle developed for any species from which a marker gene can be cloned. For example, monoclonal antibodies have been recently developed for lampreys, enabling the study of novel lymphocytic lineages (Hirano et al. 2013) at the base of the vertebrates' phylogenetic tree. However, the high costs and technical challenges associated with the development of such tools for less studied species, in particular outside of human and mouse, significantly limit characterization of immune cells types. In the absence of appropriate antibodies, the use of fluorescent reporter lines has greatly contributed to the characterization of immune cells in zebrafish (Udvadia and Linney 2003). Many lines are available where fluorescent proteins are expressed under the control of immune cells' specific promoters, such as *mpx* for neutrophils (Renshaw et al. 2006), *lck* for T cells and NK cells (Carmona et al. 2017; Langenau et al. 2004), *foxp3* for T regs (Kasheta et al. 2017), and *rag2* for B cells (Page et al. 2013). Isolation of these fluorescent cells of a particular lineage from whole organs by FACS enables population-based gene profiling (Rougeot et al. 2014) or targeted single-cell gene expression analysis [using multiplex single-cell RT-PCR (Moore et al. 2016)]. Thanks to the availability of reporter lines like these, it has been also possible to study transcriptional changes of immune cells during the course of infection (Saraceni et al. 2016).

However, not all immune cell types are clearly defined based on expression of a single gene, and therefore, antibodies or transgenic lines for prospective isolation of defined cell populations have only been developed for a handful of immune cell types in zebrafish and are almost absent in other non-mammalian vertebrate species.

Recently, the development of sensitive and accurate single-cell RNA-sequencing (scRNA-seq) protocols has profoundly transformed our ability to comprehensively analyze immune cells and deconvolve their heterogeneity with virtually no need for cell sorting based on specific markers, opening the possibility to study immune cells in any vertebrate species.

2   Single-Cell RNA-Seq in Immunology

Over the past years, numerous scRNA-seq protocols have been developed. Currently, all scRNA-seq protocols consist of first retro-transcription of mRNA in each cell into cDNA, second amplification of cDNA by polymerase chain reaction (PCR) or in vitro transcription, and finally sequencing of cDNA libraries, enabling quantification of the expression level of individual transcripts. Kolodziejczyk et al. (2015) provide a comprehensive review of individual scRNA-seq protocols and their relative strengths and weaknesses. Different techniques can be classified based on the sequencing coverage of the transcripts. Methods such as Smart-Seq2 (Picelli et al. 2014) and CEL-seq2 (Hashimshony et al. 2016) are based on full-length mRNA sequencing, allowing for isoform quantification and de novo transcriptome assembly. Droplet-based methods such as 10X-Genomics (Zheng et al. 2017) and InDrops (Zilionis et al.

2016) generate only 3′ terminal-end short sequencing reads, due to the integration of a sequencing priming site on the oligo(dT) primer used for reverse transcription.

Droplet-based methods have a high throughput (tens of thousands of cells). Plate-based methods such as Smart-seq2 have lower throughput (hundreds of cells) but can be coupled with FACS index sorting, simultaneously producing light-scattering and protein expression measurements along with single-cell transcriptomes.

Moreover, full-length mRNA sequencing protocols such as Smart-seq2 allow for coupled T- and B-cell receptor reconstruction and transcriptome quantification in the same cell, for species where VDJ loci have been annotated, opening the door for high-throughput simultaneous analysis of T- and B-cell heterogeneity and clonality (Afik et al. 2017; Canzar et al. 2016; Eltahla et al. 2016; Stubbington et al. 2016).

Current scRNA-seq protocols amplify about ten percent of the transcripts in each cell (Islam et al. 2014). Moreover, burst-like stochastic activation of transcription (Raj and van Oudenaarden 2008) introduces additional intrinsic variability (gene 'drop-outs') in measurements of single-cell transcriptomes. This represents a limitation to detect low-abundance transcripts in a single cell. However, in practice, the large number of cells compensates this weakness and the full transcriptome of any cell type can be accurately reconstructed by computationally pooling (i.e., clustering) cells with similar transcriptomes (see later for estimations on the minimum number of cells for scRNA-seq required to sample a population of interest). As such, unsupervised clustering or other more supervised classification of cells into cell types or states is a fundamental task in scRNA-seq data analysis. The goal is to detect clusters of cells that have sufficiently similar transcriptomes with other cells of the same cluster and significantly different transcriptomes from cells in other clusters. Unsupervised approaches use the whole transcriptome or the set of most variable genes, without additional information. More supervised approaches can pre-select specific markers based on a priori knowledge. However, as mentioned earlier, the high dropout rate in scRNA-seq often results in inaccurate or incomplete classification when using only on a few markers [see example in Fig. 1 and discussion in Carmona et al. (2017)]. In addition to the classical k-means and hierarchical clustering that perform well, tens of clustering methods have been applied to the task of clustering scRNA-seq data [reviewed in Andrews and Hemberg (2018)]. Very closely related to clustering are the concepts of dimensionality reduction and visualization with methods like principal component analysis (PCA), multi-dimensional scaling (MDS), or t-distributed stochastic neighbor embedding (t-SNE), all of which are fundamental to summarize high-dimensional data produced by scRNA-seq assays and enable human interpretation of cell heterogeneity.

scRNA-seq has been used to deconvolve immune cell type heterogeneity by identifying novel distinct immune cell subsets in health and diseases (Papalexi and Satija 2017; Stubbington et al. 2017) and can provide new insights into the development and evolution of cell types (Marioni and Arendt 2017). Moreover, scRNA-seq has the potential to reconstruct or predict developmental 'trajectories' for immune cells. Developmental trajectories are now routinely profiled with scRNA-seq. Such studies have led to a deeper understanding of the regulation of early myeloid, conventional

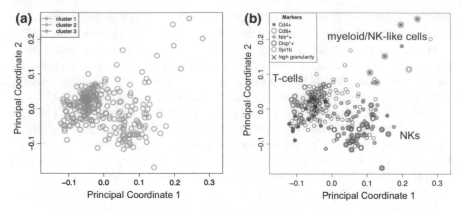

Fig. 1 Low-dimensionality projection of single-cell RNA-seq data from zebrafish immune cells [images from Carmona et al. (2017)]. **a** Multi-dimensional scaling (MDS) and clustering of lck+ cells from adult zebrafish spleens reveal three main clusters. **b** MDS projection where expression of specific marker genes revealed T cells, natural killers (NKs), and a smaller population of cells with features of both myeloid and NK cells

dendritic cells (Schlitzer et al. 2015), and megakaryocytic differentiation (Psaila et al. 2016), as well as of fate decisions along T-helper subtypes commitment during malaria infection (Lönnberg et al. 2017).

3 Single-Cell RNA-Seq of Immune Cells Beyond Mammals: Case Studies in Zebrafish

Recently, we and others have successfully applied scRNA-seq to characterize immune cells in zebrafish (Athanasiadis et al. 2017; Carmona et al. 2017; Tang et al. 2017).

To this end, we took advantage of a zebrafish transgenic line expressing GFP under the control of the lymphocyte-specific transcription factor *lck* (Langenau et al. 2004). This line was previously thought to be T-cell-specific and indeed enabled researchers to track zebrafish T-cell development in vivo. However, *lck* in mouse and human is expressed in both T and NK cells, where it has an important role in activation by phosphorylating tyrosine residues within the immunoreceptor tyrosine-based activation motifs (ITAM) of the cytoplasmic tails of the TCR-gamma chains and CD3 subunits and of NK activating receptors (Binstadt et al. 1997). Therefore, we speculated that this expression pattern could be conserved in teleosts and that a larger heterogeneity may exist among *lck*+zebrafish lymphocytes than previously thought.

To address this question, our collaborators from the Cvejic and Teichmann groups performed scRNA-seq of *lck*+zebrafish cells (Carmona et al. 2017). Briefly, spleens from adult Tg(*lck*:GFP) zebrafish were dissected and single GFP + cells were FACS (index) sorted into 96-well plates containing lysing buffer, for subsequent cDNA library preparation. Of note, index sorting allows for parameters recorded on the cytometer to be linked to each sorted cell. Single-cell cDNA libraries were generated using the Smart-seq2 protocol (Picelli et al. 2014) and sequenced using 125-bp paired-end reads on Illumina HiSeq2000.

Gene expression was quantified as transcripts per million (TPMs) in each cell using the zebrafish reference transcriptome (Howe et al. 2013, and global similarities between all pairs of cells were computed using Pearson's correlation. Unsupervised clustering analysis suggested the presence of at least three distinct cell types (Carmona et al. 2017). Given the high-dimensionality of these data (~12,000 genes × 300 cells), dimensionality reduction methods are needed to aid visual interpretation of the cellular heterogeneity. Multi-dimensional scaling (MDS) was applied to the similarity matrix to obtain a low-dimensionality projection (different colors indicate the three clusters found in the original space) (Fig. 1a). Remarkably, T-cell-specific genes such as Cd4 or Cd8 were specifically expressed in only one of the clusters (Fig. 1b). Upon analysis of the differentially expressed genes among the three clusters, we found that the second cluster specifically expressed multiple members of the NITR and DICP innate immune receptors (Fig. 1b) as well as the antimicrobial peptide NK-lysin 2 (Carmona et al. 2017). Comparative genomics and functional studies suggested that these genes might be expressed by a fish immune cell type equivalent to mammalian natural killer cells (Haire et al. 2012; Pereiro et al. 2015; Yoder et al. 2004) [see also (Moore et al. 2016) for a description of NK-lysin 4 expression based on single-cell qPCR]. These results suggest that cells in cluster 2 may correspond to a natural killer-like cell type in zebrafish. Interestingly, this NK signature was confirmed in a *rag1-/* knock out *lck*:GFP zebrafish line that fails to develop T cells (Tang et al. 2017), further supporting the identity of this lymphocytic subset.

The third and smaller cell cluster (green points Fig. 1a) presented features of both natural killer and myeloid cells, including higher granularity and specific expression of the transcription factor *Spi1b,* the granulocyte/macrophage colony-stimulating factor receptor beta (*csf2rb*), the Src family tyrosine kinase *hck* along with NK lysins 3 and 4, perforin (*prf1.7*), and chemokines *ccl35.1* and *ccl33.3* (Carmona et al. 2017). Interestingly, a very similar population was described (Tang et al. 2017), indicating that this represents likely a novel subtype of innate immune cells.

Remarkably, by clustering cells based on their transcriptional similarity and annotating the clusters based on the expression of a handful of marker genes with known or predicted function, scRNA-seq allowed us to obtain for the first time the full transcriptomes of T cells and Natural killers in an evolutionary distant vertebrate (Carmona et al. 2017). In addition, these data provided expression information about multiple cell-type-specific cytokines, cytokine receptors, and other immune-related genes. The evolutionary insights from these data are discussed later.

In larger scale experiments, Langenau and colleagues (Tang et al. 2017) used scRNA-seq to study cell heterogeneity within zebrafish kidney marrow, the equivalent of mammalian bone marrow (BM) and site of larval and adult hematopoiesis. First, they isolated cells from reporter transgenic fish lines and performed Smart-seq2 to obtain gene signatures of defined lineages: Tg(*Runx1 + 23:*GFP) for hematopoietic stem and progenitor cells (HSPCs) (Tamplin et al. 2015); Tg(*cd41:*GFP), for both HSPCs and thrombocytes (Lin 2005); Tg(*mpx:*GFP), for neutrophil/myeloid lineages (Mathias et al. 2006; Renshaw et al. 2006), Tg(*rag2:*GFP), for marrow-derived B cells (Page et al. 2013); and Tg(*lck:*GFP), for T and NK cells (Carmona et al. 2017; Langenau et al. 2004; Moore et al. 2016).

Having defined signatures for most hematopoietic cell lineages, the researchers then profiled almost 4,000 unlabeled single cells isolated from the kidney of wild-type zebrafish using InDrops scRNA-seq, obtaining an unbiased landscape of the blood and kidney cells. Finally, Tang et al. were able to study immune cell deficiencies within DNA–protein kinase catalytic subunit (*prkdc*) and interleukin-2 receptor γ a (*il2rga*) mutant fish, identifying blood cell losses in T, B, and natural killer cells. This highlights the power of scRNA-seq to address the impact of specific genetic mutants on different immune cell compartments.

Immune cells develop from multipotent stem cells through the process known as hematopoiesis. The molecular pathways governing hematopoiesis are highly conserved in vertebrates, where most (if not all) critical transcription factors involved in hematopoiesis in mammals have orthologs in fish (Davidson and Zon 2004). In the classical model of hematopoiesis, an organized hematopoietic lineage tree starts with multipotent hematopoietic stem cells and is then followed by oligopotent and unipotent progenitors. However, recent single-cell results challenged the classic model and proposed that hematopoietic progenitor cell types are very heterogeneous (Notta et al. 2016; Paul et al. 2015).

In zebrafish, Cvejic and colleagues applied single-cell RNA-seq to study thrombocyte lineage commitment and were able to reconstruct a continuous lineage differentiation pathway (Macaulay et al. 2016). More recently, their study was extended to reconstruct differentiation trajectories of most hematopoiesis lineages by performing scRNA-seq of kidney-derived blood cells from eight different zebrafish transgenic reporter lines (Athanasiadis et al. 2017). Their results indicated that the haematopoietic program is highly conserved between zebrafish and higher vertebrates.

Altogether, these different single-cell RNA-seq studies demonstrate that immune cell types can be identified by clustering cells based on transcriptional similarity in zebrafish and provided detailed phenotypic characterization of distinct immune cell populations. Although several of these studies took advantage of reporter lines, the increase in throughput of scRNA-seq techniques will make it increasingly feasible to start from unsorted cell populations to decipher the complexity of immune cell types in any vertebrate species.

4 Envisioning Immune Cell Type Identification Across Vertebrates from Blood Samples

Building upon the success of these recent single-cell RNA-seq studies in zebrafish, we envision that a powerful approach to identify and characterize immune cells in any vertebrate species could be to perform scRNA-seq on whole blood cells or peripheral blood mononuclear cells (PBMC)-enriched (e.g., by using Ficoll gradient). ScRNA-seq of PBMCs has allowed for detailed characterization of leukocytes heterogeneity in mammals from fresh (Zheng et al. 2017) and cryopreserved (Guillaumet-Adkins et al. 2017) samples. In species for which sufficient amounts of the blood are not readily accessible, like zebrafish, dissection and sampling of immune tissues such as kidney marrow might be a viable alternative to obtain a high proportion of (unlabeled) immune cells (Tang et al. 2017). Different leukocyte isolation methods from peripheral blood have been employed for different fish species (Inoue et al. 2002; Pierrard et al. 2012), indicating that although erythrocytes are nucleated in non-mammalian species, physical separation of red blood cells from other cell types found in blood is feasible in most cases.

In species for which reference genomes or transcriptomes are available for sequence mapping or transcript quantification, high-throughput droplet-based scRNA-seq methods based on 3' counting, such as 10× Genomics or InDrops, represent an attractive approach. For other organisms with limited or lack of genomic information, full-length mRNA scRNA-seq protocols, such as Smart-seq2, enables de novo transcriptome assembly using tools such as Trinity (Grabherr et al. 2011) and Velvet (Schulz et al. 2012). RNA-seq followed by de novo transcriptome assembly has been successfully conducted in common (Ji et al. 2012) and crucian carp (Liao et al. 2013; Rhee et al. 2014), leading to the identification of hundreds of immune-related genes. Of note, it has been shown that representative assemblies can be generated with as few as 20 million reads from tissue samples (Francis et al. 2013), suggesting that the amount of reads obtained in standard scRNA-seq experiments is likely enough to reconstruct the full transcriptome.

Once reads have been mapped to their respective transcripts, clustering techniques should be used to group cells based on their transcriptional similarity. Representative transcriptome for different cells types is then obtained by pooling together cells of each cluster. Cluster annotation should be done first by using existing markers derived from studies in mammalian. Given that most immune marker genes are conserved in jawed vertebrates, this approach is likely to work very well and was successfully applied in zebrafish (Carmona et al. 2017; Tang et al. 2017). Recently, we also proposed a similar approach for identifying ILCs in single-cell RNA-seq that relies on gene signatures that are conserved in most jawed vertebrates (Suffiotti et al. 2017). Whether totally new clusters will be left after annotating all those that display similarity with mammalian immune cell types is still unclear. However, it is likely that some clusters will show less striking resemblance to the mammalian counterpart, which can indicate distinct rates of evolution. In our work, for instance, we observed that many genes in the NK-like cell cluster (including predicted NK receptors NITR and DICP genes) were not conserved in mammalian species.

In Fig. 2, we delineate the main steps of this proposed strategy and the number of cells that should be considered as starting material. In the order of 10^7 leukocytes and thrombocytes per ml of the blood can be typically recovered from Asian catfish (Pierrard et al. 2012). Teleost fish typically contains comparable amounts of

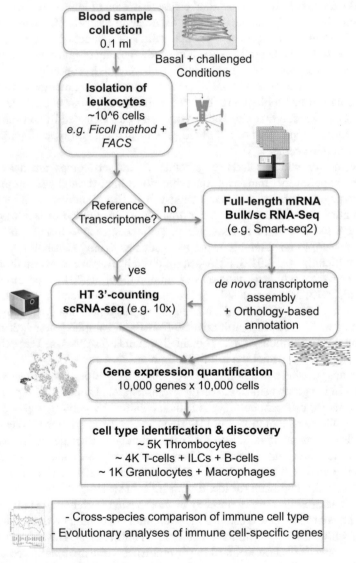

Fig. 2 General pipeline for studying immune cell types at the single-cell gene expression level, starting from whole blood samples, in any organism with enough accessible blood

thrombocytes and leukocytes, and among leukocytes, up to 90% can correspond to lymphocytes (Rey Vázquez and Guerrero 2007). Therefore, in a plausible scenario, from 100 μl of the blood of a fish with typical blood composition, we estimate that one million leukocytes and thrombocytes can be obtained. This amount is two orders of magnitude bigger than what is required to obtain 10,000 single-cell transcriptomes using 10x Genomics current technologies (for which the cell recovery rate is about 50% (Zheng et al. 2017). From these 10K cells, we would roughly obtain 5K thrombocytes, 4K lymphocytes, and 1K of granulocytes and monocytes. As a reference, in zebrafish, 3,000 single-cell transcriptomes obtained from kidney marrow were sufficient to identify HSCs, thrombocytes, macrophages, neutrophils, different subsets of T cell and Natural killer cells, B cells, and erythrocytes, among others (Tang et al. 2017).

Overall, our estimates indicate that the approach outlined in Fig. 2 is likely applicable to species from most vertebrate clades.

5 Consequences for Our Understanding of Immune Cell Type Evolution, with a Focus on Lymphocytes

Phylogenctic explorations of the adaptive immune system have relied on the identification of genes such as *TCR*, *BCR*, *RAG1/RAG2*, *MHC class I* and *class II* as the key elements of the adaptive immune system. This search has led to the identification of these genes in all jawed vertebrates that have been carefully examined, including cartilaginous fish, like sharks (Cannon et al. 2004; Cooper and Alder 2006)

However, the potential for unbiased, marker-free identification of cell types as clusters in the transcriptome space in any organism opens novel opportunities in comparative immunology that go far beyond genomic conservation of immune genes and might contribute to our understanding of the origin and evolution of cell types (Arendt et al. 2016; Marioni and Arendt 2017).

With current scRNA-seq technologies, it is now virtually possible to 'discover' de novo all of the main immune cell populations by sequencing thousands of cells and clustering them based on transcriptome similarity (Trapnell 2015). By applying scRNA-seq beyond mammals, for instance, we have identified novel types of natural killer-like cells (Carmona et al. 2017). By anchoring on expected NK features, such as expression of innate immune receptors (e.g., *nitr2a*, *dicp1.1*), cytotoxic molecules (Perforin, granzymes), antimicrobial peptides (*nkl.2* in particular), and lack of T- and B-cell features, multiple novel NK-specific molecules were identified, including cytokines, cytokine receptors, and other immune-related genes. Many of these genes have recently evolved by duplication and neofunctionalization and do not necessarily have mammalian orthologs (Carmona et al. 2017). Indeed, changes in immune gene expression patterns (a kind of neofunctionalization) are likely to reflect adaptation to different habitats (Yang et al. 2016). Interestingly, zebrafish might represent a particularly attractive model to study NKs and other ILCs, considering the robust immunity of rag1-/- mutants that lack T and B cells, but are yet able to maintain a low mortality rate (Tokunaga et al. 2017).

Finally, once immune cell type-specific genes have been identified, genomic sequence analysis, such as ortholog identification and calculation of non-synonymous to synonymous substitution rates, can inform us about cell type-specific evolutionary rates. This approach was first introduced by ourselves and collaborators when characterizing T and NK cells in zebrafish (Carmona et al. 2017). A striking result from this study was the much lower conservation of transmembrane NK-specific proteins compared to other transmembrane proteins, including those specific for T cells.

6   Conclusion

The last decade has witnessed an exponential growth of single-cell RNA-seq data (Svensson et al. 2018), but most of these data came from only a few big academic centers that had invested into this technology since the beginning. With the advent of streamlined protocols such as those available at $10\times$ Genomics or InDrops, it is very likely that scRNA-seq will become available in many more places in the next few years and will be routinely performed to analyze gene expression at the single-cell level in many different tissues and species (Regev et al. 2017).

The use of this technology to characterize immune cell populations from a broad range of vertebrate species following the pipeline proposed in Fig. 2 will likely impact our understanding of the evolution of cell types (Arendt 2008; Arendt et al. 2016; Marioni and Arendt 2017) and especially of immune cell types which are difficult to isolate in the absence of antibodies. First, it could confirm the existence of mammalian immune cell types predicted based on the conservation of known markers in other non-mammalian species. This is likely the most important insight that will come from such studies. For instance, it could provide transcriptomic evidence about the conservation of ILCs (Suffiotti et al. 2017; Vivier et al. 2016). Similarly, it will enable scientists to monitor gene expression changes in specific immune cell populations upon infections or other stimuli. Second, it could help predict new markers and study the evolution rate of immune cell types, for instance, by assessing the sequence conservation across vertebrate species or dN/dS ratio of genes specifically expressed in a given cell type in a given species (Carmona et al. 2017). Third, it will help studying neofunctionalization events (i.e., genes that are not expressed in the same cell types as their mammalian orthologs). Finally, it may potentially reveal completely novel immune cell types that have no mammalian equivalent, providing similar insights into the forces that drive the evolution of the adaptive immune system as the discovery of the adaptive immune system in jawless vertebrates (Pancer et al. 2004). For these different reasons, we anticipate that, in the next few years, comparative immunology will strongly benefit from technological developments in scRNA-seq that so far were mainly applied to mouse and human samples.

References

Afik S, Yates KB, Bi K, Darko S, Godec J, Gerdemann U, Swadling L, Douek DC, Klenerman P, Barnes EJ et al (2017) Targeted reconstruction of T cell receptor sequence from single cell RNA-seq links CDR3 length to T cell differentiation state. Nucleic Acids Res 45:e148

Andrews TS, Hemberg M (2018) Identifying cell populations with scRNASeq. Mol Aspects Med 59:114–122

Arendt D (2005) Genes and homology in nervous system evolution: comparing gene functions, expression patterns, and cell type molecular fingerprints. Theory Biosci Theor Den Biowissenschaften 124:185–197

Arendt D (2008) The evolution of cell types in animals: emerging principles from molecular studies. Nat Rev Genet 9:868–882

Arendt D, Musser JM, Baker CVH, Bergman A, Cepko C, Erwin DH, Pavlicev M, Schlosser G, Widder S, Laubichler MD et al (2016) The origin and evolution of cell types. Nat Rev Genet 17:744–757

Athanasiadis EI, Botthof JG, Andres H, Ferreira L, Lio P, Cvejic A (2017) Single-cell RNA-sequencing uncovers transcriptional states and fate decisions in haematopoiesis. Nat Commun 8

Binstadt BA, Brumbaugh KM, Leibson PJ (1997) Signal transduction by human NK cell MHC-recognizing receptors. Immunol Rev 155:197–203

Brawand D, Soumillon M, Necsulea A, Julien P, Csárdi G, Harrigan P, Weier M, Liechti A, Aximu-Petri A, Kircher M et al (2011) The evolution of gene expression levels in mammalian organs. Nature 478:343–348

Cannon JP, Haire RN, Rast JP, Litman GW (2004) The phylogenetic origins of the antigen-binding receptors and somatic diversification mechanisms. Immunol Rev 200:12–22

Canzar S, Neu KE, Tang Q, Wilson PC, Khan AA (2016) Basic: BCR assembly from single cells. Bioinformatics 631

Carmona SJ, Teichmann SA, Ferreira L, Macaulay IC, Stubbington MJT, Cvejic A, Gfeller D (2017) Single-cell transcriptome analysis of fish immune cells provides insight into the evolution of vertebrate immune cell types. Genome Res 27:451–461

Clark G, Stockinger H, Balderas R, van Zelm MC, Zola H, Hart D, Engel P (2016) Nomenclature of CD molecules from the tenth human leucocyte differentiation antigen workshop. Clin Transl Immunol 5:e57

Contreras V, Urien C, Guiton R, Alexandre Y, Vu Manh T-P, Andrieu T, Crozat K, Jouneau L, Bertho N, Epardaud M et al (2010) Existence of CD8α-like dendritic cells with a conserved functional specialization and a common molecular signature in distant mammalian species. J Immunol Baltim Md 1950(185):3313–3325

Cooper MD, Alder MN (2006) The evolution of adaptive immune systems. Cell 124:815–822

Cooper MD, Peterson RD, Good RA (1965) Delineation of the thymic and bursal lymphoid systems in the chicken. Nature 205:143–146

Cooper MD, Raymond DA, Peterson RD, South MA, Good RA (1966) The functions of the thymus system and the bursa system in the chicken. J Exp Med 123:75–102

Crozat K, Guiton R, Guilliams M, Henri S, Baranek T, Schwartz-Cornil I, Malissen B, Dalod M (2010) Comparative genomics as a tool to reveal functional equivalences between human and mouse dendritic cell subsets. Immunol Rev 234:177–198

Davidson AJ, Zon LI (2004) The "definitive" (and 'primitive') guide to zebrafish hematopoiesis. Oncogene 23:7233–7246

Eltahla AA, Rizzetto S, Pirozyan MR, Betz-Stablein BD, Venturi V, Kedzierska K, Lloyd AR, Bull RA, Luciani F (2016) Linking the T cell receptor to the single cell transcriptome in antigen-specific human T cells. Immunol Cell Biol 94:604–611

Flajnik MF, Kasahara M (2010) Origin and evolution of the adaptive immune system: genetic events and selective pressures. Nat Rev Genet 11:47–59

Francis WR, Christianson LM, Kiko R, Powers ML, Shaner NC, Haddock SH (2013) A comparison across non-model animals suggests an optimal sequencing depth for de novo transcriptome assembly. BMC Genom 14:167

Grabherr MG, Haas BJ, Yassour M, Levin JZ, Thompson DA, Amit I, Adiconis X, Fan L, Raychowdhury R, Zeng Q et al (2011) Full-length transcriptome assembly from RNA-Seq data without a reference genome. Nat Biotechnol 29:644–652

Guillaumet-Adkins A, Rodríguez-Esteban G, Mereu E, Mendez-Lago M, Jaitin DA, Villanueva A, Vidal A, Martinez-Marti A, Felip E, Vivancos A et al (2017) Single-cell transcriptome conservation in cryopreserved cells and tissues. Genome Biol 18

Haire RN, Cannon JP, O'Driscoll ML, Ostrov DA, Mueller MG, Turner PM, Litman RT, Litman GW, Yoder JA (2012) Genomic and functional characterization of the diverse immunoglobulin domain-containing protein (DICP) family. Genomics 99:282–291

Hashimshony T, Senderovich N, Avital G, Klochendler A, de Leeuw Y, Anavy L, Gennert D, Li S, Livak KJ, Rozenblatt-Rosen O et al (2016) CEL-Seq2: sensitive highly-multiplexed single-cell RNA-Seq. Genome Biol 17

Hirano M, Guo P, McCurley N, Schorpp M, Das S, Boehm T, Cooper MD (2013) Evolutionary implications of a third lymphocyte lineage in lampreys. Nature 501:435–438

Howe K, Clark MD, Torroja CF, Torrance J, Berthelot C, Muffato M, Collins JE, Humphray S, McLaren K, Matthews L et al (2013) The zebrafish reference genome sequence and its relationship to the human genome. Nature 496:498–503

Inoue T, Moritomo T, Tamura Y, Mamiya S, Fujino H, Nakanishi T (2002) A new method for fish leucocyte counting and partial differentiation by flow cytometry. Fish Shellfish Immunol 13:379–390

Islam S, Zeisel A, Joost S, La Manno G, Zajac P, Kasper M, Lönnerberg P, Linnarsson S (2014) Quantitative single-cell RNA-seq with unique molecular identifiers. Nat Methods 11:163–166

Jacobson G, Muncaster S, Mensink K, Forlenza M, Elliot N, Broomfield G, Signal B, Bird S (2017) Omics and cytokine discovery in fish: presenting the Yellowtail kingfish (Seriola lalandi) as a case study. Dev Comp Immunol 75:63–76

Ji P, Liu G, Xu J, Wang X, Li J, Zhao Z, Zhang X, Zhang Y, Xu P, Sun X (2012) Characterization of common carp transcriptome: sequencing, de novo assembly, annotation and comparative genomics. PLoS ONE 7:e35152

Kasheta M, Painter CA, Moore FE, Lobbardi R, Bryll A, Freiman E, Stachura D, Rogers AB, Houvras Y, Langenau DM et al (2017) Identification and characterization of T reg-like cells in zebrafish. J Exp Med 214:3519–3530

Kiessling R, Klein E, Wigzell H (1975) "Natural" killer cells in the mouse. I. Cytotoxic cells with specificity for mouse Moloney leukemia cells. Specificity and distribution according to genotype. Eur J Immunol 5:112–117

Kolodziejczyk AA, Kim JK, Svensson V, Marioni JC, Teichmann SA (2015) The technology and biology of single-cell RNA sequencing. Mol Cell 58:610–620

Langenau DM, Ferrando AA, Traver D, Kutok JL, Hezel J-PD, Kanki JP, Zon LI, Look AT, Trede NS (2004) In vivo tracking of T cell development, ablation, and engraftment in transgenic zebrafish. Proc Natl Acad Sci 101:7369–7374

Liao X, Cheng L, Xu P, Lu G, Wachholtz M, Sun X, Chen S (2013) Transcriptome analysis of crucian carp (Carassius auratus), an important aquaculture and hypoxia-tolerant species. PLoS ONE 8:e62308

Lin H-F (2005) Analysis of thrombocyte development in CD41-GFP transgenic zebrafish. Blood 106:3803–3810

Lönnberg T, Svensson V, James KR, Fernandez-Ruiz D, Sebina I, Montandon R, Soon MSF, Fogg LG, Nair AS, Liligeto UN et al (2017) Single-cell RNA-seq and computational analysis using temporal mixture modeling resolves T_H1/T_{FH} fate bifurcation in malaria. Sci Immunol 2:eaal2192

Macaulay IC, Svensson V, Labalette C, Ferreira L, Hamey F, Voet T, Teichmann SA, Cvejic A (2016) Single-cell RNA-sequencing reveals a continuous spectrum of differentiation in hematopoietic cells. Cell Rep 14:966–977

Manh T-PV, Alexandre Y, Baranek T, Crozat K, Dalod M (2013) Plasmacytoid, conventional, and monocyte-derived dendritic cells undergo a profound and convergent genetic reprogramming during their maturation. Eur J Immunol 43:1706–1715

Marioni JC, Arendt D (2017) How single-cell genomics is changing evolutionary and developmental biology. Annu Rev Cell Dev Biol 33:537–553

Martin SAM, Król E (2017) Nutrigenomics and immune function in fish: new insights from omics technologies. Dev Comp Immunol 75:86–98

Mathias JR, Perrin BJ, Liu T-X, Kanki J, Look AT, Huttenlocher A (2006) Resolution of inflammation by retrograde chemotaxis of neutrophils in transgenic zebrafish. J Leukoc Biol 80:1281–1288

Mebius RE, Rennert P, Weissman IL (1997) Developing lymph nodes collect CD4 + CD3-LT beta + cells that can differentiate to APC, NK cells, and follicular cells but not T or B cells. Immunity 7:493–504

Moore FE, Garcia EG, Lobbardi R, Jain E, Tang Q, Moore JC, Cortes M, Molodtsov A, Kasheta M, Luo CC et al (2016) Single-cell transcriptional analysis of normal, aberrant, and malignant hematopoiesis in zebrafish. J Exp Med 213:979–992

Morales Poole JR, Paganini J, Pontarotti P (2017) Convergent evolution of the adaptive immune response in jawed vertebrates and cyclostomes: an evolutionary biology approach based study. Dev Comp Immunol 75:120–126

Mora-Velandia LM, Castro-Escamilla O, Méndez AG, Aguilar-Flores C, Velázquez-Avila M, Tussié-Luna MI, Téllez-Sosa J, Maldonado-García C, Jurado-Santacruz F, Ferat-Osorio E et al (2017) A human Lin⁻ CD123 + CD127 low population endowed with ILC features and migratory capabilities contributes to immunopathological hallmarks of psoriasis. Front Immunol 8:176

Notta F, Zandi S, Takayama N, Dobson S, Gan OI, Wilson G, Kaufmann KB, McLeod J, Laurenti E, Dunant CF et al (2016) Distinct routes of lineage development reshape the human blood hierarchy across ontogeny. Science 351:aab2116–aab2116

Page DM, Wittamer V, Bertrand JY, Lewis KL, Pratt DN, Delgado N, Schale SE, McGue C, Jacobsen BH, Doty A et al (2013) An evolutionarily conserved program of B-cell development and activation in zebrafish. Blood 122:e1–e11

Pancer Z, Amemiya CT, Ehrhardt GRA, Ceitlin J, Larry Gartland G, Cooper MD (2004) Somatic diversification of variable lymphocyte receptors in the agnathan sea lamprey. Nature 430:174–180

Papalexi E, Satija R (2017) Single-cell RNA sequencing to explore immune cell heterogeneity. Nat Rev Immunol 18:35–45

Paul F, Arkin Y, Giladi A, Jaitin DA, Kenigsberg E, Keren-Shaul H, Winter D, Lara-Astiaso D, Gury M, Weiner A et al (2015) Transcriptional heterogeneity and lineage commitment in myeloid progenitors. Cell 163:1663–1677

Pereiro P, Varela M, Diaz-Rosales P, Romero A, Dios S, Figueras A, Novoa B (2015) Zebrafish Nk-lysins: first insights about their cellular and functional diversification. Dev Comp Immunol 51:148–159

Picelli S, Faridani OR, Björklund ÅK, Winberg G, Sagasser S, Sandberg R (2014) Full-length RNA-seq from single cells using Smart-seq2. Nat Protoc 9:171–181

Pierrard M-A, Roland K, Kestemont P, Dieu M, Raes M, Silvestre F (2012) Fish peripheral blood mononuclear cells preparation for future monitoring applications. Anal Biochem 426:153–165

Psaila B, Barkas N, Iskander D, Roy A, Anderson S, Ashley N, Caputo VS, Lichtenberg J, Loaiza S, Bodine DM et al (2016) Single-cell profiling of human megakaryocyte-erythroid progenitors identifies distinct megakaryocyte and erythroid differentiation pathways. Genome Biol 17

Raj A, van Oudenaarden A (2008) Nature, nurture, or chance: stochastic gene expression and its consequences. Cell 135:216–226

Regev A, Teichmann SA, Lander ES, Amit I, Benoist C, Birney E, Bodenmiller B, Campbell P, Carninci P, Clatworthy M et al (2017) The human cell Atlas. ELife 6

Renshaw SA, Trede NS (2012) A model 450 million years in the making: zebrafish and vertebrate immunity. Dis Model Mech 5:38–47

Renshaw SA, Loynes CA, Trushell DMI, Elworthy S, Ingham PW, Whyte MKB (2006) A transgenic zebrafish model of neutrophilic inflammation. Blood 108:3976–3978

Rey Vázquez G, Guerrero GA (2007) Characterization of blood cells and hematological parameters in Cichlasoma dimerus (Teleostei, Perciformes). Tissue Cell 39:151–160

Rhee J-S, Jeong C-B, Kim D-H, Kim I-C, Lee YS, Lee C, Lee J-S (2014) Immune gene discovery in the crucian carp Carassius auratus. Fish Shellfish Immunol 36:240–251

Robbins SH, Walzer T, Dembélé D, Thibault C, Defays A, Bessou G, Xu H, Vivier E, Sellars M, Pierre P et al (2008) Novel insights into the relationships between dendritic cell subsets in human and mouse revealed by genome-wide expression profiling. Genome Biol 9:R17

Rougeot J, Zakrzewska A, Kanwal Z, Jansen HJ, Spaink HP, Meijer AH (2014) RNA sequencing of FACS-sorted immune cell populations from zebrafish infection models to identify cell specific responses to intracellular pathogens. Methods Mol Biol Clifton NJ 1197:261–274

Saraceni PR, Romero A, Figueras A, Novoa B (2016) Establishment of infection models in zebrafish larvae (Danio rerio) to study the pathogenesis of aeromonas hydrophila. Front Microbiol 7:1219

Schlitzer A, Sivakamasundari V, Chen J, Sumatoh HRB, Schreuder J, Lum J, Malleret B, Zhang S, Larbi A, Zolezzi F et al (2015) Identification of cDC1- and cDC2-committed DC progenitors reveals early lineage priming at the common DC progenitor stage in the bone marrow. Nat Immunol 16:718–728

Schulz MH, Zerbino DR, Vingron M, Birney E (2012) Oases: robust de novo RNA-seq assembly across the dynamic range of expression levels. Bioinform Oxf Engl 28:1086–1092

Spits H, Artis D, Colonna M, Diefenbach A, Di Santo JP, Eberl G, Koyasu S, Locksley RM, McKenzie ANJ, Mebius RE et al (2013) Innate lymphoid cells—a proposal for uniform nomenclature. Nat Rev Immunol 13:145–149

Stubbington MJT, Lönnberg T, Proserpio V, Clare S, Speak AO, Dougan G, Teichmann SA (2016) T cell fate and clonality inference from single-cell transcriptomes. Nat Methods 13:329–332

Stubbington MJT, Rozenblatt-Rosen O, Regev A, Teichmann SA (2017) Single-cell transcriptomics to explore the immune system in health and disease. Science 358:58–63

Sudhagar A, Kumar G, El-Matbouli M (2018) Transcriptome analysis based on RNA-seq in understanding pathogenic mechanisms of diseases and the immune system of fish: a comprehensive review. Int J Mol, Sci 19

Suffiotti M, Carmona SJ, Jandus C, Gfeller D (2017) Identification of innate lymphoid cells in single-cell RNA-Seq data. Immunogenetics 69:439–450

Svensson V, Vento-Tormo R, Teichmann SA (2018) Exponential scaling of single-cell RNA-seq in the past decade. Nat Protoc 13:599–604

Tamplin OJ, Durand EM, Carr LA, Childs SJ, Hagedorn EJ, Li P, Yzaguirre AD, Speck NA, Zon LI (2015) Hematopoietic stem cell arrival triggers dynamic remodeling of the perivascular niche. Cell 160:241–252

Tang Q, Iyer S, Lobbardi R, Moore JC, Chen H, Lareau C, Hebert C, Shaw ML, Neftel C, Suva ML et al (2017) Dissecting hematopoietic and renal cell heterogeneity in adult zebrafish at single-cell resolution using RNA sequencing. J Exp Med 214:2875–2887

Tokunaga Y, Shirouzu M, Sugahara R, Yoshiura Y, Kiryu I, Ototake M, Nagasawa T, Somamoto T, Nakao M (2017) Comprehensive validation of T- and B-cell deficiency in rag1-null zebrafish: implication for the robust innate defense mechanisms of teleosts. Sci Rep 7

Trapnell C (2015) Defining cell types and states with single-cell genomics. Genome Res 25:1491–1498

Traver D, Paw BH, Poss KD, Penberthy WT, Lin S, Zon LI (2003) Transplantation and in vivo imaging of multilineage engraftment in zebrafish bloodless mutants. Nat Immunol 4:1238–1246

Udvadia AJ, Linney E (2003) Windows into development: historic, current, and future perspectives on transgenic zebrafish. Dev Biol 256:1–17

Villani A-C, Satija R, Reynolds G, Sarkizova S, Shekhar K, Fletcher J, Griesbeck M, Butler A, Zheng S, Lazo S et al (2017) Single-cell RNA-seq reveals new types of human blood dendritic cells, monocytes, and progenitors. Science 356

Vivier E, van de Pavert SA, Cooper MD, Belz GT (2016) The evolution of innate lymphoid cells. Nat Immunol 17:790–794

Vu Manh T-P, Marty H, Sibille P, Le Vern Y, Kaspers B, Dalod M, Schwartz-Cornil I, Quéré P (2014) Existence of conventional dendritic cells in Gallus gallus revealed by comparative gene expression profiling. J Immunol Baltim Md 1950(192):4510–4517

Vu Manh T-P, Elhmouzi-Younes J, Urien C, Ruscanu S, Jouneau L, Bourge M, Moroldo M, Foucras G, Salmon H, Marty H et al (2015) Defining mononuclear phagocyte subset homology across several distant warm-blooded vertebrates through comparative transcriptomics. Front Immunol 6:299

Warner NL, Szenberg A, Burnet FM (1962) The immunological role of different lymphoid organs in the chicken. I. Dissociation of immunological responsiveness. Aust J Exp Biol Med Sci 40:373–387

Yang J, Chen X, Bai J, Fang D, Qiu Y, Jiang W, Yuan H, Bian C, Lu J, He S et al (2016) The Sinocyclocheilus cavefish genome provides insights into cave adaptation. BMC Biol 14:1

Yoder JA, Litman RT, Mueller MG, Desai S, Dobrinski KP, Montgomery JS, Buzzeo MP, Ota T, Amemiya CT, Trede NS et al (2004) Resolution of the novel immune-type receptor gene cluster in zebrafish. Proc Natl Acad Sci 101:15706–15711

Zapata A, Amemiya CT (2000) Phylogeny of lower vertebrates and their immunological structures. Curr Top Microbiol Immunol 248:67–107

Zheng GXY, Terry JM, Belgrader P, Ryvkin P, Bent ZW, Wilson R, Ziraldo SB, Wheeler TD, McDermott GP, Zhu J et al (2017) Massively parallel digital transcriptional profiling of single cells. Nat Commun 8:14049

Zilionis R, Nainys J, Veres A, Savova V, Zemmour D, Klein AM, Mazutis L (2016) Single-cell barcoding and sequencing using droplet microfluidics. Nat Protoc 12:44–73

Iwama G, Nakanishi, T (1996) Fish physiology, vol. 15: The fish immune system: organism, pathogen and environment. Academic Press, New York

Evolutionary Impacts of Alternative Transposition

Weijia Su, Sharu Paul Sharma and Thomas Peterson

Abstract With the development of rapid DNA-sequencing techniques, more and more eukaryotic genomes are sequenced, assembled, and annotated. Now, we know that many eukaryotic genomes are large, highly repetitive, and very complex. One question that remains is what evolutionary forces generate the complexity of eukaryotic genomes? In this chapter, we discuss one of the possible answers to this question: the capacity of transposable elements to induce diverse genomic recombinations through various transposition reactions. Transposable elements (TEs or transposons) are DNA sequences that can move from one genomic location to another, and they are highly represented in most eukaryotic genomes. TEs are often called "junk DNA" because most copies are silenced and have no obvious function; however, various studies have indicated that TEs have made major contributions to shaping eukaryotic genomes and regulating gene expression. We show that TE transpositions can induce a variety of genome rearrangements including deletion, inversion, duplication, and translocation. These occur as direct products of alternative transposition pathways. Unlike standard transposition which involves only one transposon, alternative transposition involves two distinct TEs, undergoes more complicated movements, and generates more significant genome structure variations. Here, we describe various types of alternative transposition pathways, the diverse genome rearrangements they generate, and their potential role in the process of genome evolution.

1 Transposable Element and Alternative Transposition

Transposable elements (TEs), also known as transposons, are DNA sequences that can move themselves from one genomic site to another. These elements were first discovered in the 1940s by the iconic geneticist Barbara McClintock (McClintock

W. Su · S. P. Sharma · T. Peterson (✉)
Department of Genetics, Development and Cell Biology, Iowa State University,
Ames, IA 50011-3260, USA
e-mail: thomasp@iastate.edu

W. Su · S. P. Sharma · T. Peterson
Department of Agronomy, Iowa State University, Ames, IA 50011-3260, USA

© Springer International Publishing AG, part of Springer Nature 2018
P. Pontarotti (ed.), *Origin and Evolution of Biodiversity*,
https://doi.org/10.1007/978-3-319-95954-2_7

1948, 1950). She named the first transposon system as *Ac/Ds*: *Dissociation* (*Ds*) is a non-autonomous transposon and its movements are regulated by the autonomous element *Activator* (*Ac*). Since the mid-twentieth century, many more transposons have been discovered and classified. Typically, there are two major classes of TE (Pray 2008): Class I transposable elements transpose by a "copy and paste" mechanism typically involving RNA-mediated reversed transcription. As these TEs have striking similarities to retroviruses, these transposons are also called retro-elements, or retrotransposons. In contrast, Class II TEs are often called DNA elements and they transpose by a "cut-and-paste" mechanism, which does not involve an RNA intermediate. For these TEs, transposase enzyme binds to specific sequences at the transposon termini, excises the TE, makes a staggered double-strand cut at a new target site, and inserts the TE. The newly transposed TE is often flanked by a target site duplication (TSD) of 2–9 base pairs that result from fill-in repair of the staggered cut at the insertion site. With the development of genomic studies, we know that TEs occupy more than half of the DNA in most of eukaryotic genomes, specifically, around 85% of maize genomes (Schnable et al. 2009), and 50% of human genomes (Lander Eric et al. 2001).

Because standard transposition involves the 5′ and 3′ termini of the same TE, it can only change the position of one transposon within the genome. In contrast, alternative transposition acts on the ends of two TE copies, one from each TE. Because the other end of each TE remains attached to the flanking DNA, the resulting transposition reaction moves both TE copies as well as the flanking host DNA, producing larger and more complicated genome rearrangements. There are two major types of alternative transpositions: In one type, termed Sister Chromatid Transposition (SCT), the transposition-competent termini of the two TEs are located in different sister chromatids. Transposase binds to the termini of two TEs on different sister chromatids and performs transposition. In the other situation, the termini of two TEs are on the same chromatid but are present in reversed orientation with respect to the normal TE configuration. Transposase binds to these reversed 5′ and 3′ termini and performs Reversed Ends Transposition (RET). Studies have shown that by breaking and rejoining chromosomes, both SCT and RET can generate a variety of genome rearrangements such as deletions (Zhang and Peterson 2005a), duplications (Zhang et al. 2013), inversions, translocations (Zhang and Peterson 1999, 2004; Zhang et al. 2006) as well as novel compound structures (Zhang et al. 2014). In addition to genome structure variations, alternative transpositions can also regulate gene expression by forming chimeric genes (Wang et al. 2015), introducing copy number variations (Zuo et al. 2016), and shuffling regulatory elements.

Although TEs comprise a major fraction of most eukaryotic genomes, TEs are often considered as useless DNA sequences, and thus came to be called "junk DNA" or "selfish/parasitic DNA." However, a number of recent studies provide strong evidence that TEs play very important roles in genome evolution, effectively shaping genome size, diversity, and architecture. An excellent example of TEs as a source of biological innovation is found in the retrotransposon (MER20) which provides regulatory sequences to over 1000 genes important for pregnancy in placental mammals, thus effectively creating a novel gene regulatory network (Lynch et al. 2011, 2012). Similarly, human endogenous retroviral (ERV) sequences containing enhancers inducible by Interferon have been co-opted as regulators of genes important for innate immunity (Chuong et al. 2016). In plants, LTR retrotransposons are responsible for massive genome expansion and the evolution of plant disease-resistance genes in hot pepper genomes (Kim et al. 2017). DNA transposons have additional potential for genome alterations due to their ability to generate double-strand DNA breaks (DSBs) during excision and insertion reactions. For example, error-prone repair of TE-induced DSBs may lead to increased numbers of mutations around TE excision sites, as has been found in various grasses including rice, maize, wheat, and barley. Due to the preference of DNA TEs to insert into genic regions, these mutations can have profound effects on the evolution of grasses and other plants (Wicker et al. 2016). Additionally, DNA transposon-derived proteins act as anti-silencing factors in Arabidopsis (Duan et al. 2017). Finally, Class II transposable elements were identified as the original source of the recombination-activating genes *RAG1* and *RAG2* which play essential roles in V(D)J recombination and adaptive immunity in vertebrate animals (Poole et al. 2017a, b; Huang et al. 2016). These examples indicate that TEs can have comprehensive and profound effects on the process of eukaryotic genome evolution.

2 Maize *p1* and *p2* Genes

To study alternative transposition reactions, we use the maize *p1/p2* genes which offer a number of experimental advantages. *p1* and *p2* are paralogous genes which encode similar R2R3 Myb transcription factors that activate the structural genes *c2*, *chi*, and *a1* encoding chalcone synthase, chalcone isomerase, and dihydroflavonol reductase, respectively (Dooner et al. 1991). Both *p1* and *p2* are located on the short arm of maize chromosome 1, about 70 kb apart from each other. The *p1* gene is expressed mainly in kernel pericarp and cob, where it confers a deep red flavonoid pigmentation (Fig. 1). The *p2* gene is not expressed in pericarp and cob, and instead *p2* transcripts accumulate in anther and silk (style) (Zhang et al. 2000; Goettel and Messing 2009). Previous studies elucidated the molecular evolution of *p1/p2* genes by tandem duplication of an ancestor p^{pre} gene and its downstream flanking sequence. The duplication event was estimated to have occurred about 2.75 million years ago (Zhang et al. 2000). Different *p1* alleles exhibit great diversity in gene structure and phenotypic expression. Conventionally, different *p1* alleles are indicated by a two-letter suffix that represents their pericarp and cob pigmentation. The first letter stands for pericarp color and the second refers to cob color (Fig. 1).

Previous studies discovered that the evolution of organ-specific expression patterns conferred by different *p1* alleles appears to be correlated with structural changes in a distal regulatory region (Zhang and Peterson 2005b). It has been proposed that early *p1* alleles conferred red pericarp and white cob (*P1-rw* type). Subsequently, the distal enhancer acquired a cob-specific regulatory region via gene conversion, thus giving rise to the red pericarp and red cob (*P1-rr*) type.

Fig. 1 Phenotypes conferred by maize *p1* alleles; the two-letter suffix indicates color in pericarp and cob: *p1-ww* has white (colorless) kernel pericarp and white cob; *P1-rw* has red pericarp and white cob, etc. The *P1-vv* allele confers variegated (striped) pericarp and variegated cob; *p1-vv* contains transposon *Ac*, whose excision leads to sectors of red revertant tissue

The maize *p1/p2* genes serve as an ideal genetic model to analyze alternative transposition mechanisms and impacts for several reasons. First, various *p1* alleles contain insertions of transposable elements such as *Ac* and fractured *Ac* (*fAc*, a partial *Ac* element with deletion of *Ac* 5′ end) within *p1* gene or in the nearby flanking regions. Second, because *p1* expression is indicated by the formation of non-essential pigments in kernels, alternative transposition events that alter *p1* expression result in prominent visible sectors. Third, because the kernel pericarp is derived from the ovary wall and thus shares a developmental lineage with the egg mother cell, pre-meiotic mutations that alter pericarp color will also be transmitted to the embryo. Thus, the same mutations that are visible as pericarp sectors can be recovered and propagated through the kernel. Last but not least, more than 100 natural variants of *p1* alleles have been reported (Goettel and Messing 2013; Brink and Styles 1966), which provides abundant genetic resources to analyze alternative transposition events and their contribution to natural variation.

3   Alternative Transposition-Induced Genome Rearrangement

3.1   AT-Induced Duplication and Deletion

In 2013, Zhang et al. showed that transposable elements can directly generate tandem duplications via alternative transposition. They identified multiple new spontaneous duplications ranging from 8 kbp to more than 5000 kbp (Zhang et al. 2013).

The duplications were derived from a progenitor allele termed *p1-ovov454* (orange-variegated pericarp and orange-variegated cob) that contains a pair of reverse-oriented *Ac* and *fAc* insertions in the *p1* gene (Fig. 2). This configuration is competent to undergo Reversed End Transposition (RET). If RET occurs following DNA replication and the excised *Ac* and *fAc* termini insert into the sister chromatid, then one chromatid will carry a tandem direct duplication and the other chromatid will have a reciprocal deletion (Fig. 2a). These two unequal chromatids will segregate into adjacent daughter cells at the next mitotic division. Because the *p1-ovov454* allele has orange-variegated pericarp and orange-variegated cob, both loss-of-function and gain-of-function mutations of *p1* can be observed in pericarp and cob. The deletion chromosome will generate white pericarp because *p1* is no longer functional, while the duplication chromosome contains two copies of *Ac* and thus will produce fewer red/white stripes due to the *Ac* negative dosage effect (McClintock 1948, 1951). Thus, this type of RET event can be detected as visible twin sectors in the mature maize ears as shown in Fig. 2b.

In addition to tandem direct duplication, alternative transposition can also generate inverted duplications. Interestingly, several papers published in the 1980s and

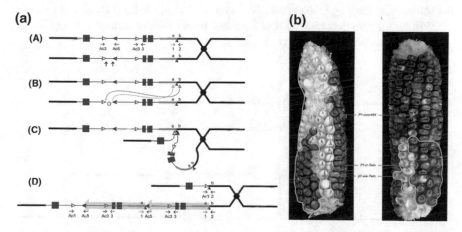

Fig. 2 **a** Mechanism of RET-generated tandem direct duplication. The blue boxes indicate *p1* exons, and red arrows indicate *fAc* and *Ac* elements (solid and open arrowheads are *Ac* 5′ and 3′ termini, respectively). Green and black triangles represent transposition target sites (*a*) and (*b*), and filled circle indicates the centromere of chromosome 1. These legends are consistent in all of the figures in this chapter. (*A*) Diagram of replicated chromosome 1 containing *p1-ovov454* allele. *Ac*-encoded transposase will cut at reverse-oriented 3′ *fAc* and 5′ *Ac* termini (short vertical arrows). (*B*) Following transposase cleavage, the inter-transposon segment of the *p1* gene forms a circle. The dotted lines show where the excised *fAc* and *Ac* termini will insert in the upper sister chromatid. (*C*) The *fAc* 3′ end joins to the proximal side (black) of the target site, while the *Ac* 5′ end joins to the distal side (green). (*D*) Following RET, the upper chromatid contains a deletion from *fAc* to the *a/b* target site, and the lower chromatid contains a direct duplication of that segment. **b** Two maize ears containing independent twin sectors (green outlines). Orange-variegated kernels have the progenitor *p1-ovov454* phenotype; white kernels contain the deletion chromosome, and adjacent orange/red kernels contain the corresponding duplication (from Zhang et al. 2013)

1990s reported structures containing inverted duplications flanked by transposable elements (Burr and Burr 1982; Courage-Tebbe et al. 1983; English et al. 1993). Weil and Wessler demonstrated the fusion of sister chromatids by chromosome-breaking *Ds* elements (Weil and Wessler 1993). Then in 1999, Zhang and Peterson identified and sequenced the junctions of reciprocal deletion and inverted duplication chromosomes generated by sister chromatid transposition (Zhang and Peterson 1999). This report included the *Ac/fAc* excision footprint and 8 bp target site duplications and thus provided strong molecular evidence for the SCT model. According to this model, transposon termini located on different chromatids are excised, and the sequences formerly flanking the TE termini are ligated together forming an *Ac/fAc* excision footprint and linking the two sister chromatids together in a chromatid bridge. The two excised *Ac/fAc* termini could insert at many potential genomic target sites; however, only insertions into a proximal site will resolve the chromatid bridge and restore chromosomal linearity. The resulting chromosome will now contain two unequal chromatids: one with an inverted duplication and the other with a reciprocal deletion (Fig. 3). These chromatids would be segregated to adjacent daughter cells in the ensuing cell division and thus could be germinally transmitted if viable. This prediction was confirmed by the isolation of a series of nested deletions flanking the *Ac/fAc* termini ranging in size from 12.5 kbp to over 4.6 Mbp (Zhang and Peterson 2005a), and a complementary set of flanking inverted duplications ranging in size from 70 kbp to 14.7 Mbp (Zuo et al. 2016).

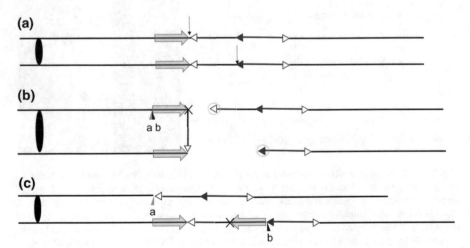

Fig. 3 Sister chromatid transposition generates inverted duplications and reciprocal deletions. The figure shows a diagram of a replicated chromosome with two identical sister chromatids. The blue-shaded arrows indicate a proximal segment that will be duplicated following SCT. **a** Transposase will cut at the 3′ terminus of *fAc* in the upper sister chromatid and the 5′ terminus of *Ac* in the lower sister chromatid (black arrows) **b** *Ac* and *fAc* termini are excised, and the two ends formerly flanking *Ac/fAc* termini are joined together at "X" to generate a chromatid bridge. **c** The excised transposon termini insert at the *a/b* target site to generate one chromatid with a deficiency (upper), and the other chromatid with a corresponding inverted duplication (lower)

3.2 AT-Induced Inversion and Translocation

In 2009, two papers reported molecular and cytological evidence that alternative transposition induced reciprocal translocations and large inversions in maize (Pulletikurti et al. 2009; Zhang et al. 2009). In Zhang et al. 2009, the authors identified a series of major chromosomal rearrangements (MCR), including 17 reciprocal translocations and two large pericentric inversions. All of these cases were derived by RET from a progenitor allele containing *Ac* and *fAc* insertions in the maize *p1* gene. As described above, these reversely oriented *Ac/fAc* termini can undergo alternative transposition in which the termini are excised and the internal *p1* sequences are joined together to form a circle. Next, if *Ac* 5′ and *fAc* 3′ insert into the same sister chromatid, either a deletion or duplication will be formed, depending on which side of the insertion site the *Ac* and *fAc* termini insert. Alternatively, if the excised transposon termini insert into another chromosome, a reciprocal translocation will be produced. This mechanism was confirmed by isolation of the *Ac/fAc* translocation junctions by PCR and Ac casting (Singh et al. 2003; Wang and Peterson 2013). The results showed that for each of the 19 MCR alleles, the 8-bp sequences flanking *Ac* and *fAc* matched each other. This indicates that the translocation and inversion junctions contained 8-bp TSDs characteristic of *Ac* transposition products, indicating that these translocations and inversions were indeed generated via RET. Finally, the MCR structures were confirmed by cytogenetic analysis using Fluorescence in situ Hybridization (FISH) of mitotic metaphase chromosomes, and propiocarmine staining of meiotic chromosomes (Fig. 4).

3.3 Composite Insertion—A Novel Structure Generated by Alternative Transposition Mediated DNA Re-replication

In the AT models described above, TE excision occurs either before DNA replication, or after the chromosomes are completely replicated. However, it is well known that some Class II "cut-and-paste" TEs transpose during DNA replication (Roberts et al. 1985; Chen et al. 1987; Peters and Craig 2001). For example, the *Ac* element has been shown to transpose during DNA replication, from a replicated donor site to an unreplicated target site. In this way, *Ac* can replicate twice in the same cell cycle and thereby increase in copy number (Greenblatt and Brink 1962; Chen et al. 1992). Therefore, it is reasonable to ask what would be the consequences if alternative transposition occurred during DNA replication.

In 2014, Zhang et al. reported a new model of alternative transposition during DNA replication. The model was developed and tested by using the *p1-ovov454* allele, the same allele described in the preceding duplication example. According to this model, the *Ac* and *fAc* insertions in the *p1* gene are replicated and then excised before the chromosome has finished replication. The excised ends are inserted into

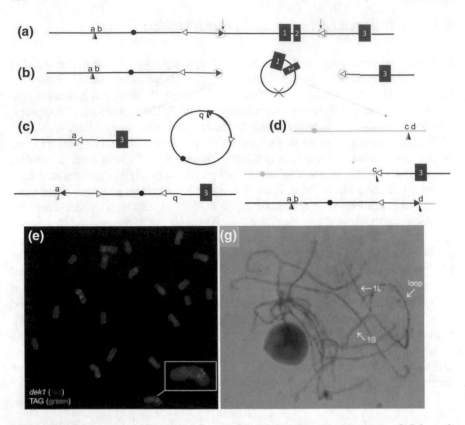

Fig. 4 Alternative transposition induced major chromosomal rearrangements. **a–d** Schematic model: **a** Configuration of *Ac* and *fAc*. Reversed transposon termini *Ac 5′* and *fAc 3′* are cleaved and excised. **b** The internal sequence between the two transposon termini forms a circle and is lost. **c** Excised transposon termini insert into the same chromosome. The upper diagram shows *Ac* insertion into a target site on opposite chromosome arm to form a ring chromosome; the lower diagram shows *Ac* insertion into distal target site and *fAc* insertion into the proximal target site, to form a pericentric inversion. **d** Transposon termini insert into another chromosome (yellow line) and produce a reciprocal translocation. **e–h**, cytogenetic analysis of MCR alleles: **e** and **f**, FISH of mitotic metaphase chromosomes. **e** Chromosome 1 pericentric inversion heterozygote (*p1-wwC30*) hybridized with *dek1* (red, detects 1S) and TAG (green, detects 1L; also 2S, 2L, and 4S), **f** Translocation chromosome T1-5 heterozygote (*p1-wwB1023*) hybridized with *dek1* (red, detects 1S), TAG (white, detects 1L; also 2S, 2L, and 4S), and *serk2-rf2e1* (green; detects 5L); **g** and **h** are propiocarmine staining of meiotic pachytene chromosomes from *p1-wwC30* and *p1-wwB1023*, respectively, heterozygous with normal. Arrows in **g** and **h** indicate sites of characteristic features. **g** Chromosome 1 pericentric inversion heterozygote (*p1-wwC30*), inversion chromosome is enlarged in the *inset*. **h** Translocation chromosome T1-5 heterozygote (*p1-wwB1023*), translocation chromosome is enlarged in the *inset* (from Pulletikurti et al. 2009; Jianbo Zhang et al. 2009)

an unreplicated target site, forming a rolling circle replicon. As DNA replication continues, *Ac*, *fAc* and their flanking sequences will be re-replicated by the rolling circle replicon. Eventually, the re-replication forks will abort spontaneously resulting

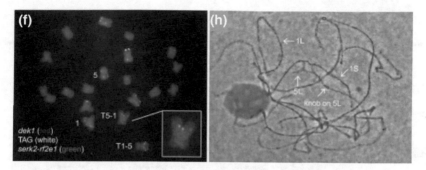

Fig. 4 (continued)

in two double-strand breaks (DSBs); fusion of these DSBs will restore the linear chromosome, forming a tandem direct duplication (TDD) and a novel structure at the insertion site termed a composite insertion (CI). This haplotype containing both TDD and CI is termed TDD-CI. If the rolling circle replication fork is released prior to DSB fusion, a solo CI will be generated without an accompanying TDD (Fig. 5). As described above, RET with insertion into a replicated target site (i.e., the sister chromatid) can also generate a TDD; however, only when the excised termini insert into unreplicated sequence will a TDD-CI or solo-CI be generated.

Using PCR-based methods such as *Ac* casting (Singh et al. 2003; Wang and Peterson 2013) and inverse PCR (iPCR) (Zhang et al. 2013) with *Ac/fAc*-specific primers, the junctions of TDDs and CIs could be isolated and sequenced. The results showed that the distance from the donor *p1* locus to the linked CI elements ranged from 13 kbp to 1.7 Mbp. Additionally, DNA gel blot experiments indicated that different CIs range up to 23 kbp, while TDDs could be over 1 Mbp. This study also showed that the DSB fusions were repaired by Homologous Recombination (HR) or Non-Homologous Ending Joining (NHEJ) at approximately equal frequencies. Because the CIs represent a novel genetic structure, these results suggest that RET-induced re-replication can duplicate and transpose genes and genetic elements and therefore may play an important role in rapid genome evolution.

4 Regulation of Gene Expression by Alternative Transposition

Recent studies showed that SCT and RET events not only generate genome rearrangements but also alter gene conformations and gene expression.

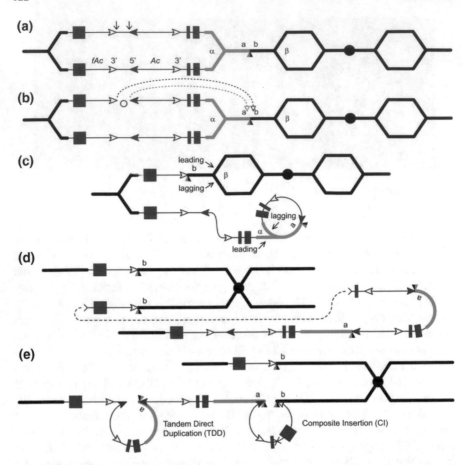

Fig. 5 RET during DNA replication generates tandem direct duplication (TDD) and composite insertion (CI). Lines indicate a replicating chromosome, and hexagons indicate replicons. Two replication forks considered here are marked α and β. **a** The locus containing *fAc/Ac* is replicated while the *a/b* target site is unreplicated. Transposase will cut at the two black arrows. **b** Transposon termini insert into unreplicated target site *a/b*. **c** Insertion of the excised transposon termini places *fAc* and *fAc*-flanking DNA ahead of replication fork β (upper chromatid), and *Ac* and *Ac*-flanking DNA ahead of replication fork α to generate a rolling circle replicon (lower chromatid). DNA replication continues. **d** Following re-replication of *fAc*, *Ac*, and a portion of the flanking sequences, DNA replication forks α and β stall and abort, resulting in chromatids terminated by broken ends (the red > or < symbol) (Michel et al. 1997). The dotted red line connects the two broken ends that will fuse together. **e** Fusion of double-strand breaks produces a chromosome with two unequal sister chromatids: The upper chromatid contains a deletion of the segment from *fAc* to the *a/b* target site. The lower chromatid contains a TDD (left-hand loop), as well as a new CI (right-hand loop). The TDD contains the DNA deleted from the upper chromatid; the CI contains the re-replicated *Ac*, *fAc*, and flanking sequences. The junction where broken chromatid ends were joined is indicated by the red × (from Zhang et al. 2014)

4.1   Novel Maize Chimeric Genes Generated by Transposition

Chimeric genes are formed through a combination of two or more coding sequences (Long and Langley 1993). These new genes can be generated by various mechanisms such as copy number variation (Rippey et al. 2013), DNA replication and repair pathways (Rogers et al. 2009) and retroposition (Wang et al. 2006). Chimeric genes are important for growth and development in plants, animals, and human. For example, chimeric genes can alter neuronal function in *Drosophila melanogaster* (Rogers et al. 2010). A recent study showed that chimeric genes arising in human deletion and duplication syndromes may be pathogenic and associated with intellectual disability (Mayo et al. 2017). In the same year, a paper published by Hao Chen stated that emergence of a novel chimeric gene in rice is responsible for multiple important traits such as grain number, plant height, and heading date (Chen et al. 2017). In the process of genome evolution, chimeric genes are important and likely one of the most common sources of developing new genes.

In 2006, Zhang et al. reported that RET events can generate novel chimeric genes (Zhang et al. 2006). Beginning with an allele containing *Ac* and *fAc* termini capable of RET, these authors derived multiple alleles termed *p1-oo* (orange pericarp and orange cob). These alleles exhibited no *Ac* activity and had deletions that spanned the *p1* and *p2* intergenic region, a distance of about 70 kbp. Fine structure mapping by PCR and sequencing of breakpoint junctions showed that the 3′ end of *fAc* had inserted into various sites in *p2* intron 2, producing a series of chimeric genes containing exons 1 and 2 of the *p2* gene, the *fAc* sequence, and exon 3 of the *p1* gene (Fig. 6). The fact that these alleles conditioned orange-colored kernel pericarp and cob indicated that they were functional. This was further confirmed by RT-PCR experiments which detected *p2-p1* fusion transcripts matching the gene structures.

In subsequent work, Wang et al. reported additional cases of chimeric genes generated by RET (Wang et al. 2015). This study was initiated with the allele *p1-ovov454* described above. Interestingly, *p1-ovov454* gave rise to a derivative allele *p1-vvD103* which had very light variegated kernel pericarp, but also exhibited semisterility when heterozygous. This semisterility is a characteristic of translocation heterozygotes, and indeed the *p1-vvD103* allele was found to be associated with a reciprocal translocation of chromosomes 1 and 10. Due to the translocation, the *p1-vvD103* allele lacks a functional *p1* gene, thus explaining the colorless pericarp. However, the appearance of infrequent orange and red sectors indicated that further mutational events could restore a functional *p1* gene or its equivalent. To test this, the authors selected whole-kernel and multi-kernel sectors with orange-red pigmentation from the *p1-vvD103* line. These alleles conferring pigmentation were analyzed and found to contain *p1-p2* chimeric genes, in surprising contrast to the *p2-p1* chimeras described above. Molecular analysis of the rearrangement junctions led to a model of their origin via RET. In this case, the reverse-oriented *Ac* and *fAc* termini in *p1-vvD103* excised and insert into *p2* on the sister chromatid. This new junction joins *p1* sequences from the donor site with the *p2* sequences at the target site. This process produces a fusion

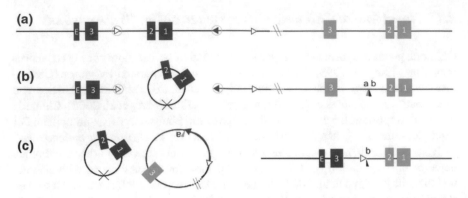

Fig. 6 RET generates chimeric gene. **a** *Ac* transposase binds to the termini of adjacent *fAc* and *Ac* insertions in the *p1* gene (blue numbered boxes indicate *p1* exons 1, 2, and 3). **b** The *Ac* and *fAc* termini are excised and the inter-transposon segment forms a circle. **c** The excised *Ac* and *fAc* transposon ends insert into target site *a/b* in intron 2 of *p2*. *Ac* 5′ end joins to *a* to form a circle, and the *fAc* 3′ end joins to *b* to generate a chimeric gene containing exons 1 and 2 of *p2* and exon 3 of *p1*

of *p1* exons 1 and 2 with exon 3 of *p2*, creating a new chimeric gene called *P1P2* (Wang et al. 2015).

Because the *P1P2* allele retains the 5′ promoter region of *p1*, it can be expressed in pericarp and produce red pigment. Expression of the chimeric gene at transcription level was confirmed by RT-PCR using primers located on exon 1 of *p1* and exon 3 of *p2*. Furthermore, the chimeric protein shared 95% identity with maize P1 protein and contains the R2R3 Myb DNA-binding region. These results indicated that the chimeric gene *P1P2* is able to restore *p1* function and activate the flavonoid biosynthetic pathway to produce red pigment in maize kernel pericarp.

These studies provide clear and convincing examples showing how alternative transposition can generate novel and functional chimeric genes. The high endogenous transposition activity of the *Ac* system enabled the isolation and identification of these chimeric genes as they arose in pedigreed materials, so the structures of progenitor, intermediate and chimeric alleles could be established with certainty. The fact that these chimeras can arise at appreciable rates in only a few generations suggests that the products of alternative transposition could have major impacts over evolutionary time.

4.2   Dosage Effects on Genes and Small RNA Expression by Alternative Transposition Induced Copy Number Alterations

Gene duplication and deletion can increase or decrease gene copy number and thereby alter or regulate gene expression (Guo et al. 1996). There are two major types of response to gene dosage changes: in dosage effect, gene expression levels will increase or decrease in proportion to the gene copy number; whereas, dosage compensation will maintain constant levels of gene expression in spite of gene copy number variation (CNV; Gupta et al. 2006). Since alternative transposition actively generates duplications and deletions in the genome, it seems valuable to investigate whether gene expression would be affected by alternative transposition induced copy number variation. In addition to changes in DNA sequences, epigenetic regulation is also crucial for proper growth and development of plants, animals and humans. Small RNAs play important roles in affecting epigenetic regulation (Holoch and Moazed 2015); however, little is known about the effects of gene copy number variation on small RNA expression.

Because alternative transposition can generate large and overlapping duplications and deletions in a known genetic background, these CNV alleles can be extremely useful for investigating dosage effects on small RNA expression. The results of such a study were published by Zuo et al. (2016). In this paper, the authors describe an allele termed *p1-ww714* which contains a 14.6 Mb inverted duplication generated by sister chromatid transposition on maize chromosome 1. This duplicated region contains approximately 300 protein-coding genes. By backcrossing with maize inbred line B73 followed by self-pollination of a heterozygote, the authors generated sibling offspring that contain two (B73/B73), three (B73/*p1-ww714*), and four (*p1-ww714/p1-ww714*) copies of the 14.6 Mbp segment. Phenotypic screening shows that plants homozygous for the duplication (*p1-ww714/p1-ww714*) exhibit delayed flowering time, reduced stature and ear length compared to sibling homozygous *B73/B73* plants, whereas heterozygous *p1-ww714/B73* plants have intermediate ear length compared to the sibling homozygotes (Fig. 7).

High-throughput transcriptome analyses via microarray and RNAseq were performed to compare gene expression levels in seedlings of the contrasting genotypes. The results show that, among 212 genes in the duplicated region with detectable expression in seedlings, 125 (~60%) were differentially expressed. Moreover, small RNA sequencing data identified a 3.3 Mb region in which small RNA transcripts are differentially expressed in proportion to gene dosage (Zuo et al. 2016). Together, these data show that alternative transposition-induced CNVs can significantly affect gene and small RNA transcript levels in maize.

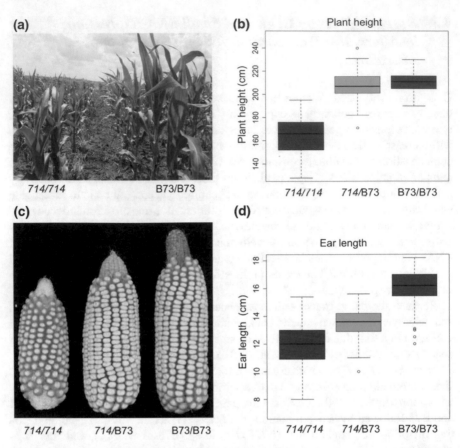

Fig. 7 Phenotypic effects of copy number variation observed in three genotypes: *p1-ww714/p1-ww714*, *p1-ww714/B73*, and *B73/B73*. Homozygous *p1-ww714/p1-ww714* plants contain four copies of a 14.6 Mbp segment and exhibit shorter plants and smaller ears (from Zuo et al. 2016)

4.3 Alternative Transposition Activates Gene Expression by Shuffling Regulatory Elements

Transcription regulatory elements such as enhancer, silencer, and insulator are crucial in controlling gene expression. Their analysis is complicated by the fact that they could be located nearby or far away from the transcription start site (TSS) of the gene they regulate (Oka et al. 2017). Enhancers are DNA elements that can be bound by activators and increase gene transcription; they are typically short, ranging from 50 to 1500 bp (Blackwood and Kadonaga 1998). Enhancers can be located more than one hundred kbp distant from the TSS, and either upstream or downstream of their target genes (Louwers et al. 2009). How enhancers regulate gene expression over long distances still remains unclear; however, it is widely believed that enhancer

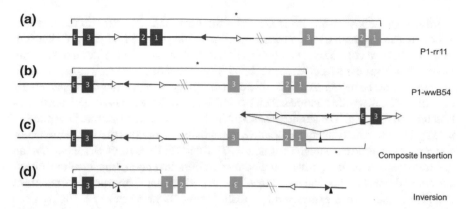

Fig. 8 Two ways to mobilize enhancer elements by alternative transposition. Numbered blue and green boxes indicate exons of *p1* and *p2* genes, respectively. The red boxes marked "E" indicate an enhancer element located 3′ of the *p1* gene (Sidorenko et al. 2000). Black triangles indicate the insertion target sites of transposable termini, asterisks, and brackets show the schematic distances between the enhancer and transcription start site of *p2*. **a** The progenitor allele *p1-rr11* contains intact *p1* and *p2* genes, with *fAc* and *Ac* insertions flanking *p1* exons 1 and 2. **b** The *p1-wwB54* allele has a deletion of *p1* exons 1 and 2 and retains *fAc/Ac* termini which can undergo further RET events to generate structures in (**c**) and (**d**). **c** Representative composite insertion (CI) allele retains the *p1-wwB54* backbone, but has gained a CI inserted into the promoter region of *p2*, thus activating *p2* expression in kernel pericarp. **d** Representative inversion allele has an inversion of *p2*, placing the *p2* promoter near the *p1* enhancer, thus activating *p2* expression in kernel pericarp

elements can interact with their target gene by looping out the sequences in between (Yadav et al. 2016).

In addition to shuffling exons, alternative transposition events can also duplicate and transpose enhancer elements and thereby alter gene expression. For example, if an enhancer element is flanking a TE involved in RET, the enhancer may be carried along with the TE to the new insertion site. This process has been observed in the case of a maize allele termed *p1-wwB54*. This allele has a deletion of exons 1 and 2 of the *p1* gene (Fig. 8b) and therefore lacks kernel pericarp pigmentation. However, *p1-wwB54* retains a nearby intact *p2* gene, although *p2* is not expressed in pericarp. As described above, RET of the *p1-wwB54* allele can create a composite insertion which can mobilize a *p1* enhancer to the vicinity of *p2*, thereby inducing expression of *p2* in kernel pericarp (Fig. 8c; Su and Peterson unpublished). Furthermore, RET can also induce inversions which place the *p2* promoter near the *p1* enhancer, thereby activating *p2* expression (Fig. 8d; Sharma and Peterson unpublished).

5 Summary

Transposable elements are well known by their ability to mobilize and increase copy number in eukaryotic genomes. Because TE transpositions often cause deleterious mutations, it's reasonable to assume that TE activity would be suppressed in order to minimize the loss of fitness incurred by the host. This can explain the fact that although TEs make up a large proportion of genomes, most TE copies are silenced; as these silenced copies have no obvious function, TEs are often referred to as "junk DNA." Yet despite the fact that the vast majority of extant TE sequences are silenced or inert, it must be recognized that these elements may have profoundly impacted the evolution of eukaryotic genomes. Yet, the full scope of TE effects on genome evolution remains unclear. The studies discussed in this chapter demonstrate how pairs of nearby TE copies can undergo alternative transposition reactions that can cause major changes in gene and chromosome structure. SCT and RET events of *hAT* elements can generate genome rearrangements including deletion, inversion, translocation, duplication, and novel compound structures. Those non-standard transposition reactions can also regulate gene expression by creating new chimeric genes, changing gene copy number and/or shuffling regulatory elements. The examples in this chapter are largely from the maize *Ac/Ds* transposon system, because these elements are highly active and are ideally located in kernel color genes allowing facile visual screens for novel rearrangements. However, we propose that the mechanisms and impacts of alternative transposition described here may also extend to other Class II transposable elements in a variety of eukaryotic genomes. These studies clearly indicate that transposable elements can readily alter genome size, increase genome diversity, and regulate gene expression in rapid events and over long-term evolutionary processes.

Acknowledgements This Research is supported by the USDA National Institute of Food and Agriculture Hatch project number IOW05282, and by State of Iowa funds.

References

Blackwood EM, Kadonaga JT (1998) Going the distance: a current view of enhancer action. Science 281(5373):60–63

Brink RA, Styles ED (1966) A collection of pericarp factors. Maize Genet Coop News Lett 40:149–160

Burr B, Burr FA (1982) Ds controlling elements of maize at the shrunken locus are large and dissimilar insertions. Cell 29(3):977–986

Chen J, Greenblatt IM, Dellaporta SL (1987) Transposition of Ac from the P locus of maize into unreplicated chromosomal sites. Genetics 117(1):109–116

Chen J, Greenblatt IM, Dellaporta SL (1992) Molecular analysis of Ac transposition and DNA replication. Genetics 130(3):665–676

Chen H, Tang Y, Liu J, Tan L, Jiang J, Wang M, Sun C (2017) Emergence of a novel chimeric gene underlying grain number in rice. Genetics 205(2):993–1002

Chuong EB, Elde NC, Feschotte C (2016) Regulatory evolution of innate immunity through co-option of endogenous retroviruses. Science 351(6277):1083–1087

Courage-Tebbe U, Döring HP, Fedoroff N, Starlinger P (1983) The controlling element Ds at the Shrunken locus in Zea mays: structure of the unstable sh-m5933 allele and several revertants. Cell 34(2):383–393

Dooner HK, Robbins TP, Jorgensen RA (1991) Genetic and developmental control of anthocyanin biosynthesis. Annu Rev Genet 25(1):173–199

Duan CG, Wang X, Xie S, Pan L, Miki D, Tang K, Wang Z (2017) A pair of transposon-derived proteins function in a histone acetyltransferase complex for active DNA demethylation. Cell Res 27(2):226

English J, Harrison K, Jones JD (1993) A genetic analysis of DNA sequence requirements for dissociation state I activity in tobacco. Plant Cell 5(5):501–514

Goettel W, Messing J (2009) Change of gene structure and function by non-homologous end-joining, homologous recombination, and transposition of DNA. PLoS Genet 5(6):e1000516

Goettel W, Messing J (2013) Epiallele biogenesis in maize. Gene 516(1):8–23

Greenblatt IM, Brink RA (1962) Twin mutations in medium variegated pericarp maize. Genetics 47(4):489–501

Guo M, Davis D, Birchler JA (1996) Dosage effects on gene expression in a maize ploidy series. Genetics 142(4):1349–1355

Gupta V, Parisi M, Sturgill D, Nuttall R, Doctolero M, Dudko OK, Oliver B (2006) Global analysis of X-chromosome dosage compensation. J Biol 5(1):3

Holoch D, Moazed D (2015) RNA-mediated epigenetic regulation of gene expression. Nat Rev Genet 16(2):71

Huang S, Tao X, Yuan S, Zhang Y, Li P, Beilinson HA, Le Petillon Y (2016) Discovery of an active RAG transposon illuminates the origins of V (D) J recombination. Cell 166(1):102–114

Kim S, Park J, Yeom SI, Kim YM, Seo E, Kim KT, Kim SB (2017) New reference genome sequences of hot pepper reveal the massive evolution of plant disease-resistance genes by retroduplication. Genome Biol 18(1):210

Lander Eric S, Linton Lauren M, Bruce B, Chad N, Zody Michael C, Jennifer B, Roel F (2001) Initial sequencing and analysis of the human genome

Long M, Langley CH (1993) Natural selection and the origin of jingwei, a chimeric processed functional gene in Drosophila. Science 260(5104):91–95

Louwers M, Bader R, Haring M, van Driel R, de Laat W, Stam M (2009) Tissue-and expression level-specific chromatin looping at Maize b1 epialleles. Plant Cell 21(3):832–842

Lynch VJ, Leclerc RD, May G, Wagner GP (2011) Transposon-mediated rewiring of gene regulatory networks contributed to the evolution of pregnancy in mammals. Nat Genet 43(11):1154

Lynch VJ, Nnamani M, Brayer KJ, Emera D, Wertheim JO, Pond SLK, Feschotte C (2012) Lineage-specific transposons drove massive gene expression recruitments during the evolution of pregnancy in mammals. arXiv:1208.4639

Mayo S, Monfort S, Roselló M, Orellana C, Oltra S, Caro-Llopis A, Martínez F (2017) Chimeric genes in deletions and duplications associated with intellectual disability. International J Genomics

McClintock B (1948) Mutable loci in maize. Carnegie Inst Wash Year Book 47:155–169

McClintock B (1950) The origin and behavior of mutable loci in maize. Proc Natl Acad Sci 36(6):344–355

McClintock B (1951). Chromosome organization and genic expression. In: Cold Spring Harbor symposia on quantitative biology, vol 16. Cold Spring Harbor Laboratory Press, pp 13–47

Michel B, Ehrlich SD, Uzest M (1997) DNA double-strand breaks caused by replication arrest. EMBO J 16(2):430–438

Oka R, Zicola J, Weber B, Anderson SN, Hodgman C, Gent JI, Stam M (2017) Genome-wide mapping of transcriptional enhancer candidates using DNA and chromatin features in maize. Genome Biol 18(1):137

Peters JE, Craig NL (2001) Tn7 recognizes transposition target structures associated with DNA replication using the DNA-binding protein TnsE. Genes Dev 15(6):737–747

Poole JRM, Huang SF, Xu A, Bayet J, Pontarotti P (2017a) The RAG transposon is active through the deuterostome evolution and domesticated in jawed vertebrates. Immunogenetics 69(6):391–400

Poole JRM, Paganini J, Pontarotti P (2017b) Convergent evolution of the adaptive immune response in jawed vertebrates and cyclostomes: An evolutionary biology approach based study. Dev Comp Immunol 75:120–126

Pray LA (2008) Transposons: The jumping genes. Nature education 1(1):204

Pulletikurti V, Yu C, Peterson T, Weber DF (2009) Cytological evidence that alternative transposition by Ac elements causes reciprocal translocations and inversions in Zea mays L. Maydica 54:457

Rippey C, Walsh T, Gulsuner S, Brodsky M, Nord AS, Gasperini M, Lee MK (2013) Formation of chimeric genes by copy-number variation as a mutational mechanism in schizophrenia. Am J Hum Genet 93(4):697–710

Roberts D, Hoopes BC, McClure WR, Kleckner N (1985) IS10 transposition is regulated by DNA adenine methylation. Cell 43(1):117–130

Rogers RL, Bedford T, Hartl DL (2009) Formation and longevity of chimeric and duplicate genes in Drosophila melanogaster. Genetics 181(1):313–322

Rogers RL, Bedford T, Lyons AM, Hartl DL (2010) Adaptive impact of the chimeric gene Quetzalcoatl in Drosophila melanogaster. Proc Natl Acad Sci 107(24):10943–10948

Schnable PS, Ware D, Fulton RS, Stein JC, Wei F, Pasternak S, Minx P (2009) The B73 maize genome: complexity, diversity, and dynamics. Science 326(5956):1112–1115

Sidorenko LV, Li X, Cocciolone SM, Chopra S, Tagliani L, Bowen B, Peterson T (2000) Complex structure of a maize Myb gene promoter: functional analysis in transgenic plants. Plant J 22(6):471–482

Singh M, Lewis PE, Hardeman K, Bai L, Rose JK, Mazourek M, Brutnell TP (2003) Activator mutagenesis of the pink scutellum1/viviparous7 locus of maize. Plant Cell 15(4):874–884

Wang D, Peterson T (2013) Isolation of sequences flanking Ac insertion sites by Ac casting. In: Plant transposable elements. Humana Press, Totowa, NJ, pp 117–122

Wang D, Yu C, Zuo T, Zhang J, Weber DF, Peterson T (2015) Alternative transposition generates new chimeric genes and segmental duplications at the maize p1 locus. Genetics 201(3):925–935

Wang W, Zheng H, Fan C, Li J, Shi J, Cai Z, Lu Z (2006) High rate of chimeric gene origination by retroposition in plant genomes. Plant Cell 18(8):1791–1802

Weil CF, Wessler SR (1993) Molecular evidence that chromosome breakage by Ds elements is caused by aberrant transposition. Plant Cell 5(5):515–522

Wicker T, Yu Y, Haberer G, Mayer KF, Marri PR, Rounsley S, Roffler S (2016) DNA transposon activity is associated with increased mutation rates in genes of rice and other grasses. Nat Commun 7:12790

Yadav VK, Rai KM, Yadav VK, Sable A (2016) An overview of transcription regulatory elements in plant. J Biol Sci Med 2(4):13–23

Zhang J, Peterson T (1999) Genome rearrangements by nonlinear transposons in maize. Genetics 153(3):1403–1410

Zhang J, Peterson T (2004) Transposition of reversed Ac element ends generates chromosome rearrangements in maize. Genetics 167(4):1929–1937

Zhang J, Peterson T (2005a) A segmental deletion series generated by sister-chromatid transposition of Ac transposable elements in maize. Genetics 171(1):333–344

Zhang F, Peterson T (2005b) Comparisons of maize pericarp color1 alleles reveal paralogous gene recombination and an organ-specific enhancer region. Plant Cell 17(3):903–914

Zhang P, Chopra S, Peterson T (2000) A segmental gene duplication generated differentially expressed myb-homologous genes in maize. Plant Cell 12(12):2311–2322

Zhang J, Zhang F, Peterson T (2006) Transposition of reversed Ac element ends generates novel chimeric genes in maize. PLoS Genet 2(10):e164

Zhang J, Yu C, Pulletikurti V, Lamb J, Danilova T, Weber DF, Peterson T (2009) Alternative Ac/Ds transposition induces major chromosomal rearrangements in maize. Genes Dev 23(6):755–765

Zhang J, Zuo T, Peterson T (2013) Generation of tandem direct duplications by reversed-ends transposition of maize Ac elements. PLoS Genet 9(8):e1003691

Zhang J, Zuo T, Wang D, Peterson T (2014) Transposition-mediated DNA re-replication in maize. Elife 3

Zuo T, Zhang J, Lithio A, Dash S, Weber DF, Wise R, Peterson T (2016) Genes and small RNA transcripts exhibit dosage-dependent expression pattern in maize copy-number alterations. Genetics 203(3):1133–1147

Allorecognition and Stem Cell Parasitism: A Tale of Competition, Selfish Genes and Greenbeards in a Basal Chordate

Anthony W. De Tomaso

Abstract The cellular branch of the vertebrate adaptive immune system was discovered via its role in transplantation, which in turn became the assay used to discover T cells and the interaction between the T cell receptor (TCR) and polymorphic MHC proteins. During this time period, multiple invertebrate species were studied and found to have similar polymorphic transplantation responses, suggesting that the underlying mechanisms may be conserved. However, while transplantation responses are similar, the molecular mechanisms are not, and perhaps one of the biggest mysteries in evolution is that the MHC and TCR are only found in jawed vertebrates, and further that there are no orthologs, or indeed any recognizable ancestral genes, in jawless fish or invertebrates. This leads to two main questions: First, if polymorphic transplantation responses are universal, do they have a common purpose? Second, what is the origin of the key molecules (MHC, TCR, BCR) of the vertebrate adaptive immune system? We are studying a natural transplantation reaction in an invertebrate chordate, called *Botryllus schlosseri*, which may lend insight into these questions. Similar to the vertebrates, transplantation reactions are controlled by a single, highly polymorphic locus. In addition, in Botryllus transplantation has a well-defined role: governing the exchange of mobile germline stem cells between individuals, and we have found that this exchange can be either beneficial, or harmful. The interplay between transplantation responses and mobile stem cells in Botryllus may provide some insight into the evolution of cellular immunity in the vertebrates.

1 Tissue Allorecognition and Transplantation

Transplantation responses, also called allorecognition, are based on the ability of an individual to discriminate its own tissues (self) from those of another individual of the same species (non-self). Allorecognition has been found in nearly all multicellular

A. W. De Tomaso (✉)
Department of Molecular, Cellular and Developmental Biology, University of California, Santa Barbara, Santa Barbara, CA 93106, USA
e-mail: detomaso@lifesci.ucsb.edu

© Springer International Publishing AG, part of Springer Nature 2018   131
P. Pontarotti (ed.), *Origin and Evolution of Biodiversity*,
https://doi.org/10.1007/978-3-319-95954-2_8

organisms, from sponges to humans (Burnet 1971). Allorecognition was initially discovered as a difference in response to a surgical transplant of tissue. For example, if skin is transplanted from one part of an individual to another (an autograft), it is usually accepted, becoming vascularized and incorporating into the existing tissue. However, if skin is transplanted between unrelated individuals (an allograft), it can be rejected, resulting in destruction of the donor tissue. This suggests a mechanism by which the non-self tissue is detected that is independent of the trauma of the surgery itself.

In addition to experimental surgical procedures, many organisms can naturally transplant tissues. For example, on crowded substrates, many sessile marine invertebrates will grow into other individuals, and this will initiate a reaction during which: (1) the two individuals will combine their tissues (fusion); or, (2) undergo a reaction whereby the two tissues stay separate. The latter rejection reaction is often accompanied by an inflammatory response followed by localized killing, which prevents tissue integration. Interestingly, this is universal, and has been described in nearly every metazoan phylum, including sponges, bryozoans, cnidarians, ascidians, and vertebrates (reviewed in Rosengarten and Nicotra 2011).

Both surgical and natural transplantation in these non-vertebrate organisms have been shown to have a genetic basis, and acceptance or rejection is based on highly polymorphic recognition systems. In other words, if two individuals are picked at random, they have a low chance of being compatible, somewhere between 5 and 10% in natural populations, alluding to recognition systems that are highly polymorphic. Overall, this level of polymorphism is unusual, and not likely due to neutral evolution (Grosberg 1988). This suggests that allorecognition is a trait subject to natural selection, but the role it plays in different species is not well understood.

Until recently, the only system in which the genetic basis of allorecognition was understood was in the vertebrates. Allorecognition is a function of the immune system, where it is controlled by the major histocompatibility complex (MHC). The MHC is a marker of self and is recognized by receptors on both T cells and natural killer (NK) cells (Davis and Bjorkman 1988; Kärre et al. 1986), and these molecules are found in all jawed vertebrates. Interestingly, allorecognition in the vertebrates is thought to be a by-product of the role the MHC plays in both T cell- and NK cell-mediated immunity. For the former, the TCR recognizes the peptide/MHC complex, and discriminates between self and non-self peptides. From a strictly immunological standpoint, different MHC alleles are not recognized as foreign MHC per se, but rather as infected self, and this is due to the fact that the TCRs are educated to self MHC/peptide complexes. The role of natural killer cells is to enforce MHC Class I expression, and they also undergo an education process to self MHC Class I alleles. NK receptors often recognize polymorphic regions of these self alleles, thus when a non-self MHC allele is present, it is not recognized.

Given these functions, what is the role of MHC polymorphism? For the T cells, MHC molecules are polymorphic because this allows them to present a more diverse repertoire of peptides. For NK cells, it has been found that viruses can go yet a step further, and often encode decoy molecules that bind NK cell receptors, mimicking binding to MHC molecules. This prevents the NK cells from detecting loss of MHC expression, and initiates an arms race, whereby NK cells recognize polymorphic epitopes on self MHC alleles, and viruses try to make decoys. In this case, MHC polymorphism serves to limit the effectiveness of the decoy molecules (reviewed in Carlyle et al. 2008).

There is also evidence of another role of the MHC, specifically in social signaling. In this case, MHC polymorphism would be used as a metric of relatedness (discussed below).

While allorecognition is universal, only in the last few years have molecules encoding allorecognition determinants been identified in invertebrates. Highly polymorphic genes have been isolated in the cnidarian, *Hydractinia symbiolongicarpus* (Nicotra et al. 2009), as well as the ascidian, *Botryllus schlosseri* (De Tomaso et al. 2005). Remarkably, the molecules identified in these different species have absolutely no evolutionary relationship to each other, or to the vertebrate MHC. In addition, orthologs of these molecules have not been found in the genomes of other species known to have allorecognition responses, for example, the sponge *Amphimedonqueenslandica* (Srivastava et al. 2010). It has also been shown that allorecognition responses between species within the same phylum can be due to separate genes: the genome of the ascidian *Ciona intestinalis* does not contain genes related to those involved in allorecognition in *B. schlosseri*, but has a well-studied allorecognition response between gametes that is dependent on completely different proteins (Harada et al. 2008). This is true even in the vertebrates: Jawless fish have an adaptive immune system that is based on proteins containing leucine-rich repeats (LRR), while their jawed cousins, including humans, use immunoglobulin-based proteins (Pancer et al. 2004). However, jawless fish do not encode any precursor molecules the jawed vertebrate MHC-based immune system, nor do jawed vertebrates contain related LRR genes in their genomes.

In summary, allorecognition responses can be found throughout the metazoa, and these systems are highly polymorphic, which is due to natural selection. However, the molecules involved in these responses are not evolutionarily related, suggesting unique origins for each system. This brings up three questions. First, what role does allorecognition play in each species? Second, how can highly discriminate recognition systems evolve so rapidly, and what are the genetic origins? And finally, why are these systems not conserved?

2   Is There a General Role of Allorecognition in the Metazoa?

The presence of allorecognition systems throughout the multicellular phyla suggests it plays an important role in some aspect of multicellular biology. In contrast, there is no evolutionary conservation of the molecules involved in the allorecognition process itself, suggesting an independent origin for each system. This is enigmatic—in complete contrast with the conservation of every other process in multicellular development and maintenance. In those cases, for example, mechanisms of cell specification and patterning, the genes utilized are highly conserved, so much so that it makes sense to study a fly to potentially gain insight into a human developmental process or disease. However, while it is easy to understand and hypothesize how polymorphism could be utilized in different species, for example immunity or social signaling, the fact that these systems have independent origins suggests that there are other processes under natural selection that we are unaware of. In general, the creation and more importantly maintenance of high levels of polymorphism seem to be a result of some sort of conflict. In other words, these unknown processes referred to above may be exploited in some manner and need to be constantly evolving. An example would be polymorphism of the MHC due to its role in immunity. As outlined above, using polymorphism to detect intruders sets up an arms race between the host T cells and NK cells versus the intruding microorganisms. This type of evolutionary conflict (also called the Red Queen hypothesis) would result in diversifying the MHC molecules (Carlyle et al. 2008; Klein et al. 1993).

Given this, in vertebrates, it could be concluded that allorecognition is merely an unintended consequence of the molecular mechanisms underlying immune recognition that it is the predictable evolutionary arms race between host and microorganisms which drives the diversification of the MHC molecules. Polymorphism at the MHC exists for immune function against microorganisms, not a role in transplantation. However, there is no evidence at all that allorecognition in *Hydractinia, Botryllus* or other species has anything to do with immune function. More importantly, the vast majority of organisms use very non-polymorphic, innate mechanisms for immune function and do quite well, including large, mobile, and long-lived creatures like giant squid. So it is clear that highly polymorphic recognition systems are not a requirement for immune function. Thus, the question becomes, are there any other generalities in multicellular biology in which allorecognition would play a role? As mentioned above, the only other hypothesized role for polymorphism is social interactions, whereby the polymorphism would be used as a metric of relatedness, for a variety of functions which presumably would be altruistic in nature. This is the basis of the 'greenbeard' concept, that is, a feature which can be recognized by related individuals, which would be used when weighing the costs and benefits of an altruistic decision (i.e., kin selection), or preventing inbreeding with a related individual (Hamilton 1964a, b; Dawkins 1976). The greenbeard would be an integral part of kin selection, and the higher the polymorphism, the higher the potential resolution for determining relatedness would be. For the vertebrate MHC, the greenbeard is thought to be an odor, detected by MHC molecules (Ruff et al. 2012). However, the exact molecular details are far from being understood.

An example of the potential general role of allorecognition may be found in botryllid ascidians. Ascidians are considered the basal chordates and have a highly polymorphic allorecognition system that is genetically analogous to the vertebrates. Initially, this was thought to indicate an evolutionary relationship (Scofield et al. Scofield et al. 1982); however, isolation of these molecules did not support this hypothesis (De Tomaso et al. 2005; Nyholm et al. 2006). Nevertheless, the overall role of allorecognition in the botryllids is very well characterized and reveals several fascinating phenomena that may indicate the origins of allorecognition in the metazoa.

3   Life History and Allorecognition in the Botryllid Ascidians

Botryllid ascidians (or sea squirts) belong to the phylum Tunicata, which are considered to be the basal chordates (Dehal et al. 2002). Botryllids grow in shallow subtidal waters throughout the world and seem to prefer still water areas, such as marinas. There are a number of species where allorecognition has been studied (Saito et al. 1994), but one of the best studied is *Botryllus schlosseri*, of which the rest of this review will focus.

The link between the chordates and tunicates is most obvious in the larval stage. Ascidians are born as tadpole larva with many chordate characteristics, including a notochord, dorsal hollow nerve chord, and a post-anal tail. Following hatching, the tadpole swims, finds a suitable substrate, and settles. Settlement initiates a dramatic metamorphosis during which most of the chordate characteristics are resorbed (e.g., the tail) and a sessile, invertebrate form emerges, called an oozooid. All ascidians undergo the transition from a motile chordate larva to an oozooid, but the botryllids belong to a subset of ascidians that are colonial and grow, not by getting bigger, but by a lifelong asexual budding process which eventually gives rise to a colony of genetically identical individuals, called zooids, united by a common circulation. Each zooid is an independent body, with a gastrointestinal tract, musculature, heart, central and peripheral nervous system, and a germline. An extracorporeal circulatory system connects all the zooids and meanders throughout the colony. In turn, the zooids and circulation are embedded within a cellulose-based tunic, which is a defining feature of the Tunicata. At the periphery of the colony, the abruptly stops and ramifies into finger-shaped projections called ampullae, and these are the site of allorecognition.

Each week, each zooid in a colony replicates itself in a process called blastogenesis, giving rise to 1–4 new zooids. This budding process has two important consequences for allorecognition. First, a colony is constantly expanding over the substrate, and second development is always occurring, as each week the budding process is regenerating all somatic and germline tissues. This constant development has major implications for the biology of allorecognition in *Botryllus*.

As colonies asexually expand outward, they often come into contact with other individuals. The first part of the colonies to come into contact is the peripheral vasculature, and contact between juxtaposed ampullae initiates the allorecognition reaction, which will result in one of two outcomes. Either the two ampullae will fuse, linking the circulation of the two individuals, or they will undergo a rejection reaction. Rejection is an inflammatory reaction during which one type of blood cell migrates to the ampullae in contact, then into the periphery between the ampullae. Upon hitting seawater, the cells burst, releasing materials that self-assembles into a melanin scar, which prevents vascular fusion. From touching to fusion or rejection takes 24–48 h and can be easily monitored and studied as it occurs outside of the body. However, one of the most intriguing characteristics of botryllid allorecognition is the genetics. Allorecognition is controlled by a single, highly polymorphic locus called the *fuhc*, for **fu**sion/**h**isto**c**ompatibility (Scofield et al. 1982). The rules of allorecognition in this system are simple: Individuals that share one or both fuhc alleles are compatible and fuse, while those sharing no alleles reject. The *fuhc* locus is extraordinarily polymorphic, with most populations having 50 to >100 alleles, which essentially restricts fusion to related individuals.

fuhc-based allorecognition is remarkably similar to the vertebrates, where responses are principally controlled by a single, highly polymorphic locus: the major histocompatibility complex (MHC). However, the rules of transplantation in the vertebrates are different, as both MHC alleles must be shared for compatibility, while individuals sharing one or none will reject. In addition, besides the MHC loci, there are hundreds of minor histocompatibility loci (mHC) in the vertebrates which also contribute to tissue compatibility. In contrast, there are no modifying loci in the botryllids, and even in crosses between wild individuals fusibility segregates in normal Mendelian ratios. However, as discussed above, despite a similarity in transmission genetics, there is no relationship between the fuhc and MHC molecules; thus, the two systems appear to have an independent evolutionary origin (De Tomaso et al. 2005).

4 Natural Stem Cell Transplantation Between Compatible Colonies

Why is there a polymorphic allorecognition system in *Botryllus*? The answer can be found by following compatible colonies following the fusion event. As described above, each individual zooid in a *Botryllus* colony is constantly regenerating new buds. Following vascular fusion, it was found that germline and somatic tissues in one parabiosed partner were actually derived from the other, and that the colonies were chimeric (Sabbadin and Zaniolo 1979). This continued to be the case even if the colonies were surgically separated and could remain so for the lifespan of the individual (>6mos; Sabbadin and Zaniolo 1979; Stoner and Weissman 1996; Stoner et al. 1999). The interpretation of these results is that following the fusion event, mobile, long-lived progenitors could transfer via the vascular connection, and once resettled could contribute to the development of germline and/or somatic tissues in the newly developing buds.

In addition, it was found that following fusion one genotype could contribute disproportionately to new germline development, often resulting in a situation where only one genotype would be represented in the gametic output of both fused individuals. This dominance of germline development by a single genotype can continue for the remainder of the lifespan of the colony, even if the fusion is surgically terminated.

In summary, when germline progenitors from two genotypes are mixed, they begin to compete for germline niches in the newly developing buds, and one genotype can dominate those from another, in a process called *germ cell parasitism* (gcp). It has been shown that gcp is a repeatable and heritable trait, and winner and loser genotypes in both lab-reared and field colonies have been identified. This demonstrates the presence of germline stem cells with a genetically determined competitive phenotype (Sabbadin and Zaniolo 1979; Stoner and Weissman 1996; Stoner et al. 1999).

In another set of experiments, natural transplantation was recapitulated by isolating cells from one colony and injecting them into the vasculature of another. It was found that experimental transplantation recapitulated natural parabiosis experiments: When cells were isolated from a gcp winner colony and transplanted into a gcp loser, they expanded and differentiated. However, when cells were isolated from a gcp loser, and transplanted to a gcp winner, no germline tissues from the donor were detected. This demonstrated that gcp competitive properties were autonomous to the cells themselves and retained upon experimental transplantation (Laird et al. 2005).

If dominant gcp genotypes exist in nature, what maintains the gcp loser genotypes in any population? The answer is the polymorphism of the fuhc locus. As each population studied carries somewhere between 50 and >100 fuhc alleles, the chances of fusion between unrelated individuals are very low, between 5 and 10%. Thus, allorecognition in *Botryllus* represents interplay between the ability to fuse coupled to the possibility of gcp, and it is the extraordinary polymorphism of the fuhc locus that regulates the natural transplantation of parasitic germline stem cells between individuals.

5 Why Is There Fusion?

The extreme polymorphism of the *fuhc* locus is not likely due to neutral evolution and implies that fusion is costly and must be regulated (Grosberg 1988). However, the costs are not consistent among all individuals. For example, a SCP winner genotype would want to fuse with any other individual and favor indiscriminate fusion, while a SCP loser would rather not fuse with any individual. Given that fusion can lead to the very high cost of loss of the germline (equivalent to being dead from an evolutionary perspective), characterizing the potential benefits of fusion is important.

Three major benefits of fusion have been hypothesized. The first is an instant increase in size. For encrusting invertebrates, almost all ecological fitness parameters are size-dependent, with larger individuals showing a significant advantage in both survivability and fecundity (reviewed in De Tomaso 2006).

There is also a relationship between size and the onset of sexual maturity. Colonial invertebrates often go through a juvenile stage, where somatic growth occurs, but the germline is not present. In *Botryllus*, this is often a 4–12 week period, although we have found that juvenile individuals contain functional germline progenitors. Thus, fusion could be beneficial to smaller or juvenile individuals. Finally, the colonial nature of *Botryllus* in theory could allow Lamarckian evolution: A chimeric individual would have increased genetic diversity that it could utilize during asexual budding in changing environmental conditions. Along those lines, it has also been shown that juvenile chimeras may have a slight advantage over individuals (Carpenter et al. 2011), but the reason for this is unknown.

The consensus is that the effective role of fuhc polymorphism is to restrict the potential of fusion to kin, whereby the individual benefits of fusion can be realized, while the potential costs of SCP are lowered (reviewed in De Tomaso 2006). The costs of SCP can be ameliorated by gains in inclusive fitness (kin selection, e.g., giving germline and somatic space to relatives); thus, the more closely related two individuals are the lower the potential costs of fusion.

6   The Origins of Fusion and Functional Aspects of Allorecognition

The origins of fusibility and evolution of allorecognition responses are intriguing in context of recent functional studies (McKitrick et al. 2011; Nyholm et al. 2006). One hypothesis is that fusion was a default process due to the colonial nature of *Botryllus*: Colonies are constantly asexually expanding over an irregular substrate, and often grow around other objects and back into themselves. The idea is that it would be costly to maintain a physical barrier against yourself, and easier to fuse then shift growth resources elsewhere. In that case, fusion came first and initially would have been indiscriminate. This would be followed by the cost of gcp, which resulted in the evolution of allorecognition polymorphism and the ability to discriminate between alleles of the *fuhc*. Another hypothesis may be that gcp winner genotypes evolved the ability to indiscriminately fuse, as they were aggressively trying to parasitize other members of the population, and polymorphic allorecognition was used to protect gcp loser genotypes.

Our initial functional studies favor the former hypothesis over the latter. We have shown that blocking expression of both ligands (unpublished) and receptors (Nyholm et al. 2006) can make an individual non-reactive. In other words, if we block expression of some of the proteins involved in allorecognition, the interacting colonies neither fuse nor reject. This suggests that this system is set-up to promote discriminate fusion. The caveat to this interpretation is that we can only transiently knock down protein expression in *Botryllus*, for approximately 2–3 weeks. It could be that these proteins play another unknown role and we did not see any effect on fitness during these experiments.

In addition, another intriguing observation on botryllid allorecognition is that in most cases, it is species-specific. There are multiple botryllid species, and each shows single-locus allorecognition as described in *B. schlosseri*. However, in most cases, individuals from different species will not interact, but will grow around each other like any other object they might encounter. In some cases where members of two species interact, rejection has been documented (reviewed in Saito et al. 1994), but this is restricted to botryllids, and ampullae do not react to inanimate objects or other species.

These species-specific interactions coupled to functional studies lend strong support for the idea that allorecognition in *B. schlosseri* did not evolve to prevent the spread of dominant gcp phenotypes, but rather to mediate desired interactions between individuals. In turn, this suggests that polymorphism of the fuhc locus originated and is maintained to promote fusion between closely related individuals.

7   Fusion Between Adults and Juveniles

If fusion is beneficial under certain conditions, as the results discussed above suggest, is there a certain stage of the life history that it is more beneficial, for example juveniles forming chimeras in order to increase their chances of survival? While chimerism between juveniles has been shown to provide a small but measurable increase in fitness (Carpenter et al. 2011), studies in both behavior and stem cell biology have indicated that it is the interaction between juveniles and adults that may be the driving force in evolution of fuhc-based histocompatibility.

The first observation was made many years ago by Grosberg and Quinn (1986), who showed that swimming larvae would preferentially settle near histocompatible individuals, both other larvae and adults. This result was the first to suggest that an individual would like to increase the opportunity to be in a group with relatives, and in turn this would of course increase the chances of fusion during that individual's lifetime. These results were corroborated when proteins within the fuhc locus were identified. Both the candidate fuhc ligand and a putative receptor, called fester, have secreted forms that are made by both the larvae and adult (De Tomaso et al. 2005; Nyholm et al. 2006). Moreover, another receptor (uncle fester) is expressed along the nerves within structures in the larvae called adhesive papillae (McKitrick et al. 2011). These are sensory organs that the larvae touch to the substrate prior to settling and undergoing metamorphosis. Thus, proteins involved in histocompatibility are expressed in a manner which is consistent with co-settlement results.

In addition, it was shown that parabiosis between a juvenile and adult resulted in a rapid and long-term germline chimerism, similar to fusion between two adults (Brown et al. 2009). This demonstrated the presence of functional germline precursors in newly metamorphosed individuals, and coupled to the potential of fuhc-based co-settlement completely changed the thinking of allorecognition responses. Prior to this, we had thought of allorecognition in two-dimensional terms, with colonies expanding over a flat substrate and growing into each other. However, the potential of a swimming larva to detect and settle near a compatible individual suggests that potential allogeneic encounters could occur orders of magnitude more frequently than previously appreciated.

Co-settlement and fusion could also provide another life history strategy for an individual. Under normal conditions, larvae settle and undergo 4–12 asexual budding cycles prior to the emergence of gametes. However, following fusion, the juvenile can immediately contribute to the next generation of gametes that will mature (Brown et al. 2009). Thus, it may be that an alternative strategy for a larva may be to fuse into an established colony and naturally transplant its' germline stem cells, skipping somatic growth completely. Even if only a single zooid in a colony made the larval germline, it could change the fecundity of that individual. Although we do not know the survival rate of larvae in most populations, it is pretty clear that the ability to harbor one's germline in an established individual and contribute to the gamete pool 2–5× faster than normal could provide an enormous potential increase in fitness.

The adult colony could have potential inclusive fitness gains from this interaction, by allowing some of its germline real estate to be used by a related individual. There may also be direct fitness gains to the adult, as somatic stem cells in the juvenile may be able to contribute to maintenance of somatic growth in the established individual. In encrusting marine communities, the limiting resource is often space; thus, it may not be surprising that a strategy has evolved whereby an individual lucky enough to establish itself would provide germline space for related individuals and potentially take advantage of the younger somatic stem cells to maintain the body and extend its own reproductive lifespan. This interaction may be reflected in the rules of allorecognition: Only a single allele needs to be shared for fusion, which ensures that larvae can fuse into both parents.

If this interaction is indeed an important part of the life history of *B. schlosseri*, it should have two implications. First, a population should consist of small patches of related individuals. Second, germline chimerism should be common. Both of these studies are underway in our laboratory now, and preliminary results suggest that chimerism is much more prevalent than previously predicted. It should be noted that if germline transfer is usually altruistic, then the fuhc locus of *B. schlosseri* is an excellent example of a 'greenbeard' which promotes co-operation and has affect on social structure. Moreover, the fact that both gcp winner and loser individuals share fuhc alleles demonstrates that the two traits are not linked, and the polymorphism at the fuhc locus is consistent with previous modeling studies (Jansen and van Baalen 2006).

8 Summary

Botryllus provides a powerful and comprehensive model to study the source, mechanisms, and broader aspects of allorecognition, from immunity and stem cell biology to ecology, that are not available in other models. However, what is not known is how general phenomenon such as stem cell parasitism is in nature. It has certainly been hypothesized that transplantation between unrelated individuals could be dangerous. Prior to the discovery of the function of the MHC, it was hard to understand why the immune system would attack and destroy a transplant. MacFarlane Burnet hypothesized that this may prevent cancer from becoming contagious (Burnet 1971), of which there are two examples, a cancer of the mouth which is rampant in Tasmanian

Devils, as well as a sexually transmitted venereal cancer in dogs (reviewed in Siddle and Kaufman 2013) In addition, pregnancy in the mammals provides an opportunity for natural transplantation of stem cells between individuals as well. Thus, it could be that stem cell parasitism is a common phenomenon in metazoans, and the main reason allorecognition has evolved. Along those lines, while this review has been focused on metazoans, allorecognition, cheating, and co-operation have also been well-studied in the transition species which cycle between unicellular and multicellular, such as social ameba (Strassmann and Queller 2011; Shaulsky and Kessing 2007).

In summary, the presence of allorecognition depends on two characteristics: protein(s) that provide a stable, lifelong definition of self, and the ability to discriminate among allelic variants. Specificity depends on polymorphism at these loci; thus, allorecognition is dependent on creating and maintaining genetic diversity. In turn, the ability to detect polymorphism can then be used for a variety of functions besides allorecognition, from preventing inbreeding to governing mutual social interactions and has also been co-opted for immunity in the vertebrates. Allorecognition seems to be a common trait, providing a metric of relatedness that can be used for a variety of functions.

References

Brown FD, Tiozzo S, Roux MM, Ishizuka K, Swalla BJ, De Tomaso AW (2009) Early lineage specification of long-lived germline precursors in the colonial ascidian *Botryllus* schlosseri. Development 136:3485–3494

Burnet FM (1971) "Self-recognition" in colonial marine forms and flowering plants in relation to the evolution of immunity. Nature 232:230–235

Carlyle JR, Mesci A, Fine JH, Chen P, Bélanger S, Tai LH, Makrigiannis AP (2008) Evolution of the Ly49 and Nkrp1 recognition systems. Semin Immunol 20:321–330

Carpenter MA, Powell JH, Ishizuka KJ, Palmeri KJ, Rendulic S, De Tomaso AW (2011) Growth and long-term somatic and germline chimerism following fusion of juvenile *Botryllus* schlosseri. Biol Bull 220:57–70

Davis MM, Bjorkman PJ (1988) T-cell antigen receptor genes and T-cell recognition. Nature 334:395–402

Dawkins R (1976) The selfish gene. Oxford University Press, Oxford

Dehal P, Satou Y, Campbell RK et al (2002) The draft genome of Ciona intestinalis: insights into chordate and vertebrate origins. Science 298:2157–2167

De Tomaso AW (2006) Allorecognition polymorphism versus parasitic stem cells. Trends Genet 22:485–490

De Tomaso AW, Nyholm SV, Palmeri KJ, Ishizuka KJ, Ludington WB, Mitchel K, Weissman IL (2005) Isolation and characterization of a protochordate histocompatibility locus. Nature 438:454–459

Grosberg RK (1988) The evolution of allorecognition specificity in clonal invertebrates. Q Rev Biol 63:377–411

Grosberg RK, Quinn JF (1986) The genetic control and consequences of kin recognition by the larvae of a colonial marine invertebrate. Nature 322:456–459

Hamilton WD (1964a) The genetical evolution of social behaviour I. J Theor Biol 7:1–16

Hamilton WD (1964b) The genetical evolution of social behaviour II. J Theor Biol 7:17–52

Harada Y, Takagaki Y, Sunagawa M, Saito T, Yamada L, Taniguchi H, Shoguchi E, Sawada H (2008) Mechanism of self-sterility in a hermaphroditic chordate. Science 320:548–550

Jansen VA, van Baalen M (2006) Altruism through beard chromodynamics. Nature 440:663–666

Kärre K, Ljunggren HG, Piontek G, Kiessling R (1986) Selective rejection of H-2-deficient lymphoma variants suggests alternative immune defence strategy. Nature 319:675–678

Klein J, Satta Y, O'hUigin C, Takahata N (1993) The molecular descent of the major histocompatiblity complex. Annu Rev Immunol 11:269–295

Laird DJ, De Tomaso AW, Weissman IL (2005) Stem cells are units of natural selection in a colonial ascidian. Cell 123:1351–1360

McKitrick TM, Muscat CC, Pierce JD, Bhattacharya D, De Tomaso AW (2011) Allorecognition in a basal chordate consists of independent activating and inhibitory pathways. Immunity 34:616–626

Nicotra ML, Powell AE, Rosengarten RD, Moreno M, Grimwood J, Lakkis FG, Dellaporta SL, Buss LW (2009) A hypervariable invertebrate allodeterminant. Curr Biol 19:583–589

Nyholm SV, Passegue E, Ludington WB, Voskoboynik A, Mitchel K, Weissman IL, De Tomaso AW (2006) Fester, a candidate allorecognition receptor from a primitive chordate. Immunity 25:163–173

Pancer Z, Amemiya CT, Ehrhardt GR, Ceitlin J, Gartland GL, Cooper MD (2004) Somatic diversification of variable lymphocyte receptors in the agnathan sea lamprey. Nature 430:174–180

Rosengarten RD, Nictora ML (2011) Model systems of invertebrate allorecognition. Curr Biol 21:R82–R92

Ruff JS, Nelson AC, Kubinak JL, Potts WK (2012) MHC signaling during social communication. Adv Exp Med Biol 738:290–313

Sabbadin A, Zaniolo G (1979) Sexual differentiation and germ cell transfer in the colonialascidian, Botryllus schlosseri. J ExpZool 207:279–301

Saito Y, Hirose E, Watanabe H (1994) Allorecognition in compound ascidians. Int J Dev Biol 38:237–247

Scofield VL, Schlumpberger JM, West LA, Weissman IL (1982) Protochordate allorecognition is controlled by a MHC-like gene system. Nature 295:499–502

Shaulsky G, Kessin RH (2007) The cold war of the social amoebae. Curr Biol 17:R684–692

Siddle HV, Kaufman J (2013) A tale of two tumours: comparison of the immune escape strategies of contagious cancers. MolImmunology 55:190–193

Srivastava M, Simakov O, Chapman J et al (2010) The Amphimedon queenslandica genome and the evolution of animal complexity. Nature 466:720–726

Stoner DS, Weissman IL (1996) Somatic and germ cell parasitism in a colonial ascidian: possible role for a highly polymorphic allorecognition system. Proc Natl Acad Sci USA 93:15254–15259

Stoner DS, Rinkevich B, Weissman IL (1999) Heritable germ and somatic cell lineagecompetitions in chimeric colonial protochordates. Proc Natl Acad Sci USA 96:9148–9153

Strassmann JE, Queller DC (2011) Evolution of cooperation and control of cheating in a social microbe. Proc Natl Acad Sci USA 108:10855–62

How to Become Selfish: Evolution and Adaptation to Self-fertilization in Plants

Natalia Wozniak and Adrien Sicard

Abstract A major trend in plant evolution is the transition from outcrossing to selfing. This transition has occurred many times independently despite some putative detrimental consequences on the evolutionary potential of individuals. In animal–pollinated flowers, this transition has often been followed by a set of changes in flower characters, sex allocation and life history. In this article, we review the evolutionary history of this transition focusing on the shift from obligate outcrossing to predominant selfing. We discuss the current knowledge on the ecological factors driving selfing evolution and the consequences on the morphology, biogeography and evolution of selfing lineages. This event constitutes an excellent model to study phenotypic evolution in plants, and several studies are starting to shed light on the underlying molecular and evolutionary mechanisms.

1 The Transition to Selfing in Plants

Reproductive systems are highly variable in flowering plants. Plants can reproduce both sexually (through cross- or self-pollination) and asexually (e.g. vegetative propagation) through different mechanisms. A majority of flowering plants (~40–60%) are unable to self-fertilize due to the existence of a self-incompatibility system and, per consequence, rely on outcrossing for their reproduction (Barrett 2013). In such system, the transfer of gametes between plants is generally mediated by pollen vectors such as animal pollinators or wind. While self-incompatibility systems appear to be predominant in plants, they have frequently broken down leading to the evolution of a large number of highly selfing lineages (~20% of flowering plants) (Barrett

N. Wozniak
Institut für Biochemie und Biologie, Universität Potsdam, Karl-Liebknecht-Str. 24-25,
14476 Potsdam-Golm, Germany

A. Sicard (✉)
Department of Plant Biology, Swedish University of Agricultural Sciences, Uppsala BioCenter,
BOX 7080, 750 07 Uppsala, Sweden
e-mail: adrien.sicard@slu.se

© Springer International Publishing AG, part of Springer Nature 2018
P. Pontarotti (ed.), *Origin and Evolution of Biodiversity*,
https://doi.org/10.1007/978-3-319-95954-2_9

143

2002). The transition from outcrossing to selfing has been so frequently observed that it is now considered as a universal feature of flowering plant evolution (Stebbins 1950, 1957, 1970, 1974). The emergence of self-compatibility does not, however, always lead to predominant selfing, but often a mixed mating system is maintained (~30%) (Vogler and Kalisz 2001). In such cases, the rate of outcrossing versus selfing may vary between populations of the same species but also within the same population when individuals are subjected to different environments (Levin 2012). Because mating strategies have profound consequences on the diversity and evolutionary potential of organisms, the causes and consequences of the recurrent transitions to selfing have been at the centre of biological debates for over a century.

1.1   The Genetic Basis of Self-compatibility Evolution

Obligate outcrossing in plants is maintained by self-incompatibility (SI) systems (Fig. 1). Many different SI systems have been described, but in most cases, they are controlled by a single polymorphic locus named the S-locus [reviewed in Fujii et al. (2016)]. Different S-haplotypes encode specific male and female determinants, and it is the molecular interaction between these determinants that controls the self- or non-self-recognition. As a result, pollen harbouring a given haplotype will not be able to fertilize the ovules of the same plant as well as of plants carrying the same S-haplotype (Fig. 1). Two main types of SI have been described in plants, the gametophytic SI (GSI) in which the male determinants are directly expressed by the haploid genotype of the pollen, and the sporophytic SI (SSI) in which it is the diploid genotype that determines the male component of the self-recognition system. GSI has been best characterized in the Solanaceae. In this genus, the female determinant encodes a glycoprotein S-RNAse, while the male determinant encodes tandem copies of the S-locus F-box (SLFs) protein whose function is to target specific proteins for degradation by the proteasome (Lee et al. 1994; Murfett et al. 1994; Mccubbin et al. 2004). The current model suggests that SLFs recognize the non-self S-RNAse and promote their degradation allowing the pollen tubes to grow and reach the ovules. In this non-self-recognition system, the SLFs would be unable to recognize the self S-RNAse and thus prevent its inhibition of pollen tube growth. An example of a well-characterized SSI is the Brassicaceae S-locus (Disorders et al. 2000; Takayama et al. 2000). This system is encoded by two tightly linked genes: a small male ligand peptide, the S-locus cysteine-rich protein (SCR) and a female S-locus receptor kinase (SRK) (Kachroo et al. 2001). In this case, the direct interaction between the SCR and the SRK will activate a phosphorylation cascade that will ultimately lead to self-pollen rejection (Fujii et al. 2016). This self-recognition system also relies on complex hierarchical dominance relations that are controlled by the interactions between multiple small RNAs and their target sites within the SCR loci (Durand et al. 2014). Self-compatibility (SC) and therefore self-fertilization evolves from the breakdown of such systems. This can occur through mutations affecting the male and/or female component, via a recombination between these two factors among

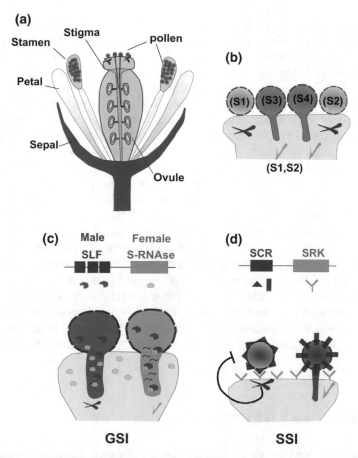

Fig. 1 Self-incompatibility systems in plants. a Schematic representation of a typical flower structure. **b** Self-pollen (in red) is rejected in self-incompatible plants whereas pollen harbouring different S-haplotypes are not (in blue). The genotypes are indicated in brackets. **c** Molecular mechanisms of Gametophytic Self-Incompatibility (GSI). The female S-RNAse is targeted for degradation by the paternal S-locus F-box protein (SLF) of foreign pollen. The SLF of self-pollen is not able to recognize the S-RNAse, which in turn degrades RNAs and inhibits pollen growth. **d** Molecular mechanisms of Sporophytic Self-Incompatibility (SSI). The S-locus cysteine-rich protein (SCR) of self-pollen grains is recognized by the female S-locus receptor kinase (SRK) leading to the inhibition of self-pollen germination. The SCR ligand of foreign pollen grains is however not recognized

different haplotypes or finally by a mutation in a modifier gene such as a downstream effector (Shimizu and Tsuchimatsu 2015).

The study of sequence variation within the S-locus in selfing lineages has received much attention, as it provides some information about the number of founding haplotypes and thus of events of selfing evolution. These analyses have revealed for instance that while some lineages have evolved from a limited number of SC

S-haplotypes, in others several independent transitions to selfing seem to have occurred. In the Brassicaceae, the speciation of *A. thaliana* from its outcrossing ancestor seems to have evolved through the fixation of three non-functional S-locus alleles; *C. rubella*, however, appears to have emerged through the fixation of a single S-locus haplotype; while in *A. lyrata* SC has evolved through multiple origins (Tsuchimatsu et al. 2010; Brandvain et al. 2013; Griffin and Willi 2014). These analyses are often complicated by the fact that once SC has evolved, the S-locus is no longer subjected to selective pressures and therefore accumulates a large number of deleterious mutations. A tendency that has, nevertheless, been observed is that SC often evolved through a mutation in the male component of the recognition system. This was proposed to relate to Bateman's principle, which states that the over-representation of male gametes over female gametes engenders a stronger mate competition between male compared to female gametes (Tsuchimatsu and Shimizu 2013). In the context of the transition to SC, a mutation in the male determinant affecting the SI recognition system renders pollen grains able to fertilize the ovules of the same plant, but also those of the plants harbouring the same haplotype group and all others in the population. A mutation in the female component is expected to have a weaker fitness advantage over other female gametophytes because they are not transported and, most importantly, they are in much-reduced number. While SC alleles seem to evolve frequently, their fixation and thus the establishment of a population of selfing individuals will depend on several ecological and evolutionary factors.

1.2 Evolutionary Triggers and Consequences

Mating systems have major implications on a population's ecology, demography and evolutionary trajectories (Charlesworth 2006). For instance, outcrossing has the benefit of maintaining a high level of heterozygosity within genomes, which allows tolerating the presence of deleterious mutations but requires a certain level of dominance for the selection of beneficial mutations. The main disadvantage is, however, that their reproductive success is highly dependent on the presence of mates and pollen vectors. Selfers, in contrast, do not depend on such environmental factors for their reproduction. Such a reproductive assurance advantage was first recognized by Darwin and since considered as one of the main factors explaining the transition to selfing (Darwin 1876). Another factor that has been proposed to promote the fixation of alleles improving self-fertilization is the fact that they confer a 3:2 transmission advantage in outcrossing populations (Fisher 1941). Once self-fertilization has evolved, it will lead to a considerable increase in homozygosity and limit the evolutionary consequences of recombination. This will, in turn, increase the expression of recessive mutations, linked selection and the influence of genetic drift (Barrett et al. 2014). As a result, shifts towards selfing and especially towards high selfing rates will be associated with a strong reduction in genetic diversity and an apparent 'population bottleneck' (Foxe et al. 2009; Busch and Delph 2012; Slotte et al. 2013;

Barrett et al. 2014). This reduced effective population size limits the efficacy of purging selection in selfing lineages and thus is expected to lead to the accumulation of deleterious mutations increasing the risk of species extinction (Glémin and Ronfort 2013; Hazzouri et al. 2013; Igic and Busch 2013; Wright et al. 2013; Slotte et al. 2013; Barrett et al. 2014). Given the consequences of selfing on genetic diversity, the main factor limiting the evolution of selfing is believed to be the abundance of recessive deleterious mutations in the ancestral outcrossing population (Lande and Schemske 1985). Indeed, through the evolution of self-fertilization, such mutations are likely to become homozygous and negatively impact plant fitness, a phenomenon known as inbreeding depression. The evolution of selfing will, therefore, depend on a balance between reproductive assurance/transmission advantage and the intensity of inbreeding depression.

1.3 Ecological Context

As outlined above, selfing genes offer both reproductive assurance and an 'automatic transmission advantage', but decrease the efficiency of natural selection in comparison to outcrossing (Cheptou 2011; Pannell 2015). These differences have generated animated discussions on the ecological contexts in which different mating systems may be favoured. The early view was that the ability of self-fertilizing plants to reproduce alone provides a selective advantage when pollen vectors or conspecific compatible mates are scarce (Darwin 1876; Baker 1955; Lloyd 1979). Such pollen limiting environments are likely to be encountered in adverse environmental conditions that would limit pollinator visitation or in low population density and fragmented habitats restricting the availability of compatible mates. Indeed, because outcrossing is often maintained by self-incompatibility systems with complex allelic dominance relationships, a lowering of genetic diversity in isolated populations may limit the probability to encounter a 'compatible allele'. As a result, the mating system is expected to evolve mostly at the margin of a species' distribution where population fragmentation and environmental challenges are often observed (Kawecki 2008; Shimizu and Tsuchimatsu 2015). In such a context, outcrossing would increase the risk of population extinction while selfing would confer an escape route through reproductive assurance. These ideas together with the observation that long-distance dispersal seems to occur preferentially in self-fertilizing species have led Baker to propose that mating system transitions will also influence the ability of populations to colonize new habitats (Baker 1955). Because such a correlation between reproductive systems and species' geographical range was not only observed in plants, it was later suggested to represent a general feature in population distribution elevating this idea to the status of 'Baker's law' (Longhurst 1955; Stebbins 1957). The existence of several counter examples has since indicated that the flexibility of breeding systems and the dynamic nature of the colonisation process make evolutionary predictions on species expansion more complex (Cheptou 2011; Pannell 2015).

A difficulty in identifying the ecological contexts driving the transition to self-fertilization is that it requires determining the direction of causality between ecology and selfing evolution (Barrett et al. 2014). This seems difficult to achieve only through correlative biogeographic analysis of mating system distribution, especially because the evolution of selfing could itself lead to species expansion and fragmentation at margin-areas where pollinators are scarce. It may, therefore, require directly testing the influence of different ecological factors on the selection of higher selfing rates. Such experiment has been conducted in the plant genus *Clarkia* by comparing the impact of population size and pollen limitation on the selection of flower traits promoting selfing (Moeller and Geber 2005). This study demonstrates that the selection of selfing occurs in a context-dependent manner with small populations or low pollinator availability favoring the selection of traits promoting selfing. The selection of these traits was, however, weaker in large populations with abundant congeners. The results of this common garden experiment were in agreement with the geographical distribution of the different mating types, suggesting an important role of reproductive assurance in promoting the transition to selfing in this genus. Furthermore, recent experiments of artificial pollinator-driven selection in *Brassica rapa* indicated that the composition of pollinator's communities also influences the evolution of plant traits and mating systems (Gervasi and Schiestl 2017). These experiments measured the effect of different pollinators on the selection of plant phenotypes, floral scent and mating system traits. This work demonstrates that pollen limitation due to less efficient pollinators results in the selection of traits increasing autogamous selfing (e.g. pistil length) as well as in a reduction in the emission of several flower scent compounds. This, further, supported the importance of reproductive assurance in driving selfing evolution. Since these are the only examples available so far, additional experiments of this type will be needed to fully appreciate the role of reproductive assurance and identify the main ecological factors triggering the transition to selfing. This change of mating strategies will, however, impose new constraints on flower phenotypes that will no longer need to optimize pollen transfer but rather to improve self-fertilization.

2 Mechanism of Adaptation to Selfing

In hermaphrodite flowers, the transition to selfing is generally associated with a characteristic set of changes in the morphology and function of flowers (Darwin 1876; Ornduff 1969). The similarities in the morphological evolution following the transition to selfing are such that it has elevated this set of phenotypic changes to the term of *selfing syndrome* (Lloyd 1979; Sicard and Lenhard 2011). It generally corresponds to a reduction of flower size and opening, a shortening of the distance between stamens and stigma, a reallocation of sexual resources from male to female gamete production as well as a decrease in the production of nectar and scent. A key question, here, is whether the mutations inducing these changes were fixed with a fitness benefit or as the result of the reduced efficiency of purging selection in selfers while potentially even incurring a fitness disadvantage. Identifying the genetic basis underlying these changes may answer this question but also provide key insights into the demographic contexts and the selective mechanisms involved in the emergence of selfing lineages (Barrett et al. 2014). For instance, if reproductive assurance has driven the evolution of self-fertilization, gene flow was most likely limited at its origin and, thus, new mutations rather than capture from standing genetic variation would be expected to underlie the evolution of the selfing syndrome. Nevertheless, an important consideration here maybe to discuss the adaptive value associated with the evolution of these different traits. Indeed, once a high selfing rate has evolved, the limited contribution of exogenous pollen renders a contribution of standing genetic variation less probable. Therefore, only the genetic basis of traits that act as mating system modifiers improving the selfing rate in the early steps of adaptation will be informative with regards to the demographic context of selfing evolution.

2.1 Adaptive Values of Selfing Syndrome Traits

One of the first arguments to support an adaptive value of the selfing syndrome may be the extent of convergence in the flower morphological changes observed between independent transitions to selfing, especially when this does not seem to correlate with the time of selfing evolution. As outlined above a long-standing question is whether the evolution of the selfing syndrome results from the relaxation of selective pressures imposed on flower traits for pollinator attraction or from the adaptive advantage of resource allocation. Theoretical studies modelling the rate of phenotypic changes under different scenarios seem to suggest that the rapid evolution of the selfing syndrome observed in some lineages (such as *C. rubella* and *I. alabamica*) is inconsistent with the relaxation of constraints and rather support an evolution under selective pressures (Foxe et al. 2009; Guo et al. 2009; Busch et al. 2011; Glémin and Ronfort 2013). Indeed, the fact that selfing syndrome traits evolved early in *C. rubella* before its geographical spread suggests that the syndrome in itself may constitute an important early step in the evolution of highly selfing lineages (Sicard et al. 2011). If this is true, it would be expected that individual selfing syndrome traits or the interaction between them would act as mating system modifiers improving the ability of plants to self-pollinate. In agreement with this idea, the loss of SI alone in the *Capsella* genus was able to allow self-fertilization but with an efficiency that is only half the one of plants in which the selfing syndrome has evolved (Sicard et al. 2011). Thus, this indicates that at least part of the selfing syndrome traits have evolved to improve self-fertilization.

2.1.1 The Selfing Syndrome as a Mating System Modifier

Herkogamy, i.e. the distance between anthers and stigma, is likely to have a direct influence on the efficiency of selfing by facilitating the deposition of self-pollen on

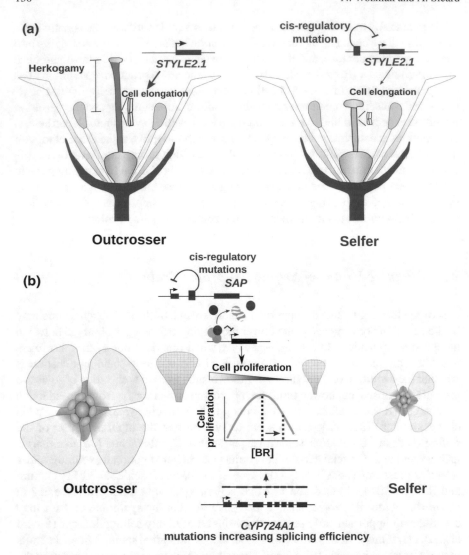

Fig. 2 Mechanisms underlying the evolution of flower morphology after the transition to selfing. a *cis*-regulatory mutation in tomato self-compatible lineage sled to a decrease in the expression level of the transcription factor STYLE2.1 which promotes cell elongation during style development. This led to a reduction of the style length and thus of herkogamy. **b** The reduction of petal size in the selfing species *C. rubella* has occurred through a reduction of cell proliferation. This has in part been achieved by *cis*-regulatory mutations affecting the expression of the F-box protein STERILE APETALA (SAP), which promotes the degradation of a repressor complex inhibiting cell proliferation. Mutations in the exon 1 and 2 of the brassinosteroid (BR) biosynthesis gene *CYP724A1* also contributed to this changes. These mutations increased its splicing efficiency, which in turn moved the BR concentrations away from the optimal quantity for growth promotion

the stigma (Fig. 2a). A correlation between the reduced stigma–anther distance and selfing rates has indeed been observed in nature and manipulating the herkogamy in *A. thaliana* was shown to have a strong impact on the efficiency of selfing (Lande and Schemske 1985; Takebayashi et al. 2006; Luo and Widmer 2013; Griffin and Willi 2014; Toräng et al. 2017). Flower size and flower opening may, likewise, contribute to improve selfing by facilitating the contact between pollen and stigma and thus indirectly influencing herkogamy. In agreement with this idea, a common genetic basis between flower size and herkogamy has been observed in *Mimulus* and genetic correlation analysis in *Capsella* indicates that both flower size and opening influence the distance between anthers and stigma (Fishman et al. 2002; Sicard et al. 2011). In the latter study, the authors detected a significant negative correlation between the efficiency of selfing and petal opening further reinforcing the importance of close anther–stigma contact for efficient self-pollination. Flower size, flower opening as well as nectar and scent production have also been shown to influence pollinator visitations (Bruce et al. 2005; Grindeland et al. 2005; Conner and Rush 2010; Heil 2011). It is, therefore, plausible that these traits also influence selfing indirectly by limiting cross-pollination (and thus the competition between self and non-self-pollen), reducing the loss of self-pollen and avoiding the damage that may be caused by insects on flowers (Sicard and Lenhard 2011).

2.1.2 Selfing Syndrome Evolution as a Result of Resource Allocation

It appears, however, less clear how changes in pollen-to-ovules (P/O) ratio could improve selfing efficiency. The negative correlation between P/O ratio and selfing rate, which has often been observed in plants, has mostly been discussed in the context of the theory of sex allocation. This theory assumes that limited reproductive resources cause an intrinsic trade-off between the investment in male and female function (Cruden 1977; Charlesworth and Charlesworth 1980; Charnov 1982). The decrease in P/O ratio in selfers is believed to be the result of local mate competition in which the investment in male or female functions will mostly be dictated by reproductive return (Charnov 1982; West 2009). In outcrossing, a larger investment in pollen number would provide higher fitness because only a small proportion of the pollen produced is likely to reach compatible pistils and also because pollen from several individuals may compete to fertilize the same ovules. In self-fertilizing plants and especially in hermaphroditic flowers, a much larger proportion of the pollen is likely to reach a compatible pistil while a lower proportion of competing pollen is present on the stigma. As a result, the large investment in male function seems unnecessary in autogamous plants which would, therefore, reallocate their resources towards the female function. Although empirical studies appear to confirm this prediction when comparing obligate outcrossers to predominant selfers, the negative correlation between P and O number is less clear when the difference in selfing and outcrossing rates are less pronounced [see Sicard and Lenhard (2011) for review]. Nevertheless, and if indeed, the reallocation of resources is the driving force of P/O reduction in selfers, it should be considered as a consequence rather than a contributor to selfing evolution. However, considering the nutritive value of pollen for pollinators, a decrease in P/O ratio may also indirectly influence selfing rates by limiting pollinator visitation (Sagili and Pankiw 2007; Mapalad et al. 2008). Unfortunately, to our knowledge, no studies have directly investigated the influence of pollen number on pollinator visitations.

The change in flower size during the evolution of the selfing syndrome has also been proposed to be a consequence of selfing evolution and the result of resource reallocation. It is believed that in self-fertilizing plants, the resources spent in building large and attractive flowers no longer provide reproductive return and would, therefore, be reallocated to other developmental processes with fitness benefit (Brunet 1992). In agreement with this hypothesis, several studies have been able to detect a trade-off between flower size and flower number or between flower size and seed size/number (Worley and Barrett 2000; Caruso 2004; Goodwillie et al. 2010). Although such correlations may suggest that reduced flower display may have an indirect influence on seed production through resource reallocation, it is difficult at this stage to determine whether this is the main driver of flower size reduction.

A difficulty in identifying the adaptive value of selfing syndrome traits is that most of these hypotheses are supported by empirical data based on correlations in highly genetically heterogeneous populations or on co-segregation analysis within bi-parental populations generated from crosses between highly phenotypically differentiated parents. While these populations may be adequate to study multi-factorial correlation, the effect of a given trait may also be confounded by the segregation of other phenotypes. Measuring the adaptive value of selfing syndrome traits may, therefore, be helped by the identification of their genetic bases. This may allow designing isogenic backgrounds in which the effects of single selfing syndrome trait on selfing rate or seed production could be directly tested. This may determine more efficiently which of these traits evolve as an adaptive response, as the relaxation of selective pressure and/or as a consequence of the increased efficiency of genetic drift in selfing lineages.

2.2   Genetic Basis of Selfing Syndrome Evolution

2.2.1   Theoretical Arguments and Empirical Quantitative Genetic Studies

Theoretical studies have made several predictions on the genetic basis of selfing syndrome evolution. These predictions mostly concern the dominance and effect size of the mutations at the origin of floral trait evolution as well as the mechanisms by which these variants become established. In general, the adaptation to new conditions such as a shift in mating system occurs through the fixation of new mutations or through the capture of segregating variants from the standing genetic variation of ancestral populations. Whether one or the other of these mechanisms underlie the evolution of selfing syndrome traits is likely to depend on the demographic and ecological context that has led to selfing evolution (Barrett et al. 2014). In pollen and mate-limiting conditions, gene flow among individuals is likely to be limited making new mutations more likely to contribute to the evolution of the selfing syndrome (Glémin and Ronfort 2013). If the transition to selfing is, however, not associated with a strong population bottleneck, standing variation would be expected to make a contribution. At the early steps of adaptation, dominant mutations are more likely to be selected because of their beneficial advantages in a heterozygous situation. As the selfing rate increases recessive mutations whose effects were masked in outcrossing populations would be expected to contribute, while no specific pattern of dominance would be predicted in highly selfing populations (Haldane 1927; Charlesworth 1992; Glémin and Ronfort 2013). When selfing evolves in a population with a high level of inbreeding depression, mutations with a large effect on the selfing rate are more likely to become fixed (Lande and Schemske 1985). Overall these theoretical works are supportive of a model of adaptation with mutations with a large effect on selfing rate followed by weaker mutations optimising selfing efficiency. Such a model would imply that few large effect mutations underlie the evolution of traits with a strong influence on the efficiency of selfing, while weaker mutations underlie changes in 'secondary traits' such as the ones allowing resources allocation or having a weaker effect on self-pollination.

Quantitative genetic studies partly support such a view with few large effect mutations underlying the breakdown of self-incompatibility or the distance between stigma and anther while numerous small to moderate effect loci appear to underlie the evolution of flower size and P/O number (Lin and Ritland 1997; Georgiady et al. 2002; Goodwillie et al. 2006; Fishman et al. 2010; Sicard and Lenhard 2011; Sicard et al. 2011; Slotte et al. 2012). It is to be noted, nevertheless, that the number of underlying mutations appears to be larger in some systems than others. Although data on dominance are very scarce, the dominance relationship of mutations seems also to conform mostly to predictions with a higher dominance level of mutations having a strong influence on the selfing rate (such as mutations at the S-locus) and a weaker dominance on the traits with less intuitive effects on selfing efficiency (Sicard and Lenhard 2011; Sicard et al. 2011; Slotte et al. 2012; Barrett et al. 2014).

In such scenario of adaptation, only mutations at an early step of selfing evolution would be expected to evolve from standing variation, yet once a high level of selfing has evolved, new mutations may be more likely to contribute. Obtaining empirical data supporting or rejecting such a scenario requires identifying the mutations underlying the evolution of selfing syndrome traits and retracing their evolutionary history. Such data are, however, still very scarce, but few studies have started to shed light on the molecular basis and the origin of these mutations.

2.2.2 Molecular Basis of Selfing Syndrome Evolution

One of the first genes underlying the evolution of a selfing syndrome trait was identified in tomato (Chen et al. 2007) (Fig. 2a). This gene encodes a helix-loop-helix transcription factor named *STYLE2.1*. Allelic variation at this locus is responsible for the reduced herkogamy that has followed the evolution of selfing in tomato culti-vars. In this species, the transition from an exserted to an inserted stigma constitutes a major adaptation to selfing, as it strongly influences the selfing rate in self-compatible lineages (Rick et al. 1978). A quantitative trait locus (QTL), termed stigma exser-tion 2.1, was found to account for most of the stigma exsertion changes between self-compatible and incompatible Solanum species (Bernacchi and Tanksley 1997; Fulton et al. 1997; Chen and Tanksley 2004). This locus was shown to encompass at least five genes among which *STYLE2.1* explained most of the phenotypic changes. The *STYLE2.1* 'short-style' allele evolved through *cis*-regulatory mutations in a ~12 kb region upstream of the transcriptional start site, decreasing the overall expression level of the gene during style development of the self-compatible species. The change in the *STYLE2.1* expression level causes a decrease in cell elongation leading to a shortening of the overall style length. Comparison of sequences between long- and short-styled species in the causative region initially identified a deletion of 450 bp at about 4 kb upstream of the transcriptional start as the causative polymorphism. Fur-ther analysis including a larger set of accessions and species failed, however, to fully correlate the presence of this deletion with the reduction of exsertion (Vosrers et al. 2014). The nature of the causative variant and its evolutionary history are therefore still uncertain.

In recent years, the small Brassicaceae genus *Capsella* has emerged as a model to study the evolution of the selfing syndrome. In this genus, two diploid species have emerged from the breakdown of the self-incompatibility system. *C. rubella* has diverged from *C. grandiflora* about 50–100 kya in a population of a potentially large number of individuals before undergoing a strong reduction in its effective popula-tion size associated with a drastic reduction in genetic diversity due to subsequent genetic drift and selection (Slotte et al. 2013; Brandvain et al. 2013). A second inde-pendent event of selfing evolution has occurred in an eastern lineage of the putative *C. grandiflora*-like outcrossing ancestor leading to the emergence of *C. orientalis* (Hurka et al. 2012). The selfing syndrome is fully established in *C. rubella* and has led to a strong reduction in flower size, flower opening, pollen to ovule ratio and flower scent. Quantitative genetic analyses revealed that most of these changes in phenotype have been caused by several mutations with a weak to moderate effect. Allele-specific expression appears to be enriched in the flower transcriptome of *C. grandiflora* × *C. rubella* flowers, suggesting a role for *cis*-regulatory mutations in the evolution of the selfing syndrome (Steige et al. 2015). Because of the potentially large number of founding individuals, both new mutations and standing variation may have contributed.

In *C. rubella*, the reduction of petal size has been caused by mutations of at least 6 loci each explaining between 6 and 10% of the selfer–outcrosser difference (Sicard et al. 2011; Slotte et al. 2012). The reduction of petal size seems to be mostly caused by a decrease in cell proliferation. Two of these QTLs have been resolved at the molecular level up to now.

One of them was shown to be caused by cis-regulatory mutations in the intron of *STERILE APETALA* (*SAP*) (Sicard et al. 2016) (Fig. 2b). *SAP* encodes an F-box protein which constitutes the specificity component of an SCF-type E3 ubiquitin ligase itself targeting specific proteins for degradation. *SAP* is now known to regulate organ growth by affecting the stability of a transcriptional repressor complex (named PEAPOD-KIX) inhibiting meristemoid proliferation (Wang et al. 2016; Li et al. 2018). In *C. rubella* mutations reducing SAP accumulation specifically during petal growth led to a reduction of the total cell number presumably through the stabilization of the cell proliferation repressor complex. The outcrosser–selfer polymorphisms contained within the causative intronic sequence were found to segregate in the present-day *C. grandiflora* populations. Association mapping indicated that two of these polymorphisms influence additively petal size in *C. grandiflora*, suggesting a scenario in which the *SAP* 'small petal' allele has evolved through the capture of a rare haplotype combining the additive effect of at least two polymorphisms from the standing genetic variation of the ancestral *C. grandiflora*-like outcrossing population. These data also illustrated the contribution of tissue-specific elements in a pleiotropic gene to enable organ-specific evolution. Furthermore, the study of nucleotidic diversity within *C. grandiflora* indicated that the *SAP* intron is under purifying selection suggesting that relaxation of selective pressure within *C. rubella* may have contributed to the *SAP* small petal allele fixation. Whether the latter has an adaptive value in a selfing context is still to be determined.

A second QTL was shown to be caused by allelic variation in a gene encoding an enzyme involved in the biosynthesis of the plant hormone brassinosteroid (BR), *CYP724A1* (Fig. 2b). In this case, two single nucleotide polymorphisms (SNPs) in exons 1 and 2 increase the splicing efficiency of all *CYP724A1* introns in *C. rubella* (Fujikura et al. 2018). The resulting increase in *CYP724A1* protein was shown to lead to higher BR levels during petal development as well as a decrease in cell proliferation. Because BR was previously known to promote organ growth, this result suggests a hyperbolic-like curve of shoot-organ response to BR concentration (Fig. 2b). Population genetic analyses indicated that the two causal mutations have evolved de novo in the *C. rubella* lineage and thus that the efficient splicing is the derived form. The effect on organ growth was not, in this case, restricted to petal size but seems to affect most plant organs. Whether such morphological change has become fixed with a selective advantage or, a disadvantage due to the reduced efficiency of purifying selection in the selfer is also still to be determined.

After the transition to selfing, *C. rubella* has lost the ability to produce volatile compounds such as benzenoids (mostly benzaldehyde (BAld)) and terpenoids (mostly trans- and cis-beta-ocimene) (Sas et al. 2016). The floral scent represents an important olfactory clue for pollinator attraction and herbivore repulsion. BAld which has been shown to attract pollinators in other species is emitted in *C. grandiflora* according to a diurnal pattern with a peak at midday, likely corresponding to the peak of its pollinators' activity. Loss of BAld production in *C. rubella* has been caused by mutations in the cinnamate: CoA ligase, CLN1 (Sas et al. 2016). In fact, based on population genetics, two independent mutations in *CLN1* leading to the absence of BAld emission seem to have occurred in *C. rubella*. Both of them affect the protein sequence either by leading to a premature stop codon or a non-synonymous mutation. These mutations could not be found in *C. grandiflora* 'old' haplotypes, indicating that these non-functional alleles have evolved through de novo mutations. Interestingly, the same enzyme was also found to underlie gain and loss of scent in pollinator-mediated speciation in *Petunia* (Amrad et al. 2016). *CLN1* appears therefore as an evolutionary hotspot for mutations causing the loss of benzenoid compounds during plant evolution. This was proposed to be explained by the fact that affecting BAld production through *CNL1* does not appear to have a pleiotropic effect on production of other phenylpropanoids, together with the advantage that affecting an early step in BAld production limits the risk of accumulating toxic intermediates. Here as well, it is still unclear whether the inactivation of *CLN1* has been fixed with a fitness advantage linked to selfing optimization, resources allocation, herbivore damages avoidance or as a result of genetic drift and the relaxation of selective pressures.

Similarly, the biosynthetic enzyme, the OCIMENE SYNTHASE (OS) was shown to underlie repeated loss of E-β-Ocimene emission in *Mimulus* (Peng et al. 2017). In this species, the ability of producing E-β-Ocimene has been only retained in *M. lewissi* which has specialized in bumblebee pollination. In contrast, this ability has been lost in hummingbird-pollinated species (such as *M. cardinalis*), small bees-pollinated species (such as *M. bicolor*) or in derived selfing species (i.e. *M. parishii*). Sequence analyses along the phylogeny of *Mimulus* revealed that several independent mutations affecting the sequence or function of the *OS* has contributed to the loss of E-β-Ocimene emission in these different lineages. In this case, these losses were hypothesized to be advantageous especially in the context of pollination syndrome in which mismatched flower visitors may cause pollen wastage. Here, as for *CLN1* the repeated 'usage' of *OS* in independent evolution of flower scent may be, at least partly, explained by the lack of pleiotropic effect associated with the inhibition of this enzyme.

These examples are, therefore, not incompatible with a model of selfing evolution in which adaptation would first occur through the capture of segregating variants and once a high rate of selfing has evolved new mutations are more likely to contribute. However, a large pitfall here is still that the adaptive value associated with these mutations has not yet been measured. This prevents determining if indeed mutations having a larger effect on selfing rate are more likely to evolve early through capture by standing variation and thus to play an important role in the establishment of selfing lineages.

3   Conclusions and Perspectives

The transition to self-fertilization has been at the centre of the attention of plant evolutionary biology for over a century leading to a large amount of theoretical studies on the causes and consequences of selfing evolution. These studies have provided plausible demographic and ecological models underlying the evolution of selfing. Genomics approaches have started to provide some support to these models demonstrating, for instance, the impact of selfing on genetic diversity and evolutionary potentials. These studies have notably highlighted the difficulty of identifying the causes of selfing evolution through the comparison of molecular data between selfer and outcrosser taxa. For instance, the apparent population bottleneck often observed in selfing lineages may simply result from the reduction of genetic diversity due to the increased effect of genetic drift and linked selection and, thus, not be suggestive of selection for reproductive assurance (Barrett et al. 2014). Additional ecological studies directly measuring the importance of several ecological factors for the selection of alleles improving selfing may be needed to fully comprehend the different scenarios leading to the selection of selfing. This may also be helped by the identification of the evolutionary history leading to the selection of mutations improving selfing efficiency.

Molecular studies have started to shed light on the molecular basis of selfing syndrome evolution. These examples indicated a contribution of both standing variation and new mutations to the evolution of the selfing syndrome. They also demonstrate that a relatively limited number of mutations with moderate (e.g. *SAP* and *CYP724A1*) to strong effect (*OS*, *CLN1* and *STYLE2.1*) contribute to phenotypic changes after selfing evolution. Four of five of the cases identified so far, involved mutations with specific functions supporting the idea that molecular features with low pleiotropy and large effects on the phenotype are more 'suitable' for phenotypic evolution (Hoekstra and Coyne 2007; Carroll 2008; Martin and Orgogozo 2013). Two out of three of the mutations inducing morphological changes led to a spatial and/or temporal change in gene expression as often observed in developmental evolution (Carroll 2008; Stern and Orgogozo 2008). While theoretical studies together with the apparent rapid changes in flower phenotypes and the level of convergence in their evolution are suggestive of their adaptive value, for many of these traits (with the exception of herkogamy) the benefit for selfing lineages has not been measured. It is, therefore, still unclear which of these traits (or mutations) were fixed as an adaptation to selfing, due to the relaxation of selective pressures or as a result of the increased effect of genetic drift. Experiments aiming to measure the fitness advantage (or disadvantage) of alleles underlying the evolution of different selfing syndrome traits are, therefore, grandly needed. The knowledge of the genetic basis underlying the evolution of the selfing syndrome in the *Capsella* genus provides a unique opportunity to address this question. Nevertheless, unravelling the genetic basis of additional independent events of selfing syndrome evolution is also important to test the existence of genetic constraints imposed on the evolution of flower phenotypes after the transition to selfing and to identify the different ecological scenarios that may lead to selfing evolution (Woźniak and Sicard 2017). Addressing these points will allow a greater understanding of the basis of phenotypic convergence after the transition to self-fertilization in plants.

References

Amrad A, Moser M, Mandel T et al (2016) Gain and loss of floral scent production through changes in structural genes during pollinator-mediated speciation. Curr Biol 26:3303–3312. https://doi.org/10.1016/j.cub.2016.10.023

Baker HG (1955) Self-compatibility and establishment after long-distance dispersal. Evolution (N Y) 9:347–349

Barrett SCH (2002) Evolution of sex: the evolution of plant sexual diversity. Nat Rev Genet 3:274–284. https://doi.org/10.1038/nrg776

Barrett SCH (2013) The evolution of plant reproductive systems: how often are transitions irreversible? Proc R Soc B-Biological Sci 280:20130913. https://doi.org/10.1098/rspb.2013.0913

Barrett SCH, Arunkumar R, Wright SI (2014) The demography and population genomics of evolutionary transitions to self-fertilization in plants. Philos Trans R Soc B Biol Sci 369:1–9

Bernacchi D, Tanksley SD (1997) An interspecific backcross of Lycopersicon esculentum X L. hirsutum: linkage analysis and a QTL study of sexual compatibility factors and floral traits. Genetics 147:861–877

Brandvain Y, Slotte T, Hazzouri KM et al (2013) Genomic identification of founding haplotypes reveals the history of the selfing species Capsella rubella. PLoS Genet 9:e1003754. https://doi.org/10.1371/journal.pgen.1003754

Bruce TJA, Wadhams LJ, Woodcock CM (2005) Insect host location: a volatile situation. Trends Plant Sci 10:269–274. https://doi.org/10.1016/j.tplants.2005.04.003

Brunet J (1992) Sex allocation in hermaphroditic plants. Trends Ecol Evol 7:79–84

Busch JW, Delph LF (2012) The relative importance of reproductive assurance and automatic selection as hypotheses for the evolution of self-fertilization. Ann Bot 109:553–562. https://doi.org/10.1093/aob/mcr219

Busch JW, Joly S, Schoen DJ (2011) Demographic signatures accompanying the evolution of selfing in leavenworthia alabamica. Mol Biol Evol 28:1717–1729. https://doi.org/10.1093/molbev/msq352

Carroll SB (2008) Evo-devo and an expanding evolutionary synthesis: a genetic theory of morphological evolution. Cell 134:25–36. https://doi.org/10.1016/j.cell.2008.06.030

Caruso C (2004) The quantitative genetics of floral trait variation in lobelia: potential constraints on adaptive evolution. Evolution (N Y) 58:732–740

Charlesworth B (1992) Evolutionary rates in partially self-fertilizing species. Am Nat AMER Nat 140:126–148

Charlesworth D (2006) Evolution of plant breeding systems. Curr Biol 16:726–735. https://doi.org/10.1016/j.cub.2006.07.068

Charlesworth D, Charlesworth B (1980) Allocation of resources to male and female functions in hermaphrodites. Biol J Linn Soc 57–74

Charnov E (1982) The theory of sex allocation. Monogr Popul Biol 18:1–355

Chen K, Cong B, Wing R et al (2007) Changes in regulation of a transcription factor lead to autogamy in cultivated tomatoes. Science (80-) 643:10–13. https://doi.org/10.1126/science.1148428

Chen KY, Tanksley SD (2004) High-resolution mapping and functional analysis of se2.1: a major stigma exsertion quantitative trait locus associated with the evolution from allogamy to autogamy in the genus lycopersicon. Genetics 168:1563–1573. https://doi.org/10.1534/genetics.103.022558

Cheptou P (2011) Clarifying Baker' s law. Ann Bot 633–641. https://doi.org/10.1093/aob/mcr127

Conner JK, Rush S (2010) Effects of flower size and number on pollinator visitation to wild radish, Raphanus raphanistrum. Oecologia 105:509–516

Cruden R (1977) Pollen-ovule ratios: a conservative indicator of breeding systems in flowering plants evolution. Biol J Linn Soc 32–46

Darwin C (1876) The effects of cross and self fertilisation in the vegetable kingdom. John Murray, London

Disorders JD, Metab E, North C et al (2000) The S receptor kinase determines self-incompatibility in Brassica stigma. 967:1998–2001

Durand E, Méheust R, Soucaze M et al (2014) Dominance hierarchy arising from the evolution of a complex small RNA regulatory network. 346:1200–1205

Fisher R (1941) Average excess and average effect of a gene substitution. Ann Eugen 11:53–63

Fishman L, Kelly AJ, Willis JH (2002) Minor quantitative trait loci underlie floral traits associated with mating system divergence in Mimulus. Evolution (N Y) 56:2138–2155. https://doi.org/10.1111/j.0014-3820.2002.tb00139.x

Fishman L, Kelly AJ, Willis JH (2010) Minor quantitative trait loci underlie floral traits associated with mating system divergence in Mimulus. Evolution (N Y) 56:2138–2155

Foxe JP, Slotte T, Stahl EA et al (2009) Recent speciation associated with the evolution of selfing in Capsella. Proc Natl Acad Sci USA 106:5241–5245. https://doi.org/10.1073/pnas.0807679106

Fujii S, Kubo K-I, Takayama S (2016) Non-self- and self-recognition models in plant self-incompatibility. Nat Plants 2:16130. https://doi.org/10.1038/nplants.2016.130

Fujikura U, Jing R, Hanada A et al (2018) Variation in splicing efficiency underlies morphological evolution in Capsella. Dev Cell 44(192–203):e5. https://doi.org/10.1016/j.devcel.2017.11.022

Fulton T, Beck-Bunn T, Emmatty D et al (1997) QTL analysis of an advanced backcross of Lycopersicon peruvianum to the cultivated tomato and comparisons with QTLs found in other wild species. Theor Appl Genet 95:881–894

Georgiady MS, Whitkus RW, Lord EM (2002) Genetic analysis of traits distinguishing outcrossing and self-pollinating forms of currant tomato, Lycopersicon pimpinellifolium (Jusl.) Mill. Genetics 161:333–344

Gervasi DDL, Schiestl FP (2017) Real-time divergent evolution in plants driven by pollinators. Nat Commun 8:14691. https://doi.org/10.1038/ncomms14691

Glémin S, Ronfort J (2013) Adaptation and maladaptation in selfing and outcrossing species: new mutations versus standing variation. Evolution (N Y) 67:225–240. https://doi.org/10.1111/j.1558-5646.2012.01778.x

Goodwillie C, Ritland C, Ritland K (2006) The genetic basis of floral traits associated with mating system evolution in Leptosiphon (Polemoniaceae): an analysis of quantitative trait loci. Evolution 60:491–504

Goodwillie C, Sargent RD, Eckert CG et al (2010) Correlated evolution of mating system and floral display traits in flowering plants and its implications for the distribution of mating system variation. New Phytol 185:311–321. https://doi.org/10.1111/j.1469-8137.2009.03043.x

Griffin PC, Willi Y (2014) Evolutionary shifts to self-fertilisation restricted to geographic range margins in North American Arabidopsis lyrata. Ecol Lett 17:484–490. https://doi.org/10.1111/ele.12248

Grindeland JM, Sletvold N, Ims RA (2005) Effects of floral display size and plant density on pollinator visitation rate in a natural population of Digitalis purpurea. Funct Ecol 19:383–390. https://doi.org/10.1111/j.1365-2435.2005.00988.x

Guo Y-L, Bechsgaard JS, Slotte T et al (2009) Recent speciation of Capsella rubella from Capsella grandiflora, associated with loss of self-incompatibility and an extreme bottleneck. Proc Natl Acad Sci USA 106:5246–5251. https://doi.org/10.1073/pnas.0808012106

Haldane J (1927) A mathematical theory of natural and artificial selection V. Selection and mutation. Math Proc Camb Philos Soc 28:838–844

Hazzouri KM, Escobar JS, Ness RW et al (2013) Comparative population genomics in collinsia sister species reveals evidence for reduced effective population size, relaxed selection, and evolution of biased gene conversion with an ongoing mating system shift. Evolution (N Y) 67:1263–1278. https://doi.org/10.1111/evo.12027

Heil M (2011) Nectar: generation, regulation and ecological functions. Trends Plant Sci 16:191–200. https://doi.org/10.1016/j.tplants.2011.01.003

Hoekstra HE, Coyne JA (2007) The locus of evolution: Evo devo and the genetics of adaptation. Evolution (N Y) 61:995–1016. https://doi.org/10.1111/j.1558-5646.2007.00105.x

Hurka H, Friesen N, German DA et al (2012) "Missing link" species Capsella orientalis and
    Capsella thracica elucidate evolution of model plant genus Capsella (Brassicaceae). Mol Ecol
    21:1223–1238. https://doi.org/10.1111/j.1365-294X.2012.05460.x
Igic B, Busch JW (2013) Is self-fertilization an evolutionary dead end? New Phytol 198:386–397.
    https://doi.org/10.1111/nph.12182
Kachroo A, Schopfer CR, Nasrallah ME (2001) Allele-specific receptor-ligand interactions in Bras-
    sica self-incompatibility. 293:1824–1827
Kawecki TJ (2008) Adaptation to marginal habitats. Annu Rev Ecol Evol Syst 39:321–342. https://
    doi.org/10.1146/annurev.ecolsys.38.091206.095622
Lande R, Schemske D (1985) The evolution of self-fertilization and inbreeding depression in plants.
    I. Genetic models. Evolution (N Y) 39:24–40
Lee HS, Huang S, Kao T-H (1994) S proteins control rejection of incompatible pollen in Petunia
    inflata. Nature 367:560–563. https://doi.org/10.1038/367560a0
Levin DA (2012) Mating system shifts on the trailing edge. Ann Bot 109:613–620. https://doi.org/
    10.1093/aob/mcr159
Li N, Liu Z, Wang Z et al (2018) STERILE APETALA modulates the stability of a repressor protein
    complex to control organ size in Arabidopsis thaliana. PLoS Genet 14:e1007218. https://doi.or
    g/10.1371/journal.pgen.1007218
Lin JZ, Ritland K (1997) Quantitative trait loci differentiating the outbreeding Mimulus guttatus
    from the inbreeding M. platycalyx. Genetics 146:1115–1121
Lloyd DG (1979) Some reproductive factors affecting the selection of self-fertilization in plants.
    Am Nat 113:67–69. https://doi.org/10.1086/283365
Longhurst AR (1955) Evolution in the notostraca. Evolution (N Y) 84–86. https://doi.org/10.2307/
    2405360
Luo Y, Widmer A (2013) Herkogamy and its effects on mating patterns in Arabidopsis thaliana.
    PLoS ONE 8:e57902. https://doi.org/10.1371/journal.pone.0057902
Mapalad KS, Leu D, Nieh JC (2008) Bumble bees heat up for high quality pollen. 2239–2242.
    https://doi.org/10.1242/jeb.016642
Martin A, Orgogozo V (2013) The Loci of repeated evolution: a catalog of genetic hotspots of
    phenotypic variation. Evolution 67:1235–1250. https://doi.org/10.1111/evo.12081
Mccubbin AG, Huang S, Kao T (2004) Identification of the pollen determinant of S-RNase-mediated
    self-incompatibility. 429:2–5. https://doi.org/10.1038/nature02545.Published
Moeller DA, Geber MA (2005) Ecological context of the evolution of self-pollination in Clarkia
    xantiana: population size, plant communities, and reproductive assurance. Evolution (N Y)
    59:786–799
Murfett J, Atherton TL, Mou B et al (1994) S-RNase expressed in transgenic Nicotiana causes
    S-allele-specific pollen rejection. Nature 367:563–566
Ornduff R (1969) Reproductive biology in relation to systematics. Taxon 18:121–133
Pannell JR (2015) Evolution of the mating system in colonizing plants. Mol Ecol 2018–2037.
    https://doi.org/10.1111/mec.13087
Peng F, Byers KJRP, Bradshaw HD (2017) Less is more: Independent loss-of-function OCIMENE
    SYNTHASE alleles parallel pollination syndrome diversification in monkeyflowers (Mimulus).
    Am J Bot 104:1055–1059. https://doi.org/10.3732/ajb.1700104
Rick CM, Holle M, Thorp RW (1978) Rates of cross-pollination in Lycopersicon pimpinellifolium:
    impact of genetic variation in floral characters. Plant Syst Evol 129:31–44. https://doi.org/10.10
    07/BF00988982
Sagili RR, Pankiw T (2007) Effects of protein-constrained brood food on honey bee (Apis mellifera
    L.) pollen foraging and colony growth. 1471–1478. https://doi.org/10.1007/s00265-007-0379-1
Sas C, Müller F, Kappel C et al (2016) Repeated inactivation of the first committed enzyme underlies
    the loss of benzaldehyde emission after the selfing transition in Capsella. Curr Biol 1–7. https://
    doi.org/10.1016/j.cub.2016.10.026

Shimizu KK, Tsuchimatsu T (2015) Evolution of selfing: recurrent patterns in molecular adaptation. Annu Rev Ecol Evol Syst 46:annurev-ecolsys-112414-054249. https://doi.org/10.1146/annurev-ecolsys-112414-054249

Sicard A, Kappel C, Lee YW et al (2016) Standing genetic variation in a tissue-specific enhancer underlies selfing-syndrome evolution in Capsella. Proc Natl Acad Sci 113:13911–13916. https://doi.org/10.1073/pnas.1613394113

Sicard A, Lenhard M (2011) The selfing syndrome: a model for studying the genetic and evolutionary basis of morphological adaptation in plants. Ann Bot. https://doi.org/10.1093/aob/mcr023

Sicard A, Stacey N, Hermann K et al (2011) Genetics, evolution, and adaptive significance of the selfing syndrome in the genus Capsella. Plant Cell 23:3156–3171. https://doi.org/10.1105/tpc.111.088237

Slotte T, Hazzouri KM, Stern D et al (2012) Genetic architecture and adaptive significance of the selfing syndrome in Capsella. Evolution 66:1360–1374. https://doi.org/10.1111/j.1558-5646.2011.01540.x

Slotte T, Hazzouri KM, Ågren JA et al (2013) The Capsella rubella genome and the genomic consequences of rapid mating system evolution. Nat Genet 45:831–835. https://doi.org/10.1038/ng.2669

Stebbins G (1950) Variation and evolution in plants. Columbia University Press, New York

Stebbins G (1957) Self-fertilization and population variability in higher plants. Am Nat 337–54

Stebbins GL (1970) Adaptive radiation of reproductive characteristics in angiosperms, I: Pollination mechanisms. Annu Rev Ecol Syst 1:307–326. https://doi.org/10.1146/annurev.es.01.110170.001515

Stebbins GL (1974) Flowering plants: evolution above the species level. Belkn Press, Cambridge, MA

Steige KA, Reimegard J, Koenig D et al (2015) Cis-regulatory changes associated with a recent mating system shift and floral adaptation in capsella. Mol Biol Evol 32:2501–2514. https://doi.org/10.1093/molbev/msv169

Stern DL, Orgogozo V (2008) The loci of evolution: how predictable is genetic evolution? Evolution 62:2155–2177. https://doi.org/10.1111/j.1558-5646.2008.00450.x

Takayama S, Shiba H, Iwano M et al (2000) The pollen determinant of self-incompatibility in Brassica campestris. Proc Natl Acad Sci 97:1920–1925

Takebayashi N, Wolf DE, Delph LF (2006) Effect of variation in herkogamy on outcrossing within a population of Gilia achilleifolia. Heredity (Edinb) 96:159–165. https://doi.org/10.1038/sj.hdy.6800780

Toräng P, Vikström L, Wunder J et al (2017) Evolution of the selfing syndrome: anther orientation and herkogamy together determine reproductive assurance in a self-compatible plant. Evolution (N Y) 71:2206–2218. https://doi.org/10.1111/evo.13308

Tsuchimatsu T, Shimizu KK (2013) Effects of pollen availability and the mutation bias on the fixation of mutations disabling the male specificity of self-incompatibility. J Evol Biol 26:2221–2232. https://doi.org/10.1111/jeb.12219

Tsuchimatsu T, Suwabe K, Shimizu-Inatsugi R et al (2010) Evolution of self-compatibility in Arabidopsis by a mutation in the male specificity gene. Nature 464:1345–1346. https://doi.org/10.1038/nature08927

Vogler DW, Kalisz S (2001) Sex among the flowers: the distribution of plant mating systems. Evolution (N Y) 55:202–204. https://doi.org/10.1554/0014-3820(2001)055[0202:SATFTD]2.0.CO;2

Vosrers S, Jewell C, Sherman N et al (2014) The timing of molecular and morphological changes underlying reproductive transitions in wild tomatoes (Solanum sect. Lycopersicon). Mol Ecol 23:1965–1978. https://doi.org/10.1111/mec.12708

Wang Z, Li N, Jiang S et al (2016) SCFSAP controls organ size by targeting PPD proteins for degradation in Arabidopsis thaliana. Nat Commun 7:11192. https://doi.org/10.1038/ncomms11192

West S (2009) Sex allocation. Princeton University Press, Princeton, NJ

Worley C, Barrett S (2000) Evolution of floral display in Eichhornia paniculata (Pontederi-aceae): direct and correlated responses to selection on flower size and number. Evolution (N Y) 54:1533–1545

Woźniak NJ, Sicard A (2017) Evolvability of flower geometry: convergence in pollinator-driven morphological evolution of flowers. Semin Cell Dev Biol. https://doi.org/10.1016/j.semcdb.201 7.09.028

Wright SI, Kalisz S, Slotte T (2013) Evolutionary consequences of self-fertilization in plants. Proc Biol Sci 280:20130133. https://doi.org/10.1098/rspb.2013.0133

Immunoglobulin-Like Domains Have an Evolutionarily Conserved Role During Gamete Fusion in *C. elegans* and Mouse

Tatsuya Tajima and Hitoshi Nishimura

Abstract The *spe-9* class genes are predominantly or exclusively expressed in the *C. elegans* male germline and play critical roles during gamete fusion. However, it is a challenge to identify mammalian orthologs that exhibit similar functions to those of the *spe-9* class, since reproductive genes evolve much faster than somatic genes. In the mouse, *Izumo1* gene encodes a sperm-specific transmembrane (TM) protein with the immunoglobulin (Ig)-like domain that indispensably acts during gamete fusion. The *C. elegans* gene *spe-45* was recently identified by forward and reverse genetic approaches. It shows male germline-enriched expression and encodes an Ig-like TM protein like IZUMO1. Worms lacking *spe-45* produce otherwise normal spermatozoa that are incapable of fusing with oocytes. Thus, *spe-45* is a new member of the *spe-9* class, and the phenotype of *spe-45* mutant worms is essentially the same as that of *Izumo1*-knockout mice. Moreover, the Ig-like domains of SPE-45 and IZUMO1 possess similar roles to each other during gamete fusion. This indicates that *C. elegans spe-45* is functionally equivalent to mouse *Izumo1* and that their roles during gamete fusion have been conserved for ~1 billion years. Intriguingly, diverged organisms also have TM proteins with Ig-like domains that are involved during gamete interactions. This suggests the evolutionarily conserved roles of the Ig-like domains during fertilization, which are presumably related to associating with *cis*- and/or *trans*-partners.

1 The *C. elegans spe-9* Class Mutants Are Useful Tools to Investigate Gamete Fusion

Among ~60 species that are comprised in the nematode genus *Caenorhabditis*, some including *C. elegans* produce self-fertile hermaphrodites and males, while the remaining other have females and males. The self-fertility of *C. elegans* hermaphrodites enables to isolate and maintain numerous mutants, resulting in that *C.*

T. Tajima · H. Nishimura (✉)
Department of Life Science, Setsunan University, Osaka 572-8508, Japan
e-mail: nishimura@lif.setsunan.ac.jp

© Springer International Publishing AG, part of Springer Nature 2018
P. Pontarotti (ed.), *Origin and Evolution of Biodiversity*,
https://doi.org/10.1007/978-3-319-95954-2_10

elegans becomes one of the excellent model animals to study a variety of biological phenomena, including reproduction (Brenner 1974). Gametogenesis, fertilization, and early development are pivotal steps during reproduction, and *C. elegans* mutants have provided important clues to clarify the molecular basis of these reproductive steps.

For example, *spe* (spermatogenesis-defective) mutants have been isolated and analyzed to access the *C. elegans* male germline functions; spermatogenesis (spermatid production via meiosis), spermiogenesis (spermatid activation into spermatozoa), and fertilization (Wolf et al. 1978; Ward et al. 1981; Ward 1986; Kimble and Ward 1988). Below we will review how *spe* genes, particularly the *spe-9* class, are involved in *C. elegans* reproduction (L'Hernault 2009; Nishimura and L'Hernault 2010, 2017; Krauchunas et al. 2016).

1.1   C. elegans Reproduction

Figure 1 shows a part of the gonad in an adult *C. elegans* hermaphrodite. Oogenesis is ongoing in a directional manner from the distal to the proximal gonad, and the fertilization-ready oocyte (indicated by "−1") resides at the most proximal region, which is connected to the spermatheca. One of the sexual features of *C. elegans* hermaphrodites is that spermatogenesis just occurs during the fourth larval (L4) stage, which leads to production of ~300 round, sessile spermatids, followed by complete switching of gametogenesis in the gonad from spermatogenesis to oogenesis (Hirsh et al. 1976). This indicates that adult *C. elegans* hermaphrodites are somatically "females carrying self-sperm."

As the first oocyte is ovulated into the spermatheca, ~300 spermatids staying at the proximal gonad (Fig. 1a) are all pushed into the spermatheca by the ovulated oocyte. Amoeboid, motile spermatozoa (self-sperm) are formed by activation of the spermatids with an unknown factor(s) in the spermatheca, the so-called spermiogenesis, and one of the self-sperm fertilizes the oocyte (Fig. 1b). The fertilized oocyte subsequently moves into the uterus by pushing the remaining self-sperm (Fig. 1c). However, in this case, the self-sperm can crawl back into the spermatheca due to their motility and wait for the next ovulated oocyte to fertilize it (Fig. 1d). This cycle is repeated in the gonad, and eventually most of the self-sperm are consumed by fertilization. This reproductive system seems to be highly effective, since an adult hermaphrodite produces ~300 self-progeny through her life, which is an almost equal number to that of self-sperm (Ward and Carrel 1979).

Like hermaphrodites, spermatogenesis occurs in the L4 male gonad, and the spermatid production continues even after the larval males become adult. The male spermatids are usually stored in the seminal vesicles and not ectopically activated into spermatozoa within the male gonad. As males sire to hermaphrodites, spermatids are ejaculated with seminal fluids that contain an activator(s) for spermiogenesis and then transform into spermatozoa in the uterus of hermaphrodites. The male sper-

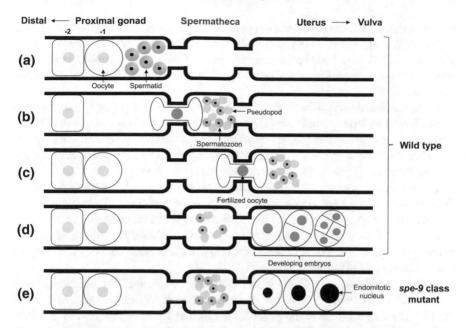

Fig. 1 Scheme of ovulation–fertilization cycles in wild-type and *spe-9* class mutant hermaphrodites. An adult *C. elegans* hermaphrodite has a U-shaped, symmetric gonad where the vulva is present at the center. Oogenesis is ongoing in a direction from the distal to the proximal gonad, and an oocyte at position "−1" is ready for being fertilized. Ovulation–fertilization cycles are shown in **a–d** for wild-type and **e** for *spe-9* class mutant hermaphrodites. **a** In the proximal gonad before the first ovulation, ~300 round spermatids that had been produced at the L4 stage are present. **b** The first oocyte at position "−1" is ovulated into the spermatheca by pushing the spermatids. In the spermatheca, spermatozoa are formed from the spermatids, and one of the self-sperm fertilizes the oocyte. **c** The fertilized oocyte moves into the uterus with pushing the remaining self-sperm. However, in this case, self-sperm can crawl back into the spermatheca to fertilize the next ovulated oocyte. **d** Self-sperms are consumed one by one in every ovulation–fertilization cycle, and eventually most of ~300 self-sperm disappear from the spermatheca. In the uterus, developing embryos are accumulated and released into the extracellular space via the vulva as they become gastrulas. **e** On the other hand, in the *spe-9* class mutants, self-sperm fails to fertilize oocytes in the spermatheca, so that unfertilized oocytes move into the uterus. The oocytes undergo endomitotic DNA replication in their nuclei and are subsequently released via the vulva

matozoa crawl into the spermatheca and fertilize oocytes even when self-sperm are already present there (Ward and Carrel 1979).

1.2   *The C. elegans spe-9 Class Genes Act During Gamete Fusion*

It is well known that *spe* genes govern many of the reproductive steps during which the *C. elegans* male germline acts. *spe* mutants can be easily found by simple criteria; *spe* mutant hermaphrodites are self-sterile but outcross-fertile after mating with wild-type males (Argon and Ward 1980; L'Hernault et al. 1988; McCarter et al. 1999). Indeed, ~60 *spe* mutants have been so far isolated, most of which are defective in either of spermatogenesis, spermiogenesis, or fertilization (L'Hernault 1997; L'Hernault and Singson 2000).

Since mutants deficient for either of *spe-9*, *spe-38*, *spe-41/trp-3*, *spe-42,* or *spe-49* all produce otherwise normal spermatozoa that are defective in fertilization, the roles of these five *spe* genes seem to be constrained to fertilization (Fig. 1e). Thus, *spe-9*, *spe-38*, *spe-41/trp-3*, *spe-42,* and *spe-49* are categorized into the *spe-9* class, which is a subgroup of *spe* genes to just participate in fertilization.

Moreover, a *C. elegans* oocyte has neither accessory cells like cumulus cells nor a thick egg coat like the zona pellucida, so that spermatozoa might directly bind to and fuse with the oocyte plasma membrane (PM) during fertilization in the spermatheca. This shows that the functional roles of the *spe-9* class genes are presumably to execute and/or regulate sperm–oocyte fusion. Each *spe-9* class gene encodes a variety of transmembrane (TM) proteins (Table 1).

SPE-9: This protein is a single-pass TM protein containing 10 epidermal growth factor (EGF)-like domains. The domain organization of SPE-9 is similar to those of the Delta family, a ligand to the Notch/LIN-12/GLP-1 family (Singson et al. 1998; Putiri et al. 2004). Thus, SPE-9 might be a *trans*-partner of an unknown oocyte surface protein. However, there is not a DSL domain in SPE-9, which is a feature of the Delta/Serrate/LAG-2 family, suggesting that SPE-9 might act in a different manner from those of the Delta family (Cordle et al. 2008).

SPE-38: This protein is a four-pass TM protein with no other significant domains (Chatterjee et al. 2005), but it has been recently elucidated that SPE-38 regulates the localization of SPE-41/TRP-3 on spermatozoa (Singaravelu et al. 2012).

SPE-41/TRP-3: This protein is a member of the transient receptor potential (TRP) canonical (TRPC) superfamily of cation channels. This protein probably forms a homo- or hetero-tetramer like other members of the TRP family (Schindl and Romanin 2007), and gamete fusion is mediated by the SPE-41/TRP-3-induced Ca^{2+} signaling (Xu and Sternberg 2003; Takayama and Onami 2016). Indeed, there is evidence for the TRPC family to play roles in lipid-processing events that might facilitate the fusion between membranes (Beech et al. 2009).

SPE-42: This protein is a six-pass TM protein containing two functional domains: the dendritic cell-specific transmembrane protein (DC-STAMP) and the C4C4-type RING finger domains (Kroft et al. 2005). Mouse DC-STAMP has been reported to play an important role in fusion of mononuclear osteoclasts (Miyamoto 2006). It is intriguing that *sneaky*, a *Drosophila* (the fruit fly) ortholog of *spe-42*, acts in breakdown of the sperm PM after fly spermatozoa enter oocytes (Wilson et al. 2006). Thus, the functional role of SPE-42 might be in sperm–oocyte fusion through the DC-STAMP domain. On the other hand, SPE-42 might form a protein complex with other sperm proteins through the C4C4-type RING finger domain, since this domain usually mediates protein-to-protein interactions (Borden 2000).

SPE-49: Quite recently, *spe-49* gene has been identified, encoding a protein of which TM topology and domain architecture are the same as those of SPE-42 (Wilson

Table 1 *C. elegans spe-9* class genes

Gene	TM[a]	Domain[b]/Feature	Localization[c]	Defect of mutant in[d]	Possible ortholog[e]
spe-9	1	EGF (10)	PM of spermatid Pseudopod of sperm	Later step after sperm–oocyte binding during fertilization	h*DLL1*, h*DLL3* h*DLL4*
spe-38	4	No obvious domain	MOs of spermatid Pseudopod of sperm	Redistribution of SPE-41/TRP-3 during spermiogenesis	Not found in mammals
spe-41/trp-3	6	TRPC family	MOs of spermatid MOs and PM of sperm	Ca^{2+} influx during fertilization	h*TRPC6* m*Trpc6*
spe-42	6	DC-STAMP (1) RING finger (1)	Not determined	Sperm–oocyte binding during fertilization	h*DCST2* m*Dcst2*
spe-49	6	DC-STAMP (1) RING finger (1)	Not determined	Fertilization	h*DCST1* m*Dcst1*

Abbreviations used are: *TM* transmembrane, *EGF* epidermal growth factor, *PM* plasma membrane, *DLL* Delta-like, *MO* membranous organelle, *TRPC* transient receptor potential-canonical, *DC-STAMP* dendritic cell-specific transmembrane protein, *DCST* DC-STAMP domain containing
This table was prepared on the basis of a published review (Nishimura and L'Hernault 2016)
[a]The number of TM domains in each predicted protein
[b]Domains that are present in each predicted protein besides the TM domain. The number of those domains is shown in parentheses
[c]Localization of each predicted protein. PM in this column means the entire cell surface
[d]For details about the defects of *spe-9* and *spe-42* mutants, see the Sect. 2.1 in the main text, Fig. 5 and Table 2
[e]Human (h) and/or mouse (m) orthologs of each *spe-9* class gene were predicted by WormBase (www.wormbase.org), except for *Izumo1*, on the basis of their nucleotide and/or predicted protein sequences

et al. 2018). Thus, SPE-49 is a paralog of SPE-42 and presumably shares a functional role(s) with SPE-42, at least partly, during gamete fusion.

Figure 2a shows how the localization of the SPE-9 class proteins is related to acquisition of the *C. elegans* sperm fertility. Unlike in many of other species, spermatogenesis once arrests in *C. elegans* after meiosis is completed. Then, by stimulation with unknown factors, spermiogenesis is initiated and round spermatids simultaneously undergo the formation and the activation of spermatozoa; pseudopods are extended, and the membranous organelles (MOs) that had been present in the cytoplasm are fused with the spermatid PM (Ward 1986). Upon the MO–PM fusion, soluble MO contents are extracellularly released, although functional roles of the contents are currently unknown. Some SPE-9 class proteins are localized on the MOs in a spermatid and then move onto the pseudopod or the entire sperm surface during spermiogen-

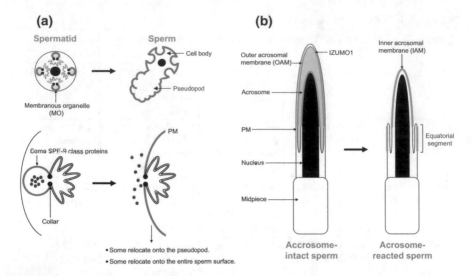

Fig. 2 Comparison of *C. elegans* spermiogenesis and mouse sperm acrosome reaction to acquire the fusogenic activity. **a** Spermiogenesis in *C. elegans*. A spermatid contains the membranous organelles (MOs) in the cytoplasm. During spermiogenesis, the pseudopod is extended to form a spermatozoon, while the MO membrane that is defined by collars, on which some SPE-9 class proteins (represented by thick pink lines) localize, fuse with the spermatid plasma membrane (PM). Soluble MO contents (shown by red dots) are extracellularly released, and the MO-localized SPE-9 class proteins relocate onto the pseudopod or the entire sperm surface. Note that the pseudopods of *C. elegans* spermatozoa are a region where gamete fusion occurs. **b** The acrosome reaction in mouse spermatozoa. IZUMO1 (represented by thick green lines), a sperm protein essential for gamete fusion, is present on the outer (OAM) and inner (IAM) acrosomal membranes. During the acrosome reaction, the OAM and the PM fuse together, soluble contents of the acrosome are released into the extracellular space, and the equatorial segment is formed, where IZUMO1 partly relocates. Again, the segment is a region where gamete fusion occurs

esis. This relocation seems to be reasonable, since *C. elegans* spermatozoa bind to and fuse with the oocyte PM via the pseudopod.

The acrosome reaction in mouse spermatozoa has similar steps to those of *C. elegans* spermiogenesis in order to acquire the sperm fusogenic activity (Fig. 2b). IZUMO1 is a sperm protein that indispensably acts during gamete fusion (for details, see the Sect. 2) (Inoue et al. 2005, 2013, 2015; Bianchi et al. 2014; Aydin et al. 2016; Ohto et al. 2016; Kato et al. 2016; Nishimura et al. 2016) and localizes on the outer and inner acrosomal membranes of acrosome-intact spermatozoa (Satouh et al. 2012). After the acrosome reaction, the equatorial segment is newly formed on acrosome-reacted spermatozoa (Florman and Fissore 2014), where IZUMO1 partly relocates (Satouh et al. 2012). Intriguingly, the segment is a region where spermatozoa fuse with the oocyte PM, like the pseudopods of *C. elegans* spermatozoa.

2   The *C. elegans spe-45* Gene Is Functionally Equivalent to the Mouse *Izumo1* Gene

Again, *C. elegans* has several advantages to study sperm–oocyte fusion; in this model organism, genetic approaches are easily available, and gamete fusion can be examined more simply than those in mammals, due to lack of accessory cells and a thick egg coat in a *C. elegans* oocyte (also described in the Sect. 1.2). Moreover, human and/or mouse homologs of the *spe-9* class genes are predicted on the basis of the domain architectures of each SPE-9 class protein (Table 1). It is unknown, however, whether those mammalian homologs are functionally related to the *C. elegans spe-9* class genes. Indeed, evolutionary rates of reproductive genes are generally faster than those of somatic genes to prevent the production of cross-hybrid species (Wyckoff et al. 2000; Swanson and Vacquier 2002; Haerty et al. 2007). Even between mammals, some mouse genes that are essentially required for reproduction are pseudogenes in the human genome (Jury et al. 1997; Grzmil et al. 2001; Shamsadin et al. 1999; Nishimura et al. 2001, 2004).

Mouse spermatozoa contain IZUMO1 (397 amino acids), a single-pass TM protein with a single immunoglobulin (Ig)-like domain (Fig. 3, upper). Gene targeting studies demonstrate that male mice lacking *Izumo1* gene are sterile and produce otherwise normal spermatozoa that have a defect in fusion with the oocyte PM (Inoue et al. 2005). Thus, the *Izumo1*-deficient male mice exhibit a similar phenotype to those of the *C. elegans spe-9* class mutants. Is there an *Izumo1*-like, *spe-9* class gene(s) in the *C. elegans* genome?

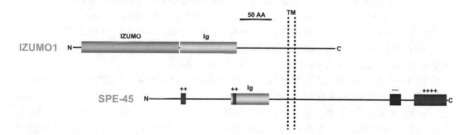

Fig. 3 Gross structures of *C. elegans* SPE-45 and mouse IZUMO1. Domain architectures of mouse IZUMO1 (upper) and *C. elegans* SPE-45 (lower) are shown. Note that *spe-45* gene used to be named *oig-7* before mutants of this gene were examined. N, the amino-terminus; IZUMO, IZUMO domain; Ig, immunoglobulin-like domain; TM, transmembrane domain; C, the carboxy-terminus; +, positively charged amino acid cluster; and −, negatively charged amino acid cluster. A scale bar indicates a 50-amino acid (AA) stretch

2.1   *spe-45 Is a New Member of the spe-9 Class*

First of all, *C. elegans* genes encoding a single-pass TM protein carrying a single-Ig-like domain were searched to identify mouse *Izumo1*-like genes (Nishimura et al. 2015; Nishimura and L'Hernault 2016). Among eight candidates, *C. elegans* genes with elevated expression in the male germline were chosen (Nishimura et al. 2015; Nishimura and L'Hernault 2016), by comparing the DNA microarray data of masculinized and feminized mutant worms (Reinke et al. 2000; Reinke et al. 2004). Following reverse transcription-PCR analysis, it was shown that *oig-7* (one Ig domain-7; this gene was later renamed *spe-45*) is a male germline-enriched gene that encodes an Ig-like, single-pass TM protein (Nishimura et al. 2015; Nishimura and L'Hernault 2016).

Figure 3 (lower) shows a domain architecture of the *oig-7* predicted protein (492 amino acids) (Nishimura et al. 2015; Singaravelu et al. 2015; Nishimura and L'Hernault 2016; Krauchunas and Singson 2016). This protein contains acidic (−) and basic (+) amino acid clusters, in addition to one Ig-like and one TM domains. The positively charged regions might associate with negatively charged substances, such as sulfated proteoglycans and phospholipids. The acidic sequence in the SPE-45 intracellular region possibly folds and binds to the cytoplasmic, basic region within a single SPE-45 protein molecule. At any rate, these structural features are probably prerequisites to regulate the function(s) and localization of SPE-45. Moreover, the entire sequence of SPE-45 protein was only 8.7% and 7.9% identical to those of mouse and human IZUMO1 proteins, respectively (Fig. 4). In particular, the sequence of the cytoplasmic tail domain in SPE-45 is highly divergent from those in mouse and human IZUMO1 (Fig. 4), implying that an intracellular event(s) which happens within *C. elegans* spermatozoa during and/or after gamete fusion might be distinguishable from those in mammalian male gametes. Thus, from a point of identity or similarity of their protein sequences, it might not be reasonable to consider that *C. elegans* SPE-45 is orthologs of mouse and human IZUMO1.

Next, the self-fertility of hermaphrodites lacking *oig-7* was examined (Nishimura et al. 2015; Singaravelu et al. 2015; Nishimura and L'Hernault 2016; Krauchunas and Singson 2016). The *oig-7* mutant laid numerous unfertilized oocytes while F1 progeny was barely produced. However, the same mutant normally produced outcross-progeny after mating with wild-type males. These results indicate a typical Spe phenotype; defects of *spe* mutants are confined to male germline functions. Thus, *oig-7* gene was categorized into *spe* and renamed as *spe-45* hereafter. This gene (*spe-45/oig-7*) was also identified by screening of forward genetically produced mutants (Singaravelu et al. 2015).

After dissection of *spe-45* mutant males, released spermatids were indistinguishable from wild-type cells in number and cytology, indicating that apparent spermatogenesis (meiosis) is normal (Nishimura et al. 2015; Singaravelu et al. 2015; Nishimura and L'Hernault 2016; Krauchunas and Singson 2016). As one of the advantage to use *C. elegans* for studies on male reproduction, *C. elegans* spermatids can activate into spermatozoa in vitro by stimulation with the bacterial protease mix-

```
C.elegans   MQA---LLYFTA-CLTFVDSKRFVLLDPDDPNLEKSASNITDNDVRSRSIKMAKFRKWLKYDVDCQFISRSELAVKISQGRC
Mouse       MGPHFTLLL-AALANCLCPGRPCIKCDQF-------VTDAL------KTFENTYLNDHLPHDIHKNVMRMVNHEVSSFGVVT
Human       MGPHFTLLC-AALAGCLLPAEGCVICDPS-------VVLAL------KSLEKDYLPGHLDAKHHKAMMERVENAVKDFQELS

C.elegans   PHESATEIKPTTTTKKPKSFFFKRKRKPLKKKDNSLFSNENKESVVRAVVGN------------NKMHEIDWTPLSFPDLEL
Mouse       SAEDSYLGAVDENTL-EQATWSFL-KDLKRITDSDL----KGELFIKELLWMLRHQKDIFNNLARQFQKEVLCPNK--CGVM
Human       LNEDAYMGVVDEATL-QKGSWSLL-KDLKRITDSDV----KGDLFVKELFWMLHLQKETFATYVARFQKEAYCPNK--CGVM

C.elegans   VSGTMVTFKCDEMNRKKKRK-----KHEEIEFVEWFVNGKRINP-SWFDWRVS---VSIDGHLGIW---------------
Mouse       SQTLIWCLKCEKQLHICRKSLDCGERHIEVHRSE------DLVLDCLLSWHRASKGLTDYSFYRVWENSSETLIAKGKEPYL
Human       LQTLIWCKNCKKEVHACRKSYDCGERNVEVPQME------DMILDCELNWHQASEGLTDYSFYRVWGNNTETLVSKGKEATL

C.elegans   --PI-GEGDGGHFECLSNGQLIASVTVTVVPIS----------------------------------------------
Mouse       TKSMVGPEDAGNYRCVLDTINQGHATVIRYDVTVLPPKHSEENQPPNIITQEEHETPVHVTPQTPPGQEPESELYPELHPEL
Human       TKPMVGPEDAGSYRCELGSVNSSPATIINFHVTVLPKMIKEEKPSPNIVTPGEATTESSISLQ------------------

C.elegans   -------------KVLVNGLFNYLFVCAIFAVATIPIGCL-LGNRNQEKKEIEVDR-MEEFLAENVFKTDQMAKEKVAGIIE
Mouse       YPELIPTVAQNPEKKMKTRLLILLTLGFVVLVASIIISVLHERKVSAKLKNA-SDEVKPT-ASDKSEATEN-----------
Human       --------PLQPEKMLASRLLGLLICGSLALITGLTFAI--E-----RRRKV-IDFIKSSLF-------------------

C.elegans   KQGVVDERQLIESKAKGNRSTIMILLQKPNTMKEKEKEDQKVSNNPAPAASASTEGATTVTEGTTAAETTVATETTATGTGT
Mouse       -----------GSKSDQSLSQQMGLKKASQADFNSDYSG-----------------------------------------
Human       -----------GLGSGAAEQTQVPKEKATDSRQQ------------------------------------------------

C.elegans   TEAAATTVTETNEEADNEEDEEEDEDDDDDGSVDGTTAGSTESKGPGTTSKSTDKGKKKKKTKKGGKKKKSGAGKGKGKKKS
Mouse       ---------------------------------------------------------------------------------
Human       ---------------------------------------------------------------------------------

C.elegans   KVSKEKKGGKKVQKKKPASKPTKKKK
Mouse       --------------------------
Human       --------------------------
```

Fig. 4 Alignment of protein sequences among *C. elegans* SPE-45 and mouse and human IZUMO1. The entire amino acid sequences of *C. elegans* SPE-45 and mouse (NP_001018013.1) and human (NP_872381.2) IZUMO1 were compared. Amino acid residues that are identical (red) to those found in *C. elegans* SPE-45 are highlighted. Blue letters indicate conserved residues among these three proteins. Regions corresponding to the immunoglobulin-like loops and the transmembrane domains are shown by green and broken squares, respectively

ture Pronase (Nelson and Ward 1980; Machaca et al. 1996). Since spermatids from *spe-45* mutant males could transform into spermatozoa by Pronase treatment at similar levels to those of wild-type male spermatids, in vitro spermiogenesis is normal. Moreover, when *spe-45* mutant males outcrossed to *fem-1* mutant hermaphrodites, which have no self-sperm and are essentially females (Doniach and Hodgkin 1984), there were many male spermatozoa in the spermatheca of the females. This suggests that mutant male spermatids can activate into spermatozoa in the uterus and crawl into the spermatheca; in vivo spermiogenesis is probably normal.

The final phenotypic analysis was to examine whether self-fertilization occurs in the spermatheca of *spe-45* mutant hermaphrodites (Nishimura et al. 2015; Nishimura and L'Hernault 2016). As described in the Sect. 1.1, a hermaphrodite limitedly produces ~300 self-sperm through her life. If *spe-45* mutant hermaphrodites produce otherwise normal self-sperm that have a defect in gamete fusion, the numbers of self-sperm would not decrease in the spermatheca even after repeated ovulation–fertilization cycles (Fig. 1). Thus, L4 hermaphrodites of wild type and *spe-45, spe-9,* and *spe-42* mutants were incubated at 20 °C for 24 or 72 h, fixed in methanol and fluorescently stained to visualize sperm nuclei (Fig. 5). Then, the numbers of self-sperm were counted in every spermatheca (Table 2).

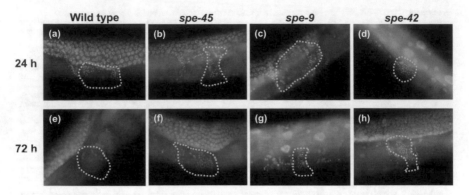

Fig. 5 *spe-9* class mutant spermatozoa are not consumed by fertilization in the spermathecae. L4 hermaphrodites of wild type (**a** and **e**) and *spe-45* (**b** and **f**), *spe-9* (**c** and **g**) and *spe-42* (**d** and **h**) mutants were incubated at 20 °C for 24 h (A-D) or 72 h (E-H), fixed in methanol at −20 °C, and fluorescently stained with 4′,6-diamidino-2-phenylindole (DAPI) to visualize sperm nuclei. Areas of the spermathecae are indicated by broken orange lines. The numbers of self-sperm in each spermatheca were counted, and Table 2 summarizes the data. Note that this figure was prepared on the basis of the previous report (Nishimura et al. 2015)

Table 2 Relative number of self-sperm in the spermatheca of *spe-45* mutant hermaphrodites after repeated ovulation

Time post L4 (h)	Genotype of tested hermaphrodite			
	Wild type	*spe-45*	*spe-9*	*spe-42*
24	100	100	100	100
72	1.4	63	52	100

This table was prepared using the reported data (Nishimura et al. 2015). L4 hermaphrodites of each genotype were incubated at 20 °C for 24 or 72 h, followed by fixation and staining with 4′,6-diamidino-2-phenylindole (DAPI) to fluorescently label sperm nuclei. This enabled to count the number of hermaphrodite-derived sperm in the spermatheca. Data shown in this table are based on the assumption that the number of self-sperm in each spermatheca at 24 h post the L4 stage is 100%

In wild type, there were numerous self-sperm in the spermatheca at 24 h post the L4 stage, and they mostly disappeared at 72 h (Fig. 5 and Table 2), indicating that approximately all of the self-sperm are consumed by fertilization in wild-type hermaphrodites. On the other hand, the self-sperm numbers of *spe-45* and *spe-9* mutants were normal in their spermathecae at 24 h, but ~63% and ~52% levels of *spe-45* and *spe-9* mutant self-sperm still stayed in the spermathecae of each mutant even at 72 h (Fig. 5 and Table 2). Unlike these two *spe* mutants, the numbers of *spe-42* mutant self-sperm at 24 and 72 h were almost equal to each other in the spermatheca (Fig. 5 and Table 2). These findings demonstrate that *spe-45* mutant worms cannot complete fertilization as well as *spe-9* and *spe-42* mutants. Thus, *spe-45* is a new member of the *spe-9* class. Moreover, *spe-45*, *spe-9,* and *spe-42* probably play distinctive roles during gamete fusion; *spe-45* and *spe-9* are functionally unrelated to the sperm binding to the oocyte PM and, rather, participate in later steps of gamete fusion. It is worth to note that mouse spermatozoa deficient for *Izumo1* can bind to, but not fuse with, the oocyte PM. Contrarily, *spe-42* is involved in the initial phase of fertilization, such as the sperm–oocyte PM binding.

2.2　The Ig-Like Domains Are Interchangeable Between C. elegans SPE-45 and Mouse IZUMO1

The phenotype of *C. elegans spe-45* mutant was similar to that of mouse *Izumo1* mutant; spermatozoa from these two mutants likely have defects in the later phase of gamete fusion. The most important question is whether or not *spe-45* and *Izumo1* are functionally related to each other. Thus, it was tested if the functional roles of the Ig-like domains are compatible between SPE-45 and IZUMO1 (Nishimura et al. 2015; Nishimura and L'Hernault 2016).

Figure 6a shows three transgenes (IgWT, IgIZUMO1, and IgIGCM3) that were used for rescue of *spe-45* mutant hermaphrodites. IgWT is a wild-type transgene, and IgIZUMO1 and IgIGCM3 encode chimeric SPE-45 proteins in which the native Ig-like domain is replaced by those of mouse IZUMO1 and *C. elegans* IGCM-3, a somatic protein that is functionally independent of reproduction.

As IgIZUMO1 was expressed in *spe-45* mutant worms, the self-fertility was res-cued to ~77% levels of those by IgWT, while significant rescue did not occur by expression of IgIGCM3 (Fig. 6b). Therefore, the rescue effect by IgIZUMO1 is pre-sumably specific to the Ig-like domain of mouse IZUMO1, indicating that the Ig-like domains have a common role between SPE-45 and IZUMO1 during gamete fusion. It is also suggested that *C. elegans spe-45* is functionally equivalent to mouse *Izumo1* and that their functional roles have been conserved for ~1 billion years.

2.3　Single-Pass TM Proteins with the Ig-Like Domains Are Involved During Gamete Interactions in Diverged Organisms

In addition to *C. elegans* SPE-45 and mouse IZUMO1, other single-pass TM proteins containing the Ig-like domains have been found in a variety of organisms to participate in gamete interactions (Fig. 7).

Mouse SPACA6: The mutant mouse line BART97b has a deletion in the *Spaca6* gene, which encodes an Ig-like TM protein, resulting in production of spermatozoa that appear to be incapable of fusing with the oocyte PM (Lorenzetti et al. 2014).

Mouse BSG: The mouse *Bsg* gene encodes a single-pass TM protein containing two Ig-like domains. Male and female mice lacking *Bsg* are both sterile, as well as the arrest of spermatogenesis in males (Igakura et al. 1998). Moreover, the BSG

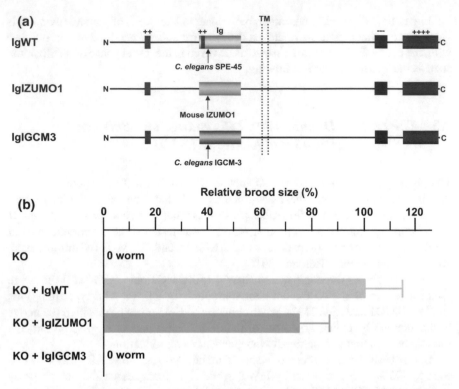

Fig. 6 Immunoglobulin-like domains are interchangeable between *C. elegans* SPE-45 and mouse IZUMO1. **a** Transgenes used for rescue of *spe-45* mutant. To rescue the self-sterility of *spe-45* mutant hermaphrodites, transgenes encoding wild-type SPE-45 (IgWT) and chimeric SPE-45 in which the native immunoglobulin (Ig)-like domain is swapped by those of mouse IZUMO1 (IgIZUMO1) or *C. elegans* IGCM-3 (IgIGCM3). IGCM-3 is a somatic protein of which function is independent of reproduction. For other structural features of each transgene-encoded protein, see Fig. 3. **b** Rescue of *spe-45* mutant (KO) by transgenes encoding chimeric SPE-45 proteins. The numbers of self-progeny produced by *spe-45* mutant hermaphrodites were examined in which either of the transgenes was expressed. The data are shown as mean ± SD, assuming that a mean of the brood sizes after rescue by IgWT is 100%. Note that this figure was prepared on the basis of the previous report (Nishimura et al. 2015)

protein is also involved during sperm interactions with the cumulus cells and the zona pellucida (Saxena et al. 2002).

Arabidopsis GEX2 and *Chlamydomonas* FUS1: GEX2 (Mori et al. 2014; Mori and Igawa 2014) and FUS1 (Ferris et al. 1996; Misamore et al. 2003) are both single-pass TM proteins with the Ig-like filamin repeat domains and involved in gamete fusion or attachment.

It still remains unclear how Ig-like domains that are found in the mouse, plant, and algal proteins act in gamete interactions. First, on the surface of mouse oocytes, IGSF8 (immunoglobulin superfamily member 8, formerly named EWI-2) associates with the tetraspanin protein CD9, which is essentially required for gamete fusion,

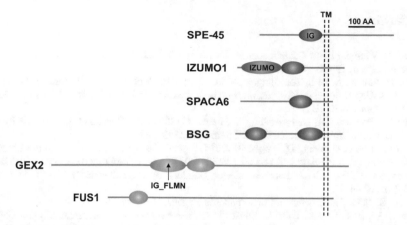

Fig. 7 Immunoglobulin-like domains are conserved in transmembrane proteins that are involved during gamete interactions in various organisms. Diverged organisms contain transmembrane (TM) proteins with the immunoglobulin-like (IG, purple) or IG-folded filamin (IG_FLMN, green) domains that play important roles in gamete interactions. A scale bar indicates a 100-amino acid (AA) stretch. IZUMO, IZUMO domain (light blue)

although the interaction is dispensable for fertilization (Inoue et al. 2012). Second, both of *Hydractinia* Alr1 and Alr2, which are TM proteins contain multiple Ig-like domains, play critical roles in recognition of self *versus* non-self (allo-recognition) through *trans*-homotypic complex formation by the Alr proteins (Karadge et al. 2015), although Alr1 and Alr2 are unlikely to function during gamete interactions. These two examples are not essentially related to fertilization, but they might provide clues to elucidate the physiological roles of the Ig-like domains during gamete interactions.

During gamete fusion in the mouse, sperm IZUMO1 associates with oocyte JUNO via the IZUMO domain (Figs. 3 and 7), but not the Ig-like domain (Inoue et al. 2013, 2015; Bianchi et al. 2014; Aydin et al. 2016; Ohto et al. 2016; Kato et al. 2016; Nishimura et al. 2016). Thus, it might be reasonable to speculate that the Ig-like domain of mouse IZUMO1 plays a role in binding to a *cis*-partner(s) rather than a *trans*-partner. Interestingly, there are *Izumo1* orthologs in birds, fish, and reptiles, but *Juno* ortholog was found only in mammals (Grayson 2015). This suggests that *Juno* gene (formerly named *flr4*, folate receptor 4) was created by a recent mammalian-specific duplication of the ancestral folate receptor gene. Since Ig-like domains acting during gamete interactions seem to be evolutionarily conserved, again, pivotal functions of those domains might not be to associate with oocyte proteins.

In the case of *C. elegans*, its genome has both male and female germline-specifically or predominantly expressed genes encoding Ig-like TM proteins besides *spe-45* (Nishimura et al. 2015). Therefore, there might be a SPE-45 partner(s) with such the domain architecture on the surface of spermatozoa and/or oocytes.

References

Argon Y, Ward S (1980) Caenorhabditis elegans fertilization-defective mutants with abnormal sperm. Genetics 96(2):413–433

Aydin H, Sultana A, Li S, Thavalingam A, Lee JE (2016) Molecular architecture of the human sperm IZUMO1 and egg JUNO fertilization complex. Nature 534(7608):562–565. https://doi.or g/10.1038/nature18595

Beech DJ, Bahnasi YM, Dedman AM, Al-Shawaf E (2009) TRPC channel lipid specificity and mechanisms of lipid regulation. Cell Calcium 45(6):583–588

Bianchi E, Doe B, Goulding D, Wright GJ (2014) Juno is the egg Izumo receptor and is essential for mammalian fertilization. Nature 508(7497):483 487. https://doi.org/10.1038/nature13203

Borden KL (2000) RING domains: master builders of molecular scaffolds? J Mol Biol 295(5):1103–1112

Brenner S (1974) The genetics of Caenorhabditis elegans. Genetics 77(1):71–94

Chatterjee I, Richmond A, Putiri E, Shakes DC, Singson A (2005) The Caenorhabditis elegans spe-38 gene encodes a novel four-pass integral membrane protein required for sperm function at fertilization. Development 132:2795–2808

Cordle J, Johnson S, Tay JZ, Roversi P, Wilkin MB, de Madrid BH, Shimizu H, Jensen S, Whiteman P, Jin B, Redfield C, Baron M, Lea SM, Handford PA (2008) A conserved face of the Jagged/Serrate DSL domain is involved in Notch trans-activation and cis-inhibition. Nat Struct Mol Biol 15(8):849–857

Doniach T, Hodgkin J (1984) A sex-determining gene, fem-1, required for both male and hermaphrodite development in Caenorhabditis elegans. Dev Biol 106(1):223–235

Ferris PJ, Woessner JP, Goodenough UW (1996) A sex recognition glycoprotein is encoded by the plus mating-type gene fus1 of Chlamydomonas reinhardtii. Mol Biol Cell 7(8):1235–1248

Florman HM, Fissore RA (2014) Fertilization in mammals. In: Plant TM, Zeleznik AJ (eds) Knobil and Neill's physiology of reproduction, 4th edn. Elsevier Academic Press, Amsterdam (Netherlands), pp 149–196

Grayson P (2015) Izumo1 and Juno: the evolutionary origins and coevolution of essential sperm-egg binding partners. R Soc Open Sci 2(12):150296. https://doi.org/10.1098/rsos.150296

Grzmil P, Kim Y, Shamsadin R, Neesen J, Adham IM, Heinlein UA, Schwarzer UJ, Engel W (2001) Human cyritestin genes (CYRN1 and CYRN2) are non-functional. Biochem J 357(Pt 2):551–556

Haerty W, Jagadeeshan S, Kulathinal RJ, Wong A, Ravi Ram K, Sirot LK, Levesque L, Artieri CG, Wolfner MF, Civetta A, Singh RS (2007) Evolution in the fast lane: rapidly evolving sex-related genes in Drosophila. Genetics 177(3):1321–1335

Hirsh D, Oppenheim D, Klass M (1976) Development of the reproductive system of Caenorhabditis elegans. Dev Biol 49(1):200–219

Igakura T, Kadomatsu K, Kaname T, Muramatsu H, Fan QW, Miyauchi T, Toyama Y, Kuno N, Yuasa S, Takahashi M, Senda T, Taguchi O, Yamamura K, Arimura K, Muramatsu T (1998) A null mutation in basigin, an immunoglobulin superfamily member, indicates its important roles in peri-implantation development and spermatogenesis. Dev Biol 194(2):152–165

Inoue N, Ikawa M, Isotani A, Okabe M (2005) The immunoglobulin superfamily protein Izumo is required for sperm to fuse with eggs. Nature 434(7030):234–238

Inoue N, Nishikawa T, Ikawa M, Okabe M (2012) Tetraspanin-interacting protein IGSF8 is dispensable for mouse fertility. Fertil Steril 98(2):465–470. https://doi.org/10.1016/j.fertnstert

Inoue N, Hamada D, Kamikubo H, Hirata K, Kataoka M, Yamamoto M, Ikawa M, Okabe M, Hagihara Y (2013) Molecular dissection of IZUMO1, a sperm protein essential for sperm-egg fusion. Development 140(15):3221–3229. https://doi.org/10.1242/dev.094854

Inoue N, Hagihara Y, Wright D, Suzuki T, Wada I (2015) Oocyte-triggered dimerization of sperm IZUMO1 promotes sperm-egg fusion in mice. Nat Commun 6:8858. https://doi.org/10.1038/nc omms9858

Jury JA, Frayne J, Hall L (1997) The human fertilin alpha gene is non-functional: implications for its proposed role in fertilization. Biochem J 321(Pt 3):577–581

Karadge UB, Gosto M, Nicotra ML (2015) Allorecognition proteins in an invertebrate exhibit homophilic interactions. Curr Biol 25(21):2845–2850. https://doi.org/10.1016/j.cub.2015.09.030

Kato K, Satouh Y, Nishimasu H, Kurabayashi A, Morita J, Fujihara Y, Oji A, Ishitani R, Ikawa M, Nureki O (2016) Structural and functional insights into IZUMO1 recognition by JUNO in mammalian fertilization. Nat Commun 7:12198. https://doi.org/10.1038/ncomms12198

Kimble J, Ward S (1988) Germ-line development and fertilization. In: Wood WB (ed) The nematode Caenorhabditis elegans. Cold Spring Harbor Laboratory Press, Cold Spring Harbor, pp 191–213

Krauchunas AR, Singson A (2016) Marriage shrines and worms impacting our understanding of mammalian fertilization. Worm 5(3):e1184389

Krauchunas AR, Marcello MR, Singson A (2016) The molecular complexity of fertilization: introducing the concept of a fertilization synapse. Mol Reprod Dev 83(5):376–386. https://doi.org/10.1002/mrd.22634

Kroft TL, Gleason EJ, L'Hernault SW (2005) The spe-42 gene is required for sperm-egg interactions during C. elegans fertilization and encodes a sperm-specific transmembrane protein. Dev Biol 286:169–181

L'Hernault SW (1997) Male germline. In: Riddle D, Blumenthal R, Meyer BJ, Priess J (eds) C. elegans II. Cold Spring Harbor Laboratory Press, Cold Spring Harbor, pp 271–294

L'Hernault SW (2009) The genetics and cell biology of spermatogenesis in the nematode C. elegans. Mol Cell Endocrinol 306(1–2):59–65. https://doi.org/10.1016/j.mce.2009.01.008

L'Hernault SW, Singson AW (2000) Developmental genetics of spermatogenesis in the nematode Caenorhabditis elegans. In: Goldberg E (ed) The testes: from stem cell to sperm function, Serono Symposium USA. Springer, New York, pp 109–119

L'Hernault SW, Shakes DC, Ward S (1988) Developmental genetics of chromosome I spermatogenesis-defective mutants in the nematode Caenorhabditis elegans. Genetics 120(2):435–452

Lorenzetti D, Poirier C, Zhao M, Overbeek PA, Harrison W, Bishop CE (2014) A transgenic insertion on mouse chromosome 17 inactivates a novel immunoglobulin superfamily gene potentially involved in sperm-egg fusion. Mamm Genome 25(3–4):141–148. https://doi.org/10.1007/s00335-013-9491-x

Machaca K, DeFelice LJ, L'Hernault SW (1996) A novel chloride channel localizes to Caenorhabditis elegans spermatids and chloride channel blockers induce spermatid differentiation. Dev Biol 176(1):1–16

McCarter J, Bartlett B, Dang T, Schedl T (1999) On the control of oocyte meiotic maturation and ovulation in Caenorhabditis elegans. Dev Biol 205(1):111–128

Misamore MJ, Gupta S, Snell WJ (2003) The Chlamydomonas Fus1 protein is present on the mating type plus fusion organelle and required for a critical membrane adhesion event during fusion with minus gametes. Mol Biol Cell 14(6):2530–2542

Miyamoto T (2006) The dendritic cell-specific transmembrane protein DC-STAMP is essential for osteoclast fusion and osteoclast bone-resorbing activity. Mod Rheumatol 16(6):341–342

Mori T, Igawa T (2014) Gamete attachment process revealed in flowering plant fertilization. Plant Signal Behav 9(12):e977715. https://doi.org/10.4161/15592324.2014.977715

Mori T, Igawa T, Tamiya G, Miyagishima SY, Berger F (2014) Gamete attachment requires GEX2 for successful fertilization in Arabidopsis. Curr Biol 24(2):170–175. https://doi.org/10.1016/j.cub.2013.11.030

Nelson GA, Ward S (1980) Vesicle fusion, pseudopod extension and amoeboid motility are induced in nematode spermatids by the ionophore monensin. Cell 19(2):457–464

Nishimura H, L'Hernault SW (2010) Spermatogenesis-defective (spe) mutants of the nematode Caenorhabditis elegans provide clues to solve the puzzle of male germline functions during reproduction. Dev Dyn 239(5):1502–1514. https://doi.org/10.1002/dvdy.22271

Nishimura H, L'Hernault SW (2016) Gamete interactions require transmembranous immunoglobulin-like proteins with conserved roles during evolution. Worm 5(3):e1197485

Nishimura H, L'Hernault SW (2017) Spermatogenesis. Curr Biol 27(18):R988–R994. https://doi.org/10.1016/j.cub.2017.07.067

Nishimura H, Cho C, Branciforte DR, Myles DG, Primakoff P (2001) Analysis of loss of adhesive function in sperm lacking cyritestin or fertilin beta. Dev Biol 233(1):204–213

Nishimura H, Kim E, Nakanishi T, Baba T (2004) Possible function of the ADAM1a/ADAM2 Fertilin complex in the appearance of ADAM3 on the sperm surface. J Biol Chem 279(33):34957–34962

Nishimura H, Tajima T, Comstra HS, Gleason EJ, L'Hernault SW (2015) The immunoglobulin-like gene spe-45 acts during fertilization in Caenorhabditis elegans like the mouse Izumo1 gene. Curr Biol 25(24):3225–3231. https://doi.org/10.1016/j.cub.2015.10.056

Nishimura K, Han L, Bianchi E, Wright GJ, de Sanctis D, Jovine L (2016) The structure of sperm Izumo1 reveals unexpected similarities with Plasmodium invasion proteins. Curr Biol 26(14):R661–R662. https://doi.org/10.1016/j.cub.2016.06.028

Ohto U, Ishida H, Krayukhina E, Uchiyama S, Inoue N, Shimizu T (2016) Structure of IZUMO1-JUNO reveals sperm-oocyte recognition during mammalian fertilization. Nature 534(7608):566–569. https://doi.org/10.1038/nature18596

Putiri E, Zannoni S, Kadandale P, Singson A (2004) Functional domains and temperature-sensitive mutations in SPE-9, an EGF repeat-containing protein required for fertility in Caenorhabditis elegans. Dev Biol 272:448–459

Reinke V, Smith HE, Nance J, Wang J, Van Doren C, Begley R, Jones SJ, Davis EB, Scherer S, Ward S, Kim SK (2000) A global profile of germline gene expression in C. elegans. Mol Cell 6(3):605–616

Reinke V, Gil IS, Ward S, Kazmer K (2004) Genome-wide germline-enriched and sex-biased expression profiles in Caenorhabditis elegans. Development 131(2):311–323

Satouh Y, Inoue N, Ikawa M, Okabe M (2012) Visualization of the moment of mouse sperm-egg fusion and dynamic localization of IZUMO1. J Cell Sci 125(Pt 21):4985–4990. https://doi.org/10.1242/jcs.100867

Saxena DK, Oh-Oka T, Kadomatsu K, Muramatsu T, Toshimori K (2002) Behaviour of a sperm surface transmembrane glycoprotein basigin during epididymal maturation and its role in fertilization in mice. Reproduction 123(3):435–444

Schindl R, Romanin C (2007) Assembly domains in TRP channels. Biochem Soc Trans 35(Pt 1):84–85

Shamsadin R, Adham IM, Nayernia K, Heinlein UA, Oberwinkler H, Engel W (1999) Male mice deficient for germ-cell cyritestin are infertile. Biol Reprod 61(6):1445–1451

Singaravelu G, Chatterjee I, Rahimi S, Druzhinina MK, Kang L, Xu XZ, Singson A (2012) The sperm surface localization of the TRP-3/SPE-41 Ca^{2+}-permeable channel depends on SPE-38 function in Caenorhabditis elegans. Dev Biol 365(2):376–383. https://doi.org/10.1016/j.ydbio.2012.02.037

Singaravelu G, Rahimi S, Krauchunas A, Rizvi A, Dharia S, Shakes D, Smith H, Golden A, Singson A (2015) Forward genetics identifies a requirement for the Izumo-like immunoglobulin superfamily spe-45 gene in Caenorhabditis elegans fertilization. Curr Biol 25(24):3220–3224. https://doi.org/10.1016/j.cub.2015.10.055

Singson A, Mercer KB, L'Hernault SW (1998) The C. elegans spe-9 gene encodes a sperm trans-membrane protein that contains EGF-like repeats and is required for fertilization. Cell 93:71–79

Swanson WJ, Vacquier VD (2002) The rapid evolution of reproductive proteins. Nat Rev Genet 3(2):137–144

Takayama J, Onami S (2016) The sperm TRP-3 channel mediates the onset of a Ca^{2+} wave in the fertilized C. elegans oocyte. Cell Rep 15(3):625–637. https://doi.org/10.1016/j.celrep.2016.03.040

Ward S (1986) Asymmetric localization of gene products during the development of Caenorhaditis elegans spermatozoa. In: Gall JG (ed) Gametogenesis and the early embryo. Alan R. Liss, Inc., New York, pp 55–75

Ward S, Carrel JS (1979) Fertilization and sperm competition in the nematode Caenorhabditis elegans. Dev Biol 73(2):304–321

Ward S, Argon Y, Nelson GA (1981) Sperm morphogenesis in wild-type and fertilization-defective mutants of Caenorhabditis elegans. J Cell Biol 91(1):26–44

Wilson KL, Fitch KR, Bafus BT, Wakimoto BT (2006) Sperm plasma membrane breakdown during Drosophila fertilization requires sneaky, an acrosomal membrane protein. Development 133(24):4871–4879

Wilson LD, Obakpolor OA, Jones AM, Richie AL, Mieczkowski BD, Fall GT, Hall RW, Rumbley JN, Kroft TL (2018) The C. elegans spe-49 gene is required for fertilization and encodes a sperm-specific transmembrane protein homologous to SPE-42. Mol Reprod Dev. https://doi.or g/10.1002/mrd.22992

Wolf N, Hirsh D, McIntosh JR (1978) Spermatogenesis in males of the free-living nematode, Caenorhabditis elegans. J Ultrastruct Res 63(2):155–169

Wyckoff GJ, Wang W, Wu CI (2000) Rapid evolution of male reproductive genes in the descent of man. Nature 403(6767):304–309

Xu XZ, Sternberg PM (2003) A C. elegans sperm TRP protein required for sperm-egg interactions during fertilization. Cell 114(3):285–297

Part III
Origin of Biodiversity

Feralisation—The Understudied Counterpoint to Domestication

R. Henriksen, E. Gering and D. Wright

Abstract Feralisation is a complex process that occurs when a domestic population is returned to the wild. It impacts species invasion biology, speciation, conservation and hybridisation and can be thought of as the reverse of domestication. Domestication has been an area of intense interest and study ever since Darwin, and useful as a model for evolution and the effects of strong directional selection. Despite domestication being used to identify genes affecting a large number of traits that change with selection, little is known about the genomic changes associated with feralisation. Much of the current work on the genetics of feralisation has focused on the detection of early hybrids (F_1 or F_2) between wild and domestic populations. Feralisation can lead to large changes in morphology, behaviour and many other traits, with the process of feralisation involving the sudden return of both natural and sexual selection. Such evolutionary forces influence predatory, foraging and mate choice decisions and exert strong effects on once domesticated, now feral, individuals. As such, feralisation provides a unique opportunity to observe the genomic and phenotypic responses to selection from a known (domesticated) standpoint and identify the genes underlying these selective targets. In this review, we summarise what is known in particular regarding the genomics of feralisation, and also the changes that feralisation has induced on brain size and behaviour.

1 Feralisation—The Understudied Counterpoint to Domestication

Domestication is a fundamental process integral to the advance of our civilisation and our development as a species. The study of domestication, at its heart, is essentially that of selection. It provides a cogent model for how evolution itself can act,

R. Henriksen · D. Wright (✉)
AVIAN Behavioural Genomics and Physiology Group, IFM Biology, Linköping University, 58183 Linköping, Sweden
e-mail: dominic.wright@liu.se

E. Gering
Department of Zoology, Michigan State University, Michigan 48824, USA

© Springer International Publishing AG, part of Springer Nature 2018
P. Pontarotti (ed.), *Origin and Evolution of Biodiversity*,
https://doi.org/10.1007/978-3-319-95954-2_11

especially when used as an artificial proxy to natural selection, with known and very strong selection pressures being applied to a population (Darwin 1859, 1868). However, this strong directional selection can also be a weakness. It is equivocal how relevant the genes and genetic architectures discovered in domestic populations are to natural populations, where selection is far more complex. If domestication is defined as a process whereby captive animals adapt to man and the environment he provides (Jensen and Wright 2014; Price 2002), feralisation is the removal of domestic animals from the domestic environment and their return to natural conditions. It can be intentional, termed de-domestication, when domestic animals, over several generations, are bred for self-sustainability (Gamborg et al. 2010), or unintentional, when a species escapes and subsequently re-adapts to its new environments. Classic examples of feral species are rabbits (*Oryctolagus cuniculus*), pigs (*Sus scrofa*) and dingos (*Canis lupus dingo*). In the case of the rabbit, multiple introductions have been made ranging through all continents and hundreds of islands (Flux and Fullagar 1992), with predominantly domestic animals being introduced. Similarly, domesticated pigs and Eurasian wild boar have been introduced to both the Old World and New World from the sixteenth and seventeenth centuries onwards. In both cases, these species have survived and thrived and are often a major problem (Statham and Middleton 1987), causing billions of dollars of damage through agricultural and environmental impacts and as disease vectors. Current research on feralisation has been largely limited to phenotypic assays, with few studies having examined how this process shapes feral gene pools and traits. Genetic studies using molecular techniques have been limited to genetic markers to broadly assess the degree of introgression between wild and feral species (Randi 2008) and the overall population structure, with studies primarily on pigs (*Sus scrofa*) (Hampton et al. 2004), wolves (*Canis lupus*) (Randi and Lucchini 2002; Verardi et al. 2006), wild cats (*Felis catus*) (Menotti-Raymond et al. 2003; Pierpaoli et al. 2003; Randi 2008), rock partridges (*Rock partridges*) and red-legged partridges (*Alectoris rufa*).

2   Feralisation and Evolution

Feralisation can relate to a multitude of evolutionary processes, including directional selection, parallel evolution and reverse evolution. Examples of parallel evolution occur when independent feral populations all redevelop ancestral characteristics, or when traits that were previously adaptive but lost during domesticated are, once again, recapitulated (known as atavism). Given that the strong directional selection of domestication should remove this ancestral variation, how does this occur? Loci that have been fixed under domestication should not be reversible unless mutation or introgression reintroduces wild alleles. Similarly, feralisation is also a model of cryptic genetic variation, with a great deal of phenotypic diversity being observed in feral populations, despite strong founder bottleneck effects indicating a low initial genetic variability (Gering et al. 2015). Speciation and biodiversification can also be a major topic of study—if wild type alleles are preserved in domestic populations, they can then be used to help re-populate, diversify or even recreate extinct species (Barlow 1999; Donlan 2005).

3 Quantitative Genetics and the Domestication/Feralisation Paradigm

Domestication has often been used a means of studying the genetic basis of phenotypic variation and in this regard acts as an excellent model for such genetic dissection. Domestication has produced some extreme effects on the phenotype of a variety of animals. In this way, it can also aid in completing the genotype–phenotype map in a variety of animals and help further our understanding of how changes in the genome can bring about alterations in both quantitative and discrete traits. For example, in regard to coat colour, genes have been identified in dogs (Schmutz and Berryere 2007), horses (Haase et al. 2009) and cattle (Hayes et al. 2010). In dogs, genes for wrinkled skin (Olsson et al. 2011), short leggedness and body size have all been discovered. In cattle and sheep, a deletion mutation has been identified in the *MSTN* gene that leads to extreme muscular development (an increase in muscle mass of 20%) (Clop et al. 2006; Grobet et al. 1997). In pigs, major genes for malignant hyperthermia (Fujii et al. 1991), glycogen content in muscles (Milan et al. 2000), ear size (Ren et al. 2011) and muscularity, back fat and heart size (Van Laere et al. 2003) have all been identified. The vast majority of the above examples however are of genes of major or 'Mendelian' effect (i.e. the trait is essentially monogenically derived). The search for minor-effect genes that affect quantitative traits has proven to be far more elusive. These genes (or more strictly polymorphisms) are far harder to locate, but will likely prove to be more ubiquitous and potentially more useful in bridging the phenotype–genotype gap. The relevance that these Mendelian mutations have to the basis of genetic variation in the wild is far more debatable, with such major mutations generally being strongly selected against in natural conditions.

While progress has been made, albeit incomplete, in understanding the genomic changes that attend domestication, the genes underlying feralisation are largely unknown. This raises several important questions whose answers are relevant for both the history and future of biodiversity evolution. For example, are the same genes that are selected during domestication once again selected upon (albeit in a different direction) during feralisation? To answer such questions, the genomic changes associated with feralisation need to be identified, which can be done both using extant populations and also, with the use of modern DNA extraction and sequencing techniques, even archaeological samples (Larson et al. 2014).

4   Hybridisation Between Wild and Domestic Populations

The majority of the work to date on the genomics of feral animals has focused on their hybridisation with their wild counterparts and is motivated by conservation concerns. In particular, many efforts have been made to assess the degree of introgression between domesticated taxa and wild, non-domesticated relatives (Randi 2008). From a conservation perspective, such hybridisation is typically regarded as harmful. Translocated and invasionary domestic populations can raise the risk of local extinction, disrupt locally adapted haplotypes or lead to outbreeding depression (Lynch 1991). Domestic animals contain genes that have been subjected to both strong artificial selection and also reduced natural selection (Jensen and Wright 2014; Price and King 1968), and the resulting alleles may then be maladaptive to the natural environment. Breeding with such individuals can disrupt locally adapted gene complexes (Allendorf et al. 2001; Rhymer and Simberloff 1996) and thereby increase extinction risk, especially when introgression occurs on large scales, swamping the natural population with maladaptive hybrids (Lynch and O'hely 2001). A classic example of this is the ongoing escape and release of farmed Atlantic salmon (*Salmo salar*), which increase both resource competition and rates of hybridisation to the detriment of native gene pools (Fleming and Einum 1997; Hutchings and Fraser 2008). Declining population persistence also occurs when these hybrids displace wild individuals as mating partners in the population, leading to decreasing population size as the overall fitness of the population falls (McGinnity et al. 2003). From a conservation perspective, the swamping of a population with hybrids can extirpate the original population entirely and is an extreme threat, for example the Przewalski horse (Dierendonck and Vries 1996), red wolf (Roy et al. 1994) and Hawaiian duck (Browne et al. 1993). Hybrids can however also increase fitness, for example, via hybrid vigour and the introgression of beneficial alleles. A probable example of increased fitness involves wild–domestic hybrids in wild boar populations. When wild boars are farmed they are frequently crossed with domestic pigs, with the resulting hybrids having increased growth and fecundity. Consequently, higher litter sizes and large population variation now occur in different populations of wild boar that have been subjected to introgressions from these animals (Gethöffer et al. 2007).

   The extent of these hybridisations in the wild can vary, though most are around a 5% frequency of recent hybrids (see below). Although these can occur fairly frequently, the extent of the hybridisations in any given population (at least that can be detected) is typically fairly low, particularly where the wild population is fairly large (Randi 2008). Wild populations of Italian wolves (~600 individuals) are sympatric with free-ranging dog populations (Verardi et al. 2006) and have a 5% hybrid introgression frequency over multiple different populations (Fabbri et al. 2007). A 3.9% hybrid introgression frequency was seen in the wild boar populations of the Netherlands and Germany (Goedbloed et al. 2013a, b), with domestic introgression coming from escaped or released hybrids used in farming. Coyotes in the USA also show some hybridisation with free-ranging domestic dogs, with 12 of 112 coyotes showing some signs of recent hybridisation (Adams et al. 2003). The wild red-legged partridge (*Alectoris rufa*) is continuously subjected to translocated hybrids that are used for hunting purposes and contain elements of Chukar partridge (*Alectoris chukar*) DNA (Baratti et al. 2005; Negro et al. 2001). An assessment of 691 individuals from a range of sites indicated hybrid Chukar mtDNA in between 5 and 6% of samples (Barilani et al. 2007; Randi and Lucchini 1998). In the case of the European wildcat, there is widespread overlap with domestic cats (McOrist and Kitchener 1994), but there are once again relatively low levels of hybridisation in Italy (3–5%) (Randi et al. 2001) and the Iberian peninsula (6.9%) (Oliveira et al. 2008). However, populations in Hungary (Pierpaoli et al. 2003) and Scotland (Beaumont et al. 2001) show far greater levels of hybridisation. Similarly, American mink (*Neovison vison*) that hybridise with domestic farmed mink also show high hybridisation levels (78 and 45% hybrid frequency in two populations), though only in regions immediately adjacent to farms, and with no hybrids in more remote wild populations (Kidd et al. 2009).

These patterns imply that introgression and hybridisation are regulated by both the genetic relatedness and natural selection acting on a given pair of domesticated/feral and wild taxa. Reproductive isolation and other evolved barriers to hybridisation, as well as poorly adapted hybrids, therefore act to limit the total degree of introgression. However, the caveat with almost all of these studies is that they are typically based on only a handful of microsatellite or SNP markers (Randi 2008). As such, they can only detect very recent introgressions (Randi 2008), and more ancient episodes will typically be missed. It is therefore possible that some of these events may have contributed advantageous alleles to the wild gene pool, but that these would require a far higher genomic resolution to detect.

5   The Hawaiian Feral Chicken

Very few studies have been performed that have assessed the genomic characteristics of a feral population in sufficient depth to identify specific genomic regions that are undergoing selection. One exception to this is the population of Hawaiian chickens on Kauai (see Fig. 1), which are proving to be a valuable resource for studying how feral genomes evolve. In terms of their origin, archaeological evidence indicates that chickens were first introduced to the Hawaiian Island chain (including Kauai) by AD1200 via human migration into the eastern Pacific (Thomson et al. 2014; Wilmshurst et al. 2011). Their sources were most likely Red Junglefowl (*Gallus gallus*) transported from the western Pacific by Polynesian settlers (Thomson et al. 2014). Therefore, wild Red Junglefowl have likely persisted on these islands for over 1000 years. In 1982 and 1992, tropical storm Ewa and Hurricane Iniki destroyed many of the coops containing Kauai's domestic chickens, releasing their occupants into local forests, and spurred large-scale species invasions. Phenotypic assessments of mitochondrial, vocalisation and plumage-based analyses concur that the contemporary population is a hybrid with both domestic and Red Junglefowl origins (Gering et al. 2015). Thus, birds inhabiting Kauai today exhibit characteristics of both the original Red

Fig. 1 Phenotypic changes associated with feralisation in the Kauai chicken. Figure taken from Callaway, E. Nature (2016) 529: 270–273 and used with permission

Junglefowl founder strain and more recently derived European domestics, and these characteristics may be involved in adaptation to feral environments.

Whole-genome re-sequencing of multiple individuals allows the identification of selective sweeps in a similar manner to that which has been performed previously with domestic chickens and Red Junglefowl (Rubin et al. 2010). These selective sweeps are caused where selection acts on a specific beneficial mutation or haplotype, reducing genetic variation at the causal loci and the surrounding region. Using this technique on 25 of the Kauai birds, Johnsson et al. identified 37 sweeps of 40 kb intervals, with seven of these representing ones found previously in the domestic chicken, thereby resulting in 30 unique sweeps that are unique to this hybrid, feral population. This was the first whole-genome study of feralisation to date (Johnsson et al. 2016a).

The sweeps Johnsson et al. identified in the Kauai genomes contained a total of 91 genes (Johnsson et al. 2016a). To help ascertain the function of these genes, Johnsson et al. used a laboratory intercross between wild (Red Junglefowl) birds and domestic chickens as a proxy for the hybrid Kauai population (Johnsson et al. 2016a). This intercross population had already been used to map gene expression as expression quantitative trait loci (eQTL) and phenotypic traits for comb mass, fecundity and other domestication-related characteristics as a means to identify the genes associated with domestication-related traits (Johnsson et al. 2012, 2014, 2015a, b, 2016b, 2018a, b; Wright et al. 2010, 2012). By overlapping these sweep regions with the results of the previous eQTL and QTL studies for comb size in the AIL, Johnsson et al. (2016a) found that two of the genes discovered in the sweep regions, *STK32A* and *DPYSL3*, strongly correlated with comb size in the laboratory cross (correlating gene expression with comb size) and are strong candidates for a QTL effecting comb mass on chromosome 13. Similarly, two of four QTL for broodiness (incubation behaviour—lacking in the domestic bird to increase egg production) were also found to overlap the selective sweep regions. The Kauai population appears to show signatures of recent sexual selection, with sweep regions indicating sexual ornament genes are under selection.

6 Feralisation, Brain Size and Brain Composition

Domestication has led to large changes in both brain size and composition (Jensen and Wright 2014). The significant changes in brain size and composition that have occurred during domestication provide a general model for studying evolution in brain mass and brain composition, especially the allometric relationship between body size and brain size, as well as the relative sizes of individual brain regions (Gonda et al. 2013). It is classically believed that domestication leads to a reduction in overall brain size, due to the decreased proportional brain size that is observed in multiple different domestic species (ranging from pigeons to chickens, pigs and mink (Kruska 2005) compared to their wild progeny). The results supporting this conclusion, however, are misleading, as the use of relative brain size is complicated by the selection for growth in many of these species during domestication. If the genetic architecture for body growth is independent from that for brain growth, then artificial selection for body size can in fact mask concomitant changes (or conservation) of brain size. This has been shown to be the case in chickens, where loci controlling variation in brain mass and body mass have separate genetic architectures and are therefore not pleiotropically constrained (Henriksen et al. 2016). As a result of recent artificial selection, the common layer breed White Leghorn (WL) has almost doubled their body size ~85%, whereas brain mass has only increased by ~15% compared to their wild progenitor, the Red Junglefowl (RJF). Using an advanced intercross between WL and RJF, it was demonstrated that domestication acted on separate loci to increase brain mass and body mass; thus, increased brain size may have arisen during domestication independently of body size, at least to some degree, and vice versa. This increase in absolute brain size seen in the chicken was accompanied by an increase in the proportional size of the cerebellum, which was made possible by differences in the genomic mechanisms controlling the development of various brain regions (Hager et al. 2012; Henriksen et al. 2016). During domestication, there can therefore be increased selection on some brain regions over others, and while it might be difficult to hypothesise about what selection pressure resulted in enlarge cerebellum in chickens, the fact that domesticated geese, turkeys and pigeons have also developed a larger proportional cerebellum than their wild progeny (Ebinger and Röhrs 1994; Ebinger and Löhmer 1986, Ebinger and Löhmer 1984) suggests that alteration in brain composition may have played an important role during avian domestication.

The brains of feral species offer a unique opportunity to study how the size and composition of the brain change when animals return to the wild. What happens to the brain when animals are introduced to a more enriched environment, but also potentially more challenging environment where they will encounter biological situations not existing or of minor importance in their previous domesticated settings? To date, there is only sparse information concerning brain size and composition in feral animals, since only very few studies have measured the brain (or brain case as an indirect measure of brain mass) of feral animals. All of these measurements have been done on mammals [such as pigs (Kruska and Röhrs 1974), cats (Derenne and Mougin 1976), mink (Kruska and Sidorovich 2003) and dogs (Schultz 1969)], and most studies have relied on cranial measurements as an indirect proxy for brain size. One of the only studies where the actual brain size and brain composition were measured involved feral pigs of the Galapagos Islands. Pigs were released on the Galapagos Islands about 100–150 B.P. to serve as a meat reserve. These feral pigs showed brain sizes within the normal distribution of domesticated forms, and distinct from that of the wild boar (Kruska and Röhrs 1974). Further measurements, however, revealed that their brain composition was slightly different from modern European domesticated races, indicating genetic and/or environmental effects of feralisation on brain structure. Feral mink and feral cats, on the other hand, have slightly smaller brain cavities and crania, respectively (Kruska and Sidorovich 2003), than their domesticated counterparts. Thus, in these species, feralisation has led to a decrease in brain size. The dingo is derived from early domesticated dogs and is probably among the animals that have been feral for the longest period of time (since circa 3000–8600 B.P.). Dingo's body sizes and brain sizes fall within the domestic distribution of dogs and not of the wolf (Schultz 1969).

Taken together, the available evidence suggests that feralisation affects brain size to a smaller degree than domestication. This may reflect evolutionary constraints imposed by the depletion of genetic variation in feral population's domesticated sources, and/or differences in the antiquities and effective population sizes of feral versus domestic populations. From an evolutionary perspective, it is interesting that the brain of successfully feralised species has not had to increase in size, which calls into question the hypothesis that a more challenging environment requires a bigger brain.

Although feralisation seems to have resulted in much smaller alteration to the brain than domestication, the small changes in brain composition that have been reported in feral pigs suggest that feralisation may involve behavioural adaptation. Very few studies have compared the behaviour of feral animals with their domesticated counterpart, and unlike studies on brain measurement which has mainly been done on mammals, the studies that have compared the behaviour of feral animals with their domestic counterpart have been done chiefly using birds. Rose and colleagues (Rose et al. 1985), for example, found that feral cockerels performed higher levels of agonistic behaviour and greater initial avoidance of a novel object than did domestic cockerels when reared and tested in laboratory settings. When measuring the behaviour of feral and domestic Japanese quails in a semi-natural condition (Nichols 1991), feral males were reported to crow less but with more variability than domestic males. This may reflect strategic responses to both sexual selection and the exposure to predation that crowing might entail. This study also found that feral quails were more alert to birds flying overhead and sudden noise and that they formed stronger pair bonds and had stronger associations with their chicks. These studies suggest some degree of behavioural adaption by these feral birds, since vigilance and parental care would likely affect survivability and fitness in wild habitats. The feral quails mentioned in the study above originated from a domestic population of domestic Japanese quails that were released on several of the Hawaiian Islands in the 1920s, while the behavioural study was carried out in the mid-1980s; thus, it is at least feasible that sufficient time had elapsed for genetically based adaptation.

From current data on the effects of feralisation on brain size and composition, it must be concluded that no actual return to the brain size of the former wild ancestor has occurred. Adaptation to a wild lifestyle and ecological niche are not necessarily connected with the evolution of a larger brain. The few behavioural studies suggest some degree of alteration in behaviour in order to re-adapt to a wild environment, but whether the behavioural changes are due to the small differences in the relative sizes of various brain regions size, or other levels of brain anatomy that have yet to be measured, and remain to be investigated.

7 Summary

Whilst domestication has been the focus of a huge amount of research, the potential for feralisation to expand our knowledge on a range of processes is relatively untapped. In this review, we have touched on what is known regarding the genomics changes associated with feralisation and how brain size and behaviour may have been affected. However, many other subjects present themselves as fruitful for further enquiry. The role of sexual selection in feral populations is one such example. Domestic populations often have highly restricted and controlled breeding schemes that limit or remove the potential for sexual selection; however, with a return to natural conditions, animals are free to once again select their own mates. How these populations change in terms of gene frequency can tell us a great deal about which genes are most responsive to this selection. Some evidence for this has been shown in chickens, but much more remains to be done. Similarly, resource allocation can change greatly for animals moving from domestic to feral environments, with domestic animals frequently selected for prioritising growth and fecundity. By researching more on feral populations, not only can these questions be answered, but we will also be able to develop more tools for the control and conservation of the wild animals that interact with these feral individuals.

Acknowledgements The research was carried out within the framework of the Linköping University Neuro-network. The project was supported by grants from the Swedish Research Council (VR), the European Research Council (advanced research grant GENEWELL 322206, consolidator grant FERALGEN 772874) and the National Science Foundation under Cooperative Agreement No. DBI-0939454. Any opinions, findings, and conclusions or recommendations expressed in this material are those of the author(s) and do not necessarily reflect the views of the National Science Foundation.

References

Adams J, Leonard J, Waits L (2003) Widespread occurrence of a domestic dog mitochondrial DNA haplotype in southeastern US coyotes. Mol Ecol 12:541–546

Allendorf FW, Leary RF, Spruell P, Wenburg JK (2001) The problems with hybrids: setting conservation guidelines. Trends Ecol Evol 16:613–622

Baratti M, Ammannati M, Magnelli C, Dessì-Fulgheri F (2005) Introgression of chukar genes into a reintroduced red-legged partridge (*Alectoris rufa*) population in central Italy. Anim Genet 36:29–35

Barilani M, Bernard-Laurent A, Mucci N, Tabarroni C, Kark S, Garrido JAP, Randi E (2007) Hybridisation with introduced chukars (*Alectoris chukar*) threatens the gene pool integrity of native rock (*A. graeca*) and red-legged (A. rufa) partridge populations. Biol Conserv 137:57–69

Barlow C (1999) Rewilding for evolution. Wild Earth 9:53–56

Beaumont M, Barratt E, Gottelli D, Kitchener A, Daniels M, Pritchard J, Bruford M (2001) Genetic diversity and introgression in the Scottish wildcat. Mol Ecol 10:319–336

Browne RA, Griffin CR, Chang PR, Hubley M, Martin AE (1993) Genetic divergence among populations of the Hawaiian Duck, Laysan Duck, and Mallard. Auk 49–56

Clop A et al (2006) A mutation creating a potential illegitimate microRNA target site in the myostatin gene affects muscularity in sheep. Nat Genet 38:813–818

Darwin C (1859) The origin of species. Mentor, New York

Darwin C (1868) The variation of animals and plants under domestication. John Murray, London

Derenne P, Mougin J (1976) DONNÉES CRANIOMÉTRIQUES SUR LE LAPIN ET LE CHAT HARET DE L' ILE AUX COCHONS, ARCHIPEL CROZET (46° 06′ S, 50° 14′ E). Mammalia 40:495–516

Dierendonck MC, Vries MF (1996) Ungulate reintroductions: experiences with the takhi or Przewalski horse (*Equus ferus przewalskii*) in Mongolia. Conserv Biol 10:728–740

Donlan J (2005) Re-wilding north America. Nature 436:913–914

Ebinger P, Löhmer R (1984) Comparative quantitative investigations on brains of rock doves, domestic and urban pigeons (Columba 1. livia) 1. J Zool Syst Evol Res 22:136–145

Ebinger P, Löhmer R (1986) A volumetric comparison of brains between greylag geese (Anser anser L.) and domestic geese. J Hirnforsch 28:291–299

Ebinger P, Röhrs M (1994) Volumetric analysis of brain structures, especially of the visual system in wild and domestic turkeys (*Meleagris gallopavo*). J Hirnforsch 36:219–228

Fabbri E et al (2007) From the Apennines to the Alps: colonization genetics of the naturally expanding Italian wolf (*Canis lupus*) population. Mol Ecol 16:1661–1671

Fleming I, Einum S (1997) Experimental tests of genetic divergence of farmed from wild Atlantic salmon due to domestication. ICES J Mar Sci 54:1051–1063

Flux JE, Fullagar PJ (1992) World distribution of the Rabbit Oryctolagus funiculus on islands. Mamm Rev 22:151–205

Fujii J et al (1991) Identification of a mutation in porcine ryanodine receptor associated with malignant hyperthermia. Science 253:448–451

Gamborg C, Gremmen B, Christiansen SB, Sandøe P (2010) De-domestication: ethics at the intersection of landscape restoration and animal welfare. Environ Values 57–78

Gering E, Johnsson M, Willis P, Getty T, Wright D (2015) Mixed-ancestry and admixture in Kauai's feral chickens: invasion of domestic genes into ancient Red Junglefowl reservoirs. Mol Ecol 24:2112–2124

Gethöffer F, Sodeikat G, Pohlmeyer K (2007) Reproductive parameters of wild boar (Sus scrofa) in three different parts of Germany. Eur J Wildl Res 53:287–297

Goedbloed D et al (2013a) Genome-wide single nucleotide polymorphism analysis reveals recent genetic introgression from domestic pigs into Northwest European wild boar populations. Mol Ecol 22:856–866

Goedbloed DJ et al (2013b) Reintroductions and genetic introgression from domestic pigs have shaped the genetic population structure of Northwest European wild boar. BMC Genet 14:43

Gonda A, Herczeg G, Merilä J (2013) Evolutionary ecology of intraspecific brain size variation: a review. Ecol Evol 3:2751–2764

Grobet L et al (1997) A deletion in the bovine myostatin gene causes the double-muscled phenotype in cattle. Nat Genet

Haase B et al (2009) Seven novel KIT mutations in horses with white coat colour phenotypes. Anim Genet 40:623–629

Hager R, Lu L, Rosen GD, Williams RW (2012) Genetic architecture supports mosaic brain evolution and independent brain–body size regulation. Nat Commun 3:1079

Hampton JO et al (2004) Molecular techniques, wildlife management and the importance of genetic population structure and dispersal: a case study with feral pigs. J Appl Ecol 41:735–743

Hayes BJ, Pryce J, Chamberlain AJ, Bowman PJ, Goddard ME (2010) Genetic architecture of complex traits and accuracy of genomic prediction: coat colour, milk-fat percentage, and type in Holstein cattle as contrasting model traits. PLoS Genet 6:e1001139

Henriksen R, Johnsson M, Andersson L, Jensen P, Wright D (2016) The domesticated brain: genetics of brain mass and brain structure in an avian species. Sci Rep 6:p.34031. https://doi.org/10.1038/srep34031

Hutchings JA, Fraser DJ (2008) The nature of fisheries-and farming-induced evolution. Mol Ecol 17:294–313

Jensen P, Wright D (2014) Behavioral genetics and animal domestication. In: Grandin T, Deesing MJ (eds) Genetics and behavior of domestic animals. Academic Press, London, pp 41–80

Johnsson M et al (2012) A sexual ornament in chickens is affected by pleiotropic alleles at HAO1 and BMP2, selected during domestication. PLoS Genet 8:e1002914. https://doi.org/10.1371/journal.pgen.1002914

Johnsson M et al (2014) The role of pleiotropy and linkage in genes affecting a sexual ornament and bone allocation in the chicken. Mol Ecol 23:2275–2286

Johnsson M, Jonsson KB, Andersson L, Jensen P, Wright D (2015a) Genetic regulation of bone metabolism in the chicken: similarities and differences to mammalian systems. PLoS Genet 11:e1005250. https://doi.org/10.1371/journal.pgen.1005250

Johnsson M, Jonsson KB, Andersson L, Jensen P, Wright D (2015b) Quantitative trait locus and genetical genomics analysis identifies putatively causal genes for fecundity and brooding in the chicken. G3: Genes|Genomes|Genetics. https://doi.org/10.1534/g3.115.024299

Johnsson M et al (2016a) Feralisation targets different genomic loci to domestication in the chicken. Nat Commun 7:12950. https://doi.org/10.1038/ncomms12950. http://www.nature.com/articles/ncomms12950-supplementary-information

Johnsson M, Williams MJ, Jensen P, Wright D (2016b) Genetical genomics of behavior: a novel chicken genomic model for anxiety behavior. Genetics 202:327–340

Johnsson M, Henriksen R, Fogelholm J, Höglund A, Jensen P, Wright D (2018a) Genetics and genomics of social behavior in a chicken model. Genetics. https://doi.org/10.1534/genetics.118. 300810

Johnsson M, Henriksen R, Höglund A, Fogelholm J, Jensen P, Wright D (2018b) Genetical genomics of growth in a chicken model. BMC Genom 19:72. https://doi.org/10.1186/s12864-018-4441-3

Kidd A, Bowman J, Lesbarreres D, Schulte-Hostedde A (2009) Hybridization between escaped domestic and wild American mink *(Neovison vison)*. Mol Ecol 18:1175–1186

Kruska D (2005) On the evolutionary significance of encephalization in some eutherian mammals: effects of adaptive radiation, domestication and feralization. Brain Behav Evol 65:73–108

Kruska D, Röhrs M (1974) Comparative-quantitative investigations on brains of feral pigs from the Galapagos Islands and of European domestic pigs. Zeitschrift für Anatomie und Entwicklungs-geschichte 144:61–73

Kruska D, Sidorovich V (2003) Comparative allometric skull morphometrics in mink (Mustela vison Schreber, 1777) of Canadian and Belarus origin; taxonomic status. Mamm Biol-Zeitschrift für Säugetierkunde 68:257–276

Larson G et al (2014) Current perspectives and the future of domestication studies. Proc Natl Acad Sci 111:6139–6146

Lynch M (1991) The genetic interpretation of inbreeding depression and outbreeding depression. Evolution 45:622–629

Lynch M, O'hely M (2001) Captive breeding and the genetic fitness of natural populations. Conserv Genet 2:363–378

McGinnity P et al (2003) Fitness reduction and potential extinction of wild populations of Atlantic salmon, *Salmo salar*, as a result of interactions with escaped farm salmon. Proc R Soc Lond B: Biol Sci 270:2443–2450

McOrist S, Kitchener AC (1994) Current threats to the European wildcat, Felis silvestris, in Scotland. Ambio (Sweden)

Menotti-Raymond M et al (2003) Second-generation integrated genetic linkage/radiation hybrid maps of the domestic cat *(Felis catus)*. J Hered 94:95–106

Milan D et al (2000) A mutation in PRKAG3 associated with excess glycogen content in pig skeletal muscle. Science 288:1248–1251

Negro J, Torres M, Godoy J (2001) RAPD analysis for detection and eradication of hybrid partridges *(Alectoris rufa × A. graeca)* in Spain. Biol Cons 98:19–24

Nichols CR (1991) A comparison of the reproductive and behavioural differences in feral and domestic Japanese quail. University of British Columbia

Oliveira R, Godinho R, Randi E, Ferrand N, Alves PC (2008) Molecular analysis of hybridisation between wild and domestic cats *(Felis silvestris)* in Portugal: implications for conservation. Conserv Genet 9:1–11

Olsson M et al (2011) A novel unstable duplication upstream of HAS2 predisposes to a breed-defining skin phenotype and a periodic fever syndrome in Chinese Shar-Pei dogs. PLoS Genet 7:e1001332

Pierpaoli M et al (2003) Genetic distinction of wildcat (Felis silvestris) populations in Europe, and hybridization with domestic cats in Hungary. Mol Ecol 12:2585–2598

Price EO (2002) Animal domestication and behaviour. CABI Publishing, Wallingford

Price EO, King JA (1968) Domestication and adaptation. In: Hafez ESE (ed) Adaptation of domestic animals. Lea and Febiger, Philadelphia, pp 34–45

Randi E (2008) Detecting hybridization between wild species and their domesticated relatives. Mol Ecol 17:285–293

Randi E, Lucchini V (1998) Organization and evolution of the mitochondrial DNA control region in the avian genus Alectoris. J Mol Evol 47:449–462

Randi E, Lucchini V (2002) Detecting rare introgression of domestic dog genes into wild wolf *(Canis lupus)* populations by Bayesian admixture analyses of microsatellite variation. Conserv Genet 3:29–43

Randi E, Pierpaoli M, Beaumont M, Ragni B, Sforzi A (2001) Genetic identification of wild and domestic cats (*Felis silvestris*) and their hybrids using Bayesian clustering methods. Mol Biol Evol 18:1679–1693

Ren J et al (2011) A missense mutation in PPARD causes a major QTL effect on ear size in pigs. PLoS Genet 7:e1002043

Rhymer JM, Simberloff D (1996) Extinction by hybridization and introgression. Annu Rev Ecol Syst 27:83–109

Rose KM, Wodzicka-Tomaszewska M, Cumming R (1985) Agonistic behaviour, responses to a novel object and some aspects of maintenance behaviour in feral-strain and domestic chickens. Appl Anim Behav Sci 13:283–294

Roy MS, Geffen E, Smith D, Ostrander EA, Wayne RK (1994) Patterns of differentiation and hybridization in North American wolflike canids, revealed by analysis of microsatellite loci. Mol Biol Evol 11:553–570

Rubin C-J et al (2010) Whole-genome resequencing reveals loci under selection during chicken domestication. Nature 464:587–591

Schmutz S, Berryere T (2007) Genes affecting coat colour and pattern in domestic dogs: a review. Anim Genet 38:539–549

Schultz W (1969) Zur Kenntnis des Hallstrom-hunds (*Canis hallstromi, 1957*). Zool Anz 183:47–72

Statham M, Middleton M (1987) Feral pigs on Flinders Island. In: Papers and proceedings of the Royal Society of Tasmania, pp 121–124

Thomson VA et al (2014) Using ancient DNA to study the origins and dispersal of ancestral Polynesian chickens across the Pacific. Proc Natl Acad Sci 111:4826–4831

Van Laere A-S et al (2003) A regulatory mutation in IGF2 causes a major QTL effect on muscle growth in the pig. Nature 425:832–836

Verardi A, Lucchini V, Randi E (2006) Detecting introgressive hybridization between free-ranging domestic dogs and wild wolves (Canis lupus) by admixture linkage disequilibrium analysis. Mol Ecol 15:2845–2855

Wilmshurst JM, Hunt TL, Lipo CP, Anderson AJ (2011) High-precision radiocarbon dating shows recent and rapid initial human colonization of East Polynesia. Proc Natl Acad Sci 108:1815–1820. https://doi.org/10.1073/pnas.1015876108

Wright D et al (2010) The genetic architecture of domestication in the chicken: effects of pleiotropy and linkage. Mol Ecol 19:5140–5156

Wright D et al (2012) Onset of sexual maturity in female chickens is genetically linked to loci associated with fecundity and a sexual ornament. Reprod Domest Anim 47:31–36. https://doi.org/10.1111/j.1439-0531.2011.01963.x

Postglacial Colonization of Northern Europe by Reptiles

J. L. Horreo and P. S. Fitze

Abstract During the Last Glacial Maximum (LGM; 20–14 Kya ago), Northern Europe was covered by ice and permafrost and the distribution of many organisms contracted into glacial refugia. After the LGM, species started to colonize areas from which ice and permafrost retracted and Northern Europe was recolonized. The LGM affected past and present distributions of many species. Different mechanisms led to the currently observed distributions and phylogeographic patterns. However, little evidence exists for their importance in determining the currently existing phylogeographic structuring. Here, we compare the post-LGM colonization patterns of four terrestrial reptile species: two lizards (*Zootoca vivipara, Lacerta agilis*) and two snakes (*Vipera berus, Zamenis longissimus*). All four species exhibit large natural current distributions in Europe and colonized areas covered by ice and/or permafrost during LGM. The results show that the most important parameters promoting fast and large post-LGM colonisations are: (i) adaptations to cooler temperatures (including the evolution of viviparity), (ii) absence of physical or climatic barriers during expansion from the refugia, and (iii) low competition with other species/subspecies during expansion, i.e. a refugium at the edge of a species distribution that allows first colonization of newly available habitat.

1 Pleistocene Glaciations

During the Late Quaternary, repeated cycles of climatic warming and cooling existed in the Northern Hemisphere, producing in Europe the contraction and expansion of the Arctic and Alpine ice caps (Svendsen et al. 2004). Two major cold events (ice ages) produced enormous ice shields that extended from the North to Central Europe, covering large areas. South of these ice shields, glaciers covered several European mountain regions, namely Pyrenees, Alps, Carpathians, and Dinaric Alps (Ehlers

J. L. Horreo (✉) · P. S. Fitze
Department of Biodiversity and Evolutionary Biology, National Museum of Natural Sciences (MNCN-CSIC), C/José Gutiérrez Abascal 2, 28006 Madrid, Spain
e-mail: horreojose@gmail.com

© Springer International Publishing AG, part of Springer Nature 2018
P. Pontarotti (ed.), *Origin and Evolution of Biodiversity*,
https://doi.org/10.1007/978-3-319-95954-2_12

and Gibbard 2004; Hughes and Woodward 2017; Wallis et al. 2016). Moreover, the forefront of these ice sheets was covered by permafrost or perennially cryotic ground (Harris et al. 1988).

The first of the two above-mentioned Pleistocene ice ages occurred 74–60 thousand years (Kya) ago. During the first ice age, the southern limit of the ice shield neither reached the British Islands nor Central Europe (van Andel 2003). However, both areas were covered by ice during the last big ice age, which is generally referred to as Last Glacial Maximum (LGM; Clark et al. 2009). The LGM was dated 20–14 Kya ago. During the LGM, a latitudinal gradient of temperature existed across Europe, and winter soil temperatures were 2–4 °C cooler than today in Southern Europe and 10–20 °C cooler in Central and Northern Europe (Barron and Pollard 2002). During this epoch, permafrost extended from the northern ice shields down to 45° N in Northern Europe (Fig. 1; Vandenberghe et al. 2014; Zech 2012). Its southern limit was approximately in central France, Northern Italy, Slovenia, Central Hungary, Northern Romania, and southern Ukraine (Renssen and Vandenberghe 2003). While the southern permafrost limit reached the Alpine and Carpathian glaciers, it did

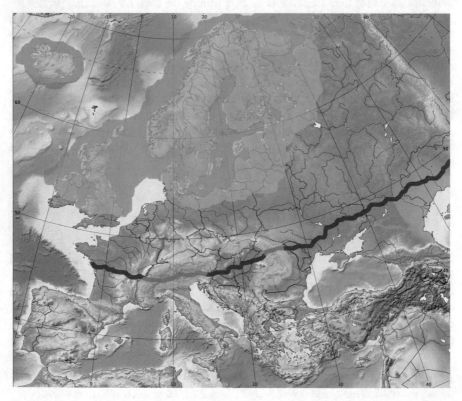

Fig. 1 Extent of continental ice sheets and mountainous glaciers (pale blue) during the Last Glacial Maximum (LGM) in Europe. The southern limit of the Last Permafrost Maximum (LPM) is shown in dark blue

not reach the Pyrenean, Apenninian, and Dynaric glaciers (Hughes and Woodward 2017).

2 LGM Refugia

The large extension of ice shields and permafrost importantly shaped species' distributions (Hewitt 2000), with many species contracting into glacial refugia. Glacial refugia are reduced species' distributions during cold periods that result from glacial expansion, which renders occupied habitat unsuitable and leads to the shrinking of the species' distribution (see Bennet and Provan 2008 for more specific details regarding the term "refugia").

During the LGM, southern European areas including parts of the Iberian Peninsula, Italy, the Pannonian Basin, the Balkans, and the Carpathian mountains remained ice-free (Ehlers and Gibbard 2004). These areas acted as refugia for many organisms that previously inhabited northern areas, and whose distributions retracted from the north due to the cooling climate and the advancing ice sheets and permafrost (e.g. Horreo et al. 2018). A great variety of animals retracted from the north. Vertebrate classes that were most affected include reptiles [e.g. the European pond terrapin, *Emys orbicularis* (Sommer et al. 2007), and the European common lizard *Zootoca vivipara* (Horreo et al. 2018)] and amphibians [e.g. the European pool frog, *Rana lessonae* (Zeisset and Beebee 2001)]. Species with higher tolerance to cold temperatures and those capable of living on permafrost did not necessarily retract into southern refugia, but remained in Northern Europe (Steward and Lister 2001). Among these were mammals (e.g. brown bear, *Ursus arctos* (Sommer and Benecke 2005) and birds [e.g. the common chaffinch *Fringilla coelebs* (Griswold and Baker 2002)]. Moreover, species inhabiting steppe habitats contracted into extrazonal refugia (Kajtoch et al. 2016). Thus, a general pattern of glacial refugia valid for all vertebrate species does not exist. The fact that tolerance to climatic cooling might explain different contraction-expansion patterns suggests that the distributions of species with similar ecological niches may contract into the same glacial refugia (e.g. Horreo et al. 2018).

However, not all species inhabiting a given biogeographic area occupied the same refugia (Hewitt 1996; Kajtoch et al. 2016; Stewart et al. 2010). This indicates that other factors such as competition among species and physical and climatic barriers are as well important (e.g. Horreo et al. 2018). These parameters may explain why some species inhabiting the same biogeographic area may exhibit the same geographic movements and share the same refugia, while movements of others species may be hindered or impeded, what results in incongruence in the location of the refugia (Anderson et al. 2002).

3   Post-LGM Colonization of Northern Europe

In the Northern Hemisphere, glaciations and inter-glacials led to south-northward contraction–expansion patterns (Hewitt 1996). During climatic cooling ice sheets and permafrost exhibited southward expansion, reducing the suitable habitat of thermophile species. Consequently, their distributions contracted into southern refugia. During inter-glacials, characterized by climatic warming, these species generally exhibited northward expansion and colonization of areas that became ice-free (Hewitt 1996). In contrast to species inhabiting temperate climates, the distributions of boreo-alpine and steppe species contracted during inter-glacials into refugia located in northern areas and extrazonal refugia, respectively, and they expanded their ranges during glacial periods (Kajtoch et al. 2016). This suggests that expansion also depended on a species' ecological niche and thus on the inhabited biogeographic area (Kajtoch et al. 2016; Stewart et al. 2010). Consequently, the current distributions of species and lineages may be the result of the combined action of the location of their refugia, the inhabited ecological niche, and the postglacial colonization patterns (Hewitt 2000; Taberlet et al. 1998).

There is evidence that postglacial colonization patterns were importantly affected by geographic barriers, including climatic, geological, and physical barriers. For example, big mountain ranges like the Pyrenees, the Alps, and the Carpathians may have acted as geographical barriers during colonization of the North (Smith et al. 2007). Such barriers left detectable genetic signatures. For example, they structured the genetic landscape of a species and they even led to speciation (e.g. *Vipera walser*; Ghielmi et al. 2016). Using genetic tools, such barriers can be traced retrospectively (Horreo et al. 2016; Wallis et al. 2016).

More specifically, the distributions of species that inhabited temperate regions contracted into different refugia during Pleistocene ice ages, and this led to reproductive isolation. Isolation promotes genetic divergence since isolation leads to the accumulation of different mutations and to differences in genetic drift, resulting in genetic differences among refugia (Hewitt 2000; Weiss and Ferrand 2007). Once the climate warmed, expansion from refugia happened and areas from which glaciers and permafrost retracted were colonized, promoting founder events that led to the biodiversity currently observed in Central and Western Europe (Wallis and Arntzen 1989; Hewitt 1996, 2000). The latitudinal retraction of central European ice shields led to latitudinal gradients of genetic diversity in species inhabiting temperate regions that were caused by founder effects and population expansion (Hewitt 2001). However, temperate species also occupied refugia in central (e.g. Austria, Slovenia, Hungary) and Eastern Europe (e.g. Deffontaine et al. 2005). During climatic warming, expansion also happened from these refugia which determined the current pattern of genetic diversity (e.g. Deffontaine et al. 2005).

4 Non-avian Sauropsids and LGM

Ectothermic species such as non-avian sauropsids (fomerly reptiles) were greatly affected by glacial contraction–expansion patterns because of their dependence on temperature. For example, the distribution of the Eurasian common lizard (*Zootoca vivipara*), the terrestrial reptile exhibiting the widest distribution in the world that also reaches farthest north (Hikida 2002), was strongly affected by glacials and inter-glacials (Horreo et al. 2018). *Zootoca vivipara* resists temperatures down to $-10\,°C$ in hibernacula (5–20 cm belowground), but it dies below this limit (Berman et al. 2016). Compared to other European reptiles, *Z. vivipara* exhibits a remarkable cold tolerance (Costanzo et al. 1995), but it cannot overwinter on permafrost, where temperatures are commonly lower than $-15\,°C$ at hibernacula depth (Berman et al. 2016). Non-avian terrestrial sauropsids are strongly affected by climatic changes (Sinervo et al. 2010) and they cannot survive on permafrost (at least not the species inhabiting Europe); thus, they are ideal organisms to understand how the LGM shaped their distributions and the current patterns of genetic diversity.

In order to provide evidence for or against different mechanisms leading to the currently observed distributional and genetic patterns, we compare four terrestrial reptile species exhibiting large distributions in Europe that include European regions that were covered by ice or permafrost during the LGM and for which the history of colonization is known. Two species are lizards with wide natural distributions ranging from Western Europe to East Asia: the Eurasian common lizard (*Zootoca vivipara*) and the sand lizard (*Lacerta agilis*). The other two are snakes: the common European viper (*Vipera berus*), which is the most widely distributed terrestrial snake species in the world, and the Aesculapian snake (*Zamenis longissimus*), which is a thermophilic species with a natural distribution ranging from Western Europe to the Caspian Sea.

4.1 The Eurasian Common Lizard (Zootoca vivipara Lichtenstein, 1823)

After the LGM, *Zootoca vivipara* colonized Northern Europe, and nowadays, it exhibits the most northerly and widest geographic distribution of any terrestrial reptile (Hikida 2002). Its natural distribution covers Eurasia and ranges from Ireland and North-Western Spain in the West to East Russia and Japan in the East and from Southern Bulgaria in the South to Northern Sweden in the North (Fig. 2). The species consists of six genetic lineages that are distributed across Eurasia [clades A to F; (Surget-Groba et al. 2006)]. Clade A (Eastern oviparous clade) is mainly located in Italy, Austria and Slovenia; clade B (Western oviparous clade) in Southern France and northern Spain; clade C (Central viviparous I clade) in Austria; clade D (Eastern viviparous clade) in Eastern Europe and Asia; clade E (Western viviparous clade) in

Fig. 2 Current natural distribution of *Zootoca vivipara* in Europe. Areas shaded with different colours represent the approximate areas inhabited by different clades (Horreo et al. 2018): green: clade A, blue: clade B, purple: clade C, orange: clade D, pink: clade E, and yellow: clade F

Western and Northern Europe; and clade F (Central viviparous II clade) in Austria and Hungary.

A recent study (Horreo et al. 2018) showed that this species inhabited different refuges during the LGM. These refuges were located in Western Europe (North, South, and West to the Pyrenees), Central Europe (Pannonian/Vienna Basin), Eastern Europe (East, West, and South of the Carpathian Mountains), North to the Black Sea, Southern Europe (Northern Italy), and probably another one in the north, south, or east of the Alps (Horreo et al. 2018). After LGM, each genetic lineage followed different colonization routes. Two general colonization patterns existed. First, the colonization of mountain areas such as the Cantabrian Mountains and the Pyrenees by clade B, the Alps by clades A, C, E, and F, and the Carpathian mountains by clades E and D. Second, two clades exhibited an enormous range expansion during which Northern Europe and Asia were colonized (Fig. 3). Clade E colonized the majority of Northern Europe, including the UK, Ireland, and the Scandinavian Peninsula. Clade D moved up North through Finland, reaching Northern Sweden, and it also exhibited an enormous expansion during which it colonized Eastern Europe, Asia,

Fig. 3 Most probable routes by which *Zootoca vivipara* colonized Northern Europe after the LGM. The Southern Last Permafrost Maximum (LPM) is shown in dark blue

and East Asia. Only two of the six clades exhibited massive northward expansion after glacial retraction and thawing of permafrost, while the other four clades colonized mountainous areas once ice coverage and permafrost disappeared.

4.2   The Common European Viper (**Vipera berus** *Linnaeus, 1758)*

The common viper is the world's most widely distributed terrestrial snake species. Its natural distribution ranges from Scotland in the West to Pacific Russia (Sakhalin) in the East, and from Northern Greece in the South to the Arctic Circle (Fig. 4). *Vipera berus* is a cold-tolerant reptile that can survive, a couple of hours in a frozen state, but not prolonged freezing (over 1 day) (Andersson and Johansson 2001), and thus, *Vipera berus* has been classified as being non-freeze tolerant. Nevertheless, it inhabits areas up to 2,600 m (Gasc et al. 1997), where precise choice of hibernation

Fig. 4 Current natural distribution of *Vipera berus* in Europe. Areas shaded with different colours represent the approximate areas inhabited by different clades (Ursenbacher et al. 2006): green: Italian clade, pink: Balkan clade, all other colours belong to the Northern clade that consists of four subclades: brown: Carpathian subclade, yellow: Central European subclade; red: Eastern subclade; and orange: Western subclade

sites is probably key to winter survival (Andersson and Johansson 2001). *Vipera berus* consists of three main genetic lineages (Ursenbacher et al. 2006): the Italian clade (Italy, northern Slovenia, Austria, south-eastern Switzerland), the Balkan clade, and the Northern clade (the northern, eastern and western Alps, Northern Europe, Asia). The Northern clade consists of four subclades: Carpathian (Romania, eastern Slovakia and south-eastern Poland), Eastern (northern Slovakia, Estonia, Finland, Russia), Central European (UK, Netherlands, Germany, Czech Republic, Denmark, Sweden Norway), and Western (Massif Central and Northern France, Switzerland, Austria).

Several LGM refugia have been detected for *Vipera berus* (Ursenbacher et al. 2006). Refugia were located in Italy, on the Balkans, in Eastern Europe (near the Carpathian Mountains, and in the east of the Carpathians), in France, and possibly in Hungary. As in the European common lizard, two of the six genetic lineages (subclades in this case) colonized the majority of Northern Europe (Fig. 5; Ursenbacher

Fig. 5 Routes by which *Vipera berus* colonized Northern Europe after the LGM. The Southern Last Permafrost Maximum (LPM) is shown in dark blue

et al. 2006) after glacial retraction and thawing of the permafrost. The Eastern sub-clade colonized the Fenno-Scandinavian Peninsula by moving north through Finland, and it colonized also Eastern Europe, Asia, and East Asia. The Central European sub-clade colonized Central and Northern Europe (including Czech Republic, Germany, the Netherlands, Great Britain, Denmark, Sweden, and Norway). The Western sub-clade also exhibits postglacial expansion, but to a much smaller geographic extent. This clade colonized northern France and Switzerland most likely from a refugium located in Southern France. The other clades mainly remained in areas that were not covered by permafrost during the LGM. One clade remained in Northern Italy and Slovenia, and two clades remained on the Balkans, one on the eastern Balkan (Romania) and the other one colonized the Dinaric Alps from the Southern Balkans.

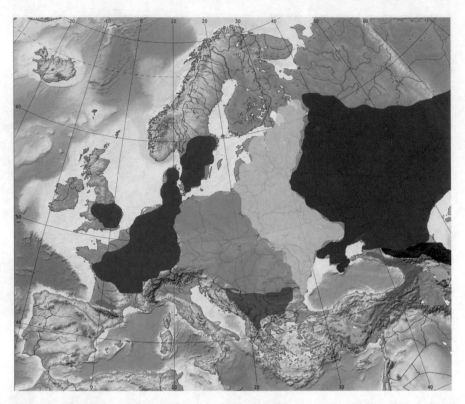

Fig. 6 Current natural distribution of *Lacerta agilis* in Europe. Areas shaded with different colours represent the approximate areas inhabited by different subspecies (Andres et al. 2014): dark blue: *L. a. agilis*, pale blue: *L. a. argus*, yellow: *L. a. chersonensis*, red: *L. a. exigua*, pink: *L. a. garzoni*, dark grey: *L. a. bosnica*, purple: *L. a. tauridica*, orange: *L. a. grusinica*, black: *L. a. boemica*, dark green: *L. a. brevicaudata*, and pale green: *L. a. iorinensis*

4.3   The Sand Lizard (Lacerta agilis Linnaeus, 1758)

The sand lizard inhabits large parts of Europe, and it is the terrestrial reptile with the second largest natural distribution (Roitberg et al. 2015). It inhabits Western France in the West, North-Western China/North-Western Mongolia in the East, Greece/Armenia in the South and Southern Sweden/South of the Russian Republic of Karelia in the North (Fig. 6). It consists of eleven subspecies (Andres et al. 2014; Bischoff 1998). *L. a. agilis, L. a. argus, L. a. chersonensis, and L. a. exigua* inhabit northern European areas (Fig. 6), while the other seven inhabit the Pyrenees, the southern Balkans, and Northern Caucasus (*L. a. garzoni, L. a. bosnica, L. a. tauridica, L. a. grusinica, L. a. boemica, L. a. brevicaudata, L. a. iorinensis*) in the South.

   In this species, LGM refugia were located in Crimea, South Caucasus, the Pannonian Basin, and the Balkans (Bischoff 1998; Zinenko et al. 2005). After the LGM,

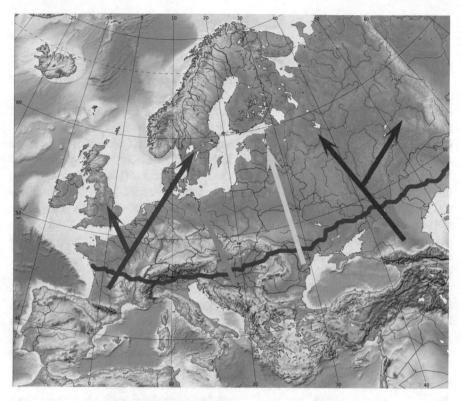

Fig. 7 Routes by which *Lacerta agilis* colonized Northern Europe after the LGM. The Southern Last Permafrost Maximum (LPM) is shown in dark blue

four subspecies colonized northern European areas from refugia located south of the Last Permafrost Limit: *L. a. agilis* colonized Western Europe, Great Britain, Denmark and Southern Sweden most likely from a refuge located in Southern France; *L. a. argus* colonized Slovakia, Czech Republic Germany, and Poland from a refuge most likely located in or close to the Pannonian Basin (Bischoff 1998; Kalyyabina-Hauf et al. 2004); *L. a. chersonensis* colonized the East of the Carpathian mountains, Belarus, Lithuania, Latvia, Estonia, and the westernmost provinces of Russia adjacent to Latvia, Estonia, and Finland; *L. a. exigua* colonized North-Eastern Europe, and Asia most likely from a refuge located in the Caucasus area (Fig. 7). The other seven subspecies inhabit areas located south of the LPM, and they did not exhibit huge range expansions.

4.4   The Aesculapian Snake (Zamenis longissimus *Laurenti,* 1768)

The Aesculapian snake is a thermophile reptile that prefers warm but not hot and moderately humid conditions. It avoids dry habitats and the limit of its southern distribution coincides with the Southern boundary of the deciduous broadleaf forest (Gomille 2002). Its current natural distribution ranges from Brittany in Western France to North-Western Iran and Azerbaijan in the East, and from Greece and North-Western Iran in the South to Northern France and central Germany in the North (Fig. 8). The Aesculapian snake consists of four clades (Musilová et al. 2010): the Western clade inhabits Northern Spain, France, Switzerland, Northern and Central Italy, Slovenia, and the westernmost stripe of the Balkans down to Greece; the Eastern clade inhabits the central and Eastern Balkans from Bulgaria to Croatia, Hungary, Slovakia, Czech Republic, and Germany); the Greek clade exclusively inhabits Greece, and the Asian clade inhabits Turkey, North-Western Iran, Azerbaijan, Georgia, and the Republic of Adygea, an adjacent Russian Oblast.

In *Z. longissimus,* two LGM refugia have been proposed: one on the Balkans and another in Western Europe (probably in the south of France) (Musilová et al. 2010). After the LGM, the Western and Eastern clades colonized important European Areas north of the LPM (Musilová et al. 2010). The Western clade exhibited northwestward expansion towards North-Western France and it colonized part of the Alps and the exact location of the refugium is unclear (Fig. 9). Three possibilities exist: the refugium was located on the Balkans, on the Italian Peninsula, or on the Spanish Peninsula. In contrast, the Eastern clade colonized northern areas most likely from a refugium located on the Balkans (e.g. in Greece). This clade expanded northwards up to Germany and probably until Denmark (Musilová et al. 2010).

5   Colonization of Northern Europe After the LGM by Reptiles

Vipera berus and *Zootoca vivipara* exhibited very similar genetic structuring in Northern Europe and the routes of colonization of northern European are highly congruent as well (Figs. 3 and 5). In both species, two different genetic clades colonized Northern Europe, Asia, and East Asia (Figs. 3 and 5). In each species, one clade colonized Central and Western Europe from the north of the Balkans, and the other clade colonized Eastern Europe, Asia, and East Asia from the north of the Black Sea. Both, *Z. vivipara* and *V. berus*, are more cold-tolerant than *L. agilis* and *Z. longissimus* and as well more cold-tolerant than other European terrestrial Sauropsids (Andersson and Johansson 2001; Voituron et al. 2002). While *Z. vivipara* exhibits freeze tolerance down to $-10°$, *V. berus* is not freeze tolerant, but survives a couple of hours in a frozen state. Moreover, in both species the clades with large North and Eastward expansions are ovoviviparous, what allows them to behaviourally heat

Fig. 8 Current natural distribution of *Zamenis longissimus* in Europe. Areas shaded with different colours represent the approximate areas inhabited by different clades (Musilová et al. 2010): red: Western clade, green: Eastern clade, blue: Greek clade, and yellow: Asian clade

their eggs by exposing themselves to warm places, a behaviour that may favour rapid expansion once winter temperatures rise to levels that they can withstand. In contrast, the distribution of *Podarcis muralis*, a species whose current distribution importantly overlaps with that of *Z. vivipara* and *V. berus*, does not reach more northern latitudes. *P. muralis* as well survives freezing over longer time scales (Claussen et al. 1990), but it is strictly oviparous and thus requires higher average daily temperatures for successful egg incubation, compared to the two viviparous species. This suggests that ovoviviparity, but not necessarily freeze tolerance was the key characteristic that allowed these two species to be the first to colonize areas from which ice and permafrost retracted after the LGM. As a consequence, they exhibited huge and fast north and eastwards range expansions, while northward expansions of more thermophile species or oviparous species were much smaller. Being the first to colonize newly available areas also means that they will colonize these areas in absence of competition with other reptiles inhabiting similar ecological niches, what increases their colonization success (Horreo et al. 2018).

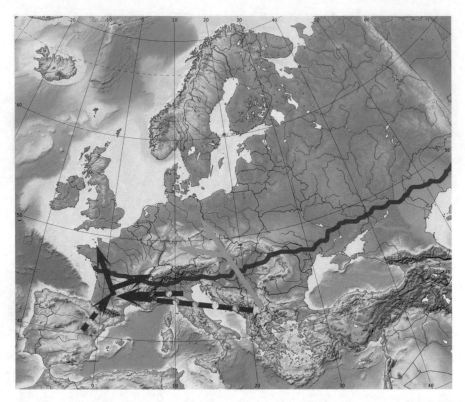

Fig. 9 Routes by which *Zamenis longissimus* colonized areas covered by ice and permafrost during the LGM. The Southern Last Permafrost Maximum (LPM) is shown in dark blue. Dashed arrows indicate potential, not confirmed refugia

The sand lizard (*Lacerta agilis*) exhibits a similar distribution as *Z. vivipara* and *V. berus*, but in contrast to those it does not inhabit Northern Europe above 61° N. *L. agilis* is not freeze tolerant since only very few specimens survive short periods slightly below 0 °C (Weigmann 1929). *L. agilis* has oviparous reproduction, suggesting again that the reproductive mode rather than freeze tolerance allow for the colonization of high northern latitudes. In contrast to *P. muralis*, *L. agilis* also exhibits important eastward expansion ranging until north-western Mongolia. The colonization patterns of north European areas and those of East Asia are very similar to those of *Z. vivipara* and *V. berus* (Figs. 3, 5 and 7). However, the current phylogeographic structure importantly differs from the other two species. While in *Z. vivipara* and *V. berus* two clades colonized Northern Europe, Asia, and East Asia, in *L. agilis* four clades colonized the same regions. The four clades inhabited four different refugia during the LGM that were most likely located in Southern France, on the Northern Balkans, North of the Black Sea and in the Caucasus area (Andres et al. 2014). Interestingly, the sand lizard, but not *V. berus* and *Z. vivipara*, inhabits the Caucasus area (Fig. 6), and it is more thermophile than the other two species.

Moreover, the sand lizard's ecological niche includes drier habitats than the ecological niche of the other two species and its origin was dated in the Late Miocene–Early Pliocene and located in the Caucasian region (Kalyyabina-Hauf et al. 2004). The clade that colonized Asia corresponds to one of the clades that had a refugia in the Caucasus area (Andres et al. 2014). This suggests that the precise location of the refugia might be a crucial determinant of the expansion success. More evidence for this theory stems from the other three *L. agilis* clades. All three clades, exhibited northward expansions that are congruent with the retraction of the ice and the permafrost. While *L. a. chersonensis* (Fig. 6 yellow area) inhabits a south–north corridor starting from the southern refugia, the expansions of *L. a. agilis* and *L. a. argus* were northwards, but as well East and Westwards, respectively. Both patterns can be explained by the lack of *L. agilis* in Italy and by the Alps that formed a barrier during colonization. In both subspecies, expansion first happened northwards until encountering suitable habitat North of the Central Cordillera of the Alps, from where longitudinal eastward and westward expansion was possible, most likely until entering into contact with the other subspecies. Competition with other subspecies also explains why *L. a. chersonensis* only expanded northwards and why *L. a. exigua* did not colonize Central and Western Europe despite its enormous success in colonizing Asia. Similar patterns also exist in *Z. vivipara* and *V. berus*, both exhibiting almost perfect South-North limits between the two clades exhibiting the most important northward expansions (Figs. 2 and 4). Consequently, it is very likely that the current distributions are shaped by intense competition among clades and subspecies.

In contrast to these species, the Aesculpian snake (*Zamenis longissimus*) inhabits warmer environments and it is also oviparous. Its colonization of Northern Europe started from a refuge located on the Balkans (Musilová et al. 2010) and most likely later than the colonization by the more cold-tolerant species. Consequently, its expansion was not as wide as that of the other species (Gomille 2002).

Here, we described how several reptiles species colonized northern European areas, Asia, and East Asia after the Last Glacial Maximum (since 21'000 Kya ago; Fig. 1). While some colonization and expansion patterns are similar among species, others differ, allowing to disentangle among hypothesis explaining colonization success. The most important detected parameters promoting fast north- and eastward range expansion are: (1) adaptations to cooler temperatures allowing to colonize further north and the evolution of viviparity that allows for successful reproduction even if substrate temperatures for egg incubation are too low (see *Z. vivipara* and *V. berus*); (2) the location of the refugia is important, since northwards expansion can be hindered by physical or climatic barriers; (3) the location of the refugia can also provide competitive advantages. A refugia from where direct northward expansion is possible without entering into competition with other clades and subspecies is key for wide geograpic expansion for wide expansions (see all described species). Simultaneously northwards expanding clades hinder longitudinal expansion, what explains why only clades in the east or west of the species' distribution to exhibit wide eastward or westward expansions. Similarly, clades that inhabited areas south of another clade or whose direct northward expansion is hindered by physical barriers (in *Z. vivipara* clade A, C, F; in *V. berus,* Italian, Balkan, and Western clade/subclade;

in *L. agilis* the following subspecies: *L. a. garzoni*, *L. a. bosnica*, *L. a. tauridica*, *L. a. grusinica*, *L. a. boemica*, *L. a. brevicaudata*, and *L. a. iorinensis*; and in *Z. longissimus,* the Greek clade) do not exhibit important range expansions, most likely due to competition with the fast expanding clades. This suggests that certain adaptations allow for the fast colonization of newly available habitat and that the precise location of the inhabited areas is crucial since range expansion follows a first-come first-served pattern.

References

Anderson RP, Peterson AT, Gómez-Laverde M (2002) Using niche-based GIS modeling to test geographic predictions of competitive exclusion and competitive release in South American pocket mice. Oikos 98:3–16

Andersson S, Johansson L (2001) Cold hardiness in the boreal adder, *Vipera berus*. Cryo Lett 22:151–156

Andres C, Franke F, Bleidorn C, Bernhard D, Schlegel M (2014) Phylogenetic analysis of the *lacerta agilis* subspecies complex. Syst Biodivers 12:43–54

Barron E, Pollard D (2002) High-resolution climate simulations of oxygen isotope stage 3 in Europe. Quat Res 28:296–309

Bennet K, Provan J (2008) What do we mean by "refugia"? Quat Sci Rev 27:2449–2455

Berman DL, Bulakhova NA, Alfimov AV, Meshcheryakova EN (2016) How the most northern lizard, *Zootoca vivipara*, overwinters in Siberia. Polar Biol 39:2411–2425

Bischoff W (1998) Zur verbreitung und systematik der zaunei-dechse, *Lacerta agilis* linnaeus, 1758. Mertensiella 1:11–30

Clark PU, Dyke AS, Shakun JD, Carlson AE, Clark J et al (2009) The last glacial maximum. Science 325:710–714

Claussen DL, Townsley MD, Bausch RG (1990) Supercooling and freeze-tolerance in the European wall lizard, *Podarcis muralis*, with a revisional history of the discovery of freeze-tolerance in vertebrates. J Comp Physiol B 160:137–143

Costanzo JP, Grenot C, Lee RE (1995) Supercooling, ice inoculation and freeze tolerance in the European common lizard, *Lacerta vivipara*. J Comp Physiol B 165:238–244

Deffontaine V, Libois R, Kotlík P, Sommer R, Nieberding C et al (2005) Beyond the Mediterranean Peninsulas: evidence of central European glacial refugia for a temperate forest mammal species, the bank vole (*Clethrionomys glareolus*). Mol Ecol 14:1727–1739

Ehlers J, Gibbard PL (2004) Quaternary glaciations-extent and chronology, Part I: Europe. Elsevier

Gasc JP, Cabela A, Crnobrnja-Isailovic J, Dolmen D, Grossenbacher K et al (1997) Atlas of amphibians and reptiles in Europe. Societas Europaea Herpetologica, Paris

Ghielmi S, Menegon M, Marsden SJ, Laddaga L, Ursenbacher S (2016) A new vertebrate for Europe: the discovery of a range-restricted relict viper in the western Italian. J Zool Syst Evolut Res 54:161–173

Gomille A (2002) Die äskulapnatter elaphe longissima – verbreitung und lebensweise in mitteleuropa. Chimaira, Frankfurt am Main, Germany

Griswold CK, Baker AJ (2002) Time to the most recent common ancestor and divergence times of populations of common chaffinches (*Fringilla coelebs*) in Europe and North Africa: insights into pleistocene refugia and current levels of migration. Evolution 56:143–153

Harris SA, French HM, Heginbottom JA, Johnston GH, Ladanyi B et al (1988) Glossary of permafrost and related ground-ice terms. Ontario, Canada

Hewitt GM (1996) Some genetic consequences of ice ages, and their role in divergence and speciation. Biol J Lin Soc 58:247–276

Hewitt GM (2000) The genetic legacy of the quaternary ice ages. Nature 405:907–913

Hewitt GM (2001) Speciation, hybrid zones and phylogeography—or seeing genes in space and time. Mol Ecol 10:537–549

Hikida H (2002) Natural history of the reptiles. Tokyo

Horreo JL, Jiménez-Valverde A, Fitze PS (2016) Ecological change predicts population dynamics and genetic diversity over 120,000 years. Glob Change Biol 22:1737–1745

Horreo JL, Peláez ML, Suárez T, Breedvled MC, Heulin B, et al (2018) Phylogeography, evolutionary history, and effects of glaciations in a species (*Zootoca vivipara*) inhabiting multiple biogeographic regions. J Biogeogr 45:1616–1627

Hughes PD, Woodward JC (2017) Quaternary glaciation in the Mediterranean mountains: a new synthesis. In: Hughes PD, Woodward JC (eds) Quaternary glaciation in the Mediterranean mountains. Geological Society, London, UK, pp 1–23

Kajtoch L, Cieslak E, Varga Z, Paul W, Mazur MA et al (2016) Phylogeographic patterns of steppe species in eastern central Europe: a review and the implications for conservation. Biodivers Conserv 25:2309–2339

Kalyyabina-Hauf SA, Milto KD, Ananjeva NB, Joger U, Kotenko TI et al (2004) Reevaluation of the status of *Lacerta agilis tauridica* suchov, 1926. Russ J Herpetol 11:65–72

Musilová R, Zavadil V, Marková S, Kotlík P (2010) Relics of the Europe's warm past: phylogeography of the Aesculapian snake. Mol Phylogenet Evol 57:1245–1252

Renssen H, Vandenberghe J (2003) Investigation of the relationship between permafrost distribution in NW Europe and extensive winter sea-ice cover in the North Atlantic Ocean during the cold phases of the last glaciation. Quat Sci Rev 22:209–223

Roitberg ES, Eplanova GV, Kotenko TI, Amat F, Carretero MA et al (2015) Geographic variation of life-history traits in the sand lizard, *Lacerta agilis*: testing Darwin's fecundity-advantage hypothesis. J Evol Biol 28:613–629

Sinervo B, Méndez-de-la-Cruz F, Miles DB, Heulin B, Bastiaans E et al (2010) Erosion of lizard diversity by climate change and altered thermal niches. Science 328:894–899

Smith W, Scott IAW, Keogh JS (2007) Molecular phylogeography of Rosenberg's goanna (reptilia:Varanidae: *Varanus rosenbergi*) and its conservation status in New South Wales. Syst Biodivers 5:361–369

Sommer RS, Benecke N (2005) The recolonization of Europe by brown bears *Ursus arctos* linnaeus, 1758 after the last glacial maximum. Mamm Rev 35:156–164

Sommer RS, Persson A, Wieseke N, Fritz U (2007) Holocene recolonization, extinction of the pond turtle, *Emys orbicularis* (l., 1758), in Europe. Quat Sci Rev 26:3099–3107

Steward JR, Lister AM (2001) Cryptic northern refugia and the origins of the modern biota. Trends Ecol Evol 16:608–613

Stewart JR, Lister AM, Barnes I, Dalen L (2010) Refugia revisited: individualistic responses of species in space and time. Proc R Soc B-Biol Sci 277:661–671

Surget-Groba Y, Heulin B, Guillaume CP, Puky M, Semenov D et al (2006) Multiple origins of viviparity, or reversal from viviparity to oviparity? The European common lizard (*Zootoca vivipara*, lacertidae) and the evolution of parity. Biol J Lin Soc 87:1–11

Svendsen JI, Alexanderson H, Astakhov VI, Demidov I, Dowdeswell JA et al (2004) Late quaternary ice sheet history of Northern Eurasia. Quat Sci Rev 23:1229–1271

Taberlet P, Fumagalli L, Wust-Saucy AG, Cosson JF (1998) Comparative phylogeography and postglacial colonization routes in Europe. Mol Ecol 7:453–464

Ursenbacher S, Carlsson M, Helfer V, Tegelström H, Fumagalli L (2006) Phylogeography and pleistocene refugia of the adder (*Vipera berus*) as inferred from mitochondrial DNA sequence data. Mol Ecol 15:3425–3437

van Andel TH (2003) Glacial environments I: The Weichselian climate in Europe between the end of the OIS-5 interglacial and the last glacial maximum. In: van Andel TH, Davies W (eds) Neanderthals and modern humans in the European landscape during the last glaciation: archaeological results of the stage 3 project. Short Run Press, Exeter, UK

Vandenberghe J, French HM, Gorbunov A, Marchenko S, Velichko AA et al (2014) The last permafrost maximum (LPM) map of the Northern Hemisphere: permafrost extent and mean annual air temperatures, 25–17 ka bp. Boreas 43:652–666

Voituron Y, Storey J, Grenot C, Storey K (2002) Freezing survival, body ice content and blood composition of the freeze-tolerant European common lizard, *Lacerta vivipara*. J Comp Physiol B 172:71–76

Wallis GP, Arntzen JW (1989) Mitochondrial-DNA variation in the crested newt super species: limited cytoplasmic gene flow among species. Evolution 43:88–104

Wallis GP, Waters JM, Upton P, Craw D (2016) Transverse alpine speciation driven by glaciation. Trends Ecol Evol 31:916–926

Weigmann R (1929) Die wirkung starker abküihlung auf amphibien und reptilien. Zeitschrift für wissenschaftliche Zoologie 24:641–692

Weiss S, Ferrand N (2007) Phylogeography of Southern European Refugia. Springer, Dordretch, The Netherlands

Zech R (2012) A permafrost glacial hypothesis—permafrost carbon might help explaining the pleistocene ice ages. Quat Sci J 61:84–92

Zeisset I, Beebee TJC (2001) Determination of biogeographical range: an application of molecular phylogeography to the European pool frog *Rana lessonae*. Proc R Soc B-Biol Sci 268:933–938

Zinenko OI, Drabkin PL, Rudyk OM (2005) Contact zone between two subspecies of the sand lizard: *Lacerta agilis exigua* eichw., 1831 and *Lacerta agilis chersonensis* andr., 1832 in three regions of the Left-Bank Ukraine. Russ J Herpetol 23:109–112

The Relative Roles of Selection and Drift in Phenotypic Variation: Some Like It Hot, Some Like It Wet

David S. Jacobs and Gregory L. Mutumi

Abstract Acoustic signals mediate important functions, e.g. orientation, foraging and communication, that impact on the survival and reproduction of animals. The propagation of acoustic signals is also known to be influenced by habitat, particularly differences in climate. It is therefore likely that the environment would exert significant influence on such signals and that selection rather than drift would be largely responsible for geographic variation in acoustic signals. We investigated the role of selection and drift in geographic variation in the echolocation of two species of horseshoe bats *Rhinolophus damarensis* and *R. clivosus* (Rhinolophidae) with wide geographic distributions across the arid and mesic biomes of southern Africa. In both species, selection was found to be the dominant evolutionary process influencing phenotypic variation; however, there was evidence of drift in *R. clivosus*. Furthermore, selection was not differentially exerted across populations because there was no change in the results when localities were excluded one at a time. Population divergence appeared to be mediated by selection on traits associated with manoeuvrability, detection and size in both species despite their disparate distributions. However, the climatic factor that best explained geographic variation in echolocation was dependent on the biomes occupied by the species. Temperature was the dominant climatic factor in *R. damarensis*, a species with a largely arid distribution. In *R. clivosus*, a species with distributions across both mesic and arid biomes, temperature and relative humidity together explained variation in echolocation.

D. S. Jacobs (✉) · G. L. Mutumi
Department of Biological Sciences, University of Cape Town, Cape Town, South Africa
e-mail: david.jacobs@uct.ac.za

Present Address
G. L. Mutumi
School of Natural Sciences, University of California, Merced, CA, USA
e-mail: gmutumi@gmail.com

© Springer International Publishing AG, part of Springer Nature 2018
P. Pontarotti (ed.), *Origin and Evolution of Biodiversity*,
https://doi.org/10.1007/978-3-319-95954-2_13

1 Introduction

Species with wide geographic distributions that span several different habitats and biomes may be subjected to a variety of evolutionary processes (e.g. selection, drift) and may experience varying degrees of isolation. Phenotypic divergence among populations in such widely distributed species may be the result of selection, founder effect or drift or a combination of some or all of these evolutionary processes acting separately, simultaneously or sequentially. The effects of these processes may also be synergistic. For example, increased isolation, i.e. gene flow is restricted, would enhance the effects of both adaptation and drift, especially when founder populations are small, accelerating the rate at which populations diverge (Wright 1943; Malhotra and Thorpe 2000; Millstein 2002; Morrone 2009). If we are to understand phenotypic evolution and the consequent lineage divergence which generates biodiversity, the relative contributions of these processes to divergence must be teased apart (Orr and Smith 1998; Coyne and Orr 2004). Most studies on phenotypic divergence conclude that selection is responsible (e.g. Weaver et al. 2007; Mutumi et al. 2016; Maluleke et al. 2017) but evidence for drift is gradually accumulating (e.g. Ackermann and Cheverud 2002, 2004; Weaver et al. 2007; Betti et al. 2010; Smith 2011; de Azevedo et al. 2015). Surprisingly, few studies have attempted to consider the relative roles of drift and selection on phenotypic divergence within the same system. This is most likely due to the controversy around both the significance of drift to biological diversification and whether or not it can be distinguished from adaptation (Brandon and Carson 1996; Millstein 2002; Brandon 2005). However, the dearth of studies that simultaneously investigate the influence of both drift and selection is of concern because the influence of selection on phenotypic divergence can be overestimated if the effects of drift are not considered (Marroig and Cheverud 2004; Betti et al. 2010).

Nevertheless, phenotypic traits that perform crucial survival and reproductive functions (Mutumi et al. 2017) and whose function is intimately integrated with environmental factors (e.g. climate) are more likely to be influenced by selection than drift (however see Betti et al. 2010).

Acoustic signals have severe fitness consequences because they are involved in several important life-history functions including orientation, foraging and communication. Furthermore, the propagation of acoustic signals is also known to be influenced by environmental factors, particular climate (Mutumi et al. 2016; Jacobs et al. 2017). It is therefore likely that the environment would exert significant influence on such signals and that selection rather than drift would be largely responsible for divergence in acoustic signals (Kirschel et al. 2011; Sun et al. 2013; Mutumi et al. 2016, 2017).

Although the most direct evidence for random genetic drift could be obtained using genetic approaches (Lande 1976; Leinonen et al. 2008; Rogell et al. 2010; Sun et al. 2013), drift should also be evident in the phenotypes of species. Several methods have been developed to test for selection within phenotypic variation against the null model of drift, for example, the rate test (Turelli 1988) and an adaptation of

Lande's (1976) quantitative genetic model (Ackermann and Cheverud 2002, 2004; de Azevedo et al. 2015; Mutumi et al. 2017).

Echolocation in bats is an acoustic-based orientation and prey capture system (Griffin 1958) that may also be used as a communication signal to discriminate conspecifics from heterospecifics (Schuchmann and Siemers 2010; Bastian and Jacobs 2015; Finger et al. 2017) and to choose mates (Puechmaille et al. 2014). Echolocation also forms adaptive complexes with other traits (Norberg and Rayner 1987) involved in flight (wings) and the detection, capture and mastication of prey (skulls). Such adaptive complexes are evident in both mammals and birds (Freeman and Lemen 2010; Jacobs et al. 2014) and are indicative of selection being the main determinant of phenotypic divergence. However, drift was much more important than climate in the evolution of skull variation in humans (Marroig and Cheverud 2004; Betti et al. 2010), as well as in the evolution of acoustic signals in Neotropical singing mice (Campbell et al. 2010), anurans (Ohmer et al. 2009) and in birds (Irwin et al. 2008). The single study (Mutumi et al. 2017) on two species of bats, *R. simulator* and *R. swinnyi*, that considered the influence of both drift and selection on phenotypic variation found no support for drift and concluded that selection was the dominant evolutionary process shaping variation in traits that impact heavily on fitness. However, this might not be true for all species of bats or for all populations within a species. For example, although drift generally exerted more influence on the evolution of human crania, in colder climates, selection was the dominant process (Betti et al. 2010). It is therefore possible that in widely distributed bat species drift may nevertheless be evident in some, if not all, populations.

Although bats are volant, their dispersal ability may be limited by distance or barriers to dispersal such as mountain ranges, water bodies and extensive human development and gene flow among populations of a species may be restricted (Moussy et al. 2013). Furthermore, in combination with such isolation, many bat species have populations that consist of tens or hundreds of individuals and drift may therefore play a role in the evolution of phenotypic traits in such small isolated populations (Whitlock 2000) even if those traits have fitness implications.

Southern African horseshoe bats (Rhinolophidae) have wide geographic distributions across spatially heterogeneous environments (Csorba et al. 2003; Monadjem et al. 2010). They also vary in population size from relatively small (tens of individuals) to relative large (thousands of individuals) as well as in body size, dispersal ability, degree of philopatry (Kunz and Parsons 2009) and wing and echolocation parameters (Jacobs et al. 2016). They are therefore ideal for testing the relative roles of drift and selection in phenotypic variation.

We investigated the relative contributions of selection and drift in phenotypic divergence associated with wing, skull (head length) and echolocation characteristics in two species of horseshoe bats, *Rhinolophus damarensis* and *R. clivosus* which differ in size, echolocation frequency and geographic distribution: *R. damarensis* had a largely arid distribution in the western half of southern Africa, and *R. clivosus* had a largely mesic distribution in the eastern half of southern Africa but extended into the dry western part of South Africa (Fig. 1). Using Lande's model (Lande 1976) adapted by Ackermann and Cheverud (2002) for phenotypic traits, we tested

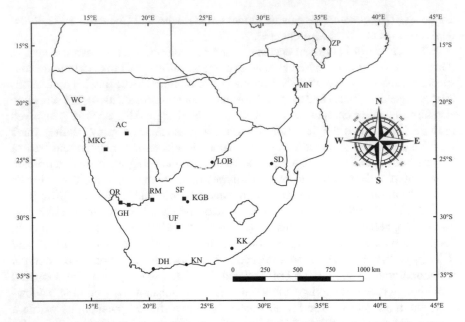

Fig. 1 Sampling localities for *Rhinolophus damarensis* and *R. clivosus* in southern Africa. Key to abbreviations: *R. damarensis* localities: WC = Wondergat Cave (20.51° S, 14.37° E), AC = Arnhem Cave (22.7° S, 18.1° E), MCK = Märcker Cave (24.0° S, 16.28° E), OR = Orange River (28.7° S, 17.54° E), GH = Goodhouse (28.9° S, 18.25° E), RM = Riemvasmaak (28.47° S, 20.29° E), SF = Soetfontein (28.38° S, 23.05° E), and UF = Untjiesburg Farm (30.83° S, 22.54° E). *R. clivosus* localities: ZP = Zomba Plateau (15.33° S, 35.28° E), MN = Monaci Mine (18.88° S, 32.72° E), LOB = Lobatse (25.24° S, 25.51° E), SUD = Sudwala (25.38° S, 30.69° E), KGB = , Koegelbeen (28.65° S, 23.35° E), KK = Kokstad (32.68° S, 27.19° E), KN = Knysna (34.06° S, 23.22° E), DH = De Hoop Nature Reserve (34.42° S, 20.35° E)

the hypothesis that selection rather than drift should be the predominant process in the evolution of traits associated with flight and sensory systems because, to be functional, these traits have to comply with the physical laws of aerodynamics and signal propagation. We evaluated the following prediction: Lande's model would yield signals of selection through the rejection of the null model of drift for traits associated with body size, flight and echolocation.

2 Methods

2.1 Study Animals

Rhinolophus damarensis (Jacobs et al. 2013) is a small insectivorous bat (10 g; Jacobs et al. 2016) with a relatively high echolocation frequency while at rest, its resting frequency (RF), of 85 kHz (Jacobs et al. 2013). It has a wide distribution in the western half of southern Africa stretching from western South Africa through Namibia to southwest Angola (Jacobs et al. 2013). This region is characterized by mostly arid conditions ranging from desert to Nama-Karoo and arid savanna (Jacobs et al. 2017).

Rhinolophus clivosus is a medium-sized bat with a mass of around 18 g and a RF of about 92 kHz (Jacobs et al. 2016). It is insectivorous and forages in and around dense foliage less than 1 m from the ground and vegetation (Neuweiler 1989; Jacobs et al. 2007). Although widespread across Africa and the Arabian Peninsula, populations of *R. clivosus* in eastern and southern Africa form a single clade distinct from populations to the north (Dool et al. 2016). We focused on this clade which is distributed from Kenya, along the eastern half of the continent into northern South Africa and along the coast of southern and western South Africa (Fig. 1). Its distribution covers a range of biomes including deserts, savannah woodlands and fringes of forests with climates that range from arid and tropical biomes with summer rainfall to Mediterranean biomes with winter rainfall (Csorba et al. 2003).

Like all other, rhinolphids *R. damaranesis* and *R. clivosus* uses high duty cycle (HDC) echolocation which means that the duration of their echolocation pulse is high relative to the time between the onset of successive pulses (Fenton 1999). Their echolocation pulses are typically dominated by a long constant frequency (CF) component and begin and/or end with a short frequency-modulated (FM) sweep (Jacobs et al. 2017). There is evidence that the echolocation pulses of both R. damarensis (Maluleke et al. 2017) and R. clivosus (Jacobs et al. 2017) are influenced by climate suggesting that selection may be a dominant process shaping the evolution of phenotypic variation in these species.

2.2 Sampling and Phenotypic Measurements

R. clivosus and *R. damarensis* were captured from eight different localities across their distributions in southern Africa (Fig. 1) using either hand nets during the day, or mist nets at night. We recorded the sex of each captured bat and determined female reproductive condition by palpating the abdomen and inspecting the mammae (Racey 1988). Age class was determined by examining the degree of epiphyseal/diaphyseal fusion (Anthony 1988). Only adult bats were used in subsequent analyses. For ethical reasons, juveniles, pregnant or lactating bats were immediately released at the site of capture.

We chose forearm length (FA) as a proxy for body size because body mass varies seasonally and diurnally in bats (e.g. Rughetti and Toffoli 2014). FA was measured to the nearest 0.1 mm using dial callipers. Echolocation pulses were recorded from bats held 30 cm away from the microphone of an ultrasound D1000X detector (Pettersson Elektronik AB, Uppsala, Sweden—www.batsound.se) at sampling frequencies of 384 and 500 kHz. We used resting frequency (RF), recorded from hand-held bats, as opposed to frequency measurements from flying individuals to avoid the variation

Table 1 Phenotypic variables

Variable name	Abbreviation	Description
Resting frequency	RF	Frequency at the centre of a pulse (kHz)
Inter-pulse interval	IPI	Time between successive pulses (ms)
Duration	Dur	Duration of the whole pulse (ms)
Duration of FMt	FMdur	Duration of the terminal frequency-modulated (FMt) component (ms)
Distomax	Disto	Time between the start of the pulse to the maximum amplitude of the pulse (ms)
Minimum frequency	FMmin	Minimum frequency of the FMt (kHz)
Bandwidth	BW	Bandwidth of the FMt
Sweep rate of the FMt	SR	SR = BW/Dur of FMt (kHz/ms)
Forearm	FA	Length of the forearm (cm)
Head length	HL	Condylobasal length measured from the tip of the nose to the lambda of the skull on live bats (cm)
Noseleaf width	NLW	Maximum diameter of the noseleaf (cm)
Arm-wing length	AWL	Straight-line distance from the shoulder to the base of the first digit (claw) of the outstretched wing (cm)
Hand-wing length	HWL	Straight-line distance from the base of the first digit (claw) to the tip of the outstretched wing (cm)
Arm-wing area	AWA	The combined area of the propatagium and the plagiopatagium (cm^2)
Hand-wing area	HWA	Area of the dactylopatagium (cm^2)
Wing area	WA	The combined area of the two wings, the entire tail membrane and the portion of the body between the wings (cm^2)
Wingspan	WS	The distance between the wingtips of a bat with wings extended so that the leading edges are angled slightly forward (Saunders and Barclay 1992) (cm)
Wing loading	WL	$WL = (mass \times 9.81\ ms^{-2})/WA\ (N\ m^{-2})$
Aspect ratio	AR	$AR = WS^2/WA$

in pulse frequency caused by horseshoe bats compensating for Doppler shifts in frequency during flight (Neuweiler 1989). Pulses of hand-held horseshoe bats, in contrast, have stable CF components, and the inter-pulse frequency variation is low (Neuweiler 1984).

Recorded echolocation calls from both species were analysed as described in Jacobs et al. (2017), and the measurements described in Table 1 were recorded.

The right wing of each captured bat was photographed and measured as described in Jacobs et al. (2007). The wing parameters measured and/or calculated are described in Table 1.

2.3 Statistical Analyses

Data were first transformed using mean standardization (each variable is divided by the mean for that variable) to equalize the scale of our variables without affecting the magnitude of standard deviations (Jacobs et al. 2013; Mutumi et al. 2017) in R statistics (R Development Core Team 2013).

2.3.1 Sexual Dimorphism

Sexual dimorphism was assessed using Manova (Siemers et al. 2005) with the phenotypic variables as dependent variables and sex and locality as categorical predictors.

2.3.2 Geographic Variation

We determined the degree of geographic variation among populations of each species through a standard discriminant function analysis (DFA) on the phenotypic variables. The DFA was done on the factor scores of factors extracted from a standard principal component analyses on the standardized phenotypic variables. The PCA was done to obtain a set of uncorrelated factors which were used as new variables in the DFA.

We also extracted the squared Mahalanobis matrix of phenotypic distances for the first two roots from the DFA and used these to construct a cluster diagram for each species to determine how bats from the different localities are grouped based on their phenotype differences.

Finally, the squared Mahalanobis distance matrix was also regressed against the geographic distance matrix for each species to determine whether the geographic patterning was driven by isolation by distance using the Mantel test in R statistics (R Development Core Team 2013), package Ade4 (Dray and Dufour 2007). The geographic distance matrix was constructed by using the paired straight-line distances among sites for each species. Straight-line distance between pairs of sites was calculated from the geographic coordinates for each site in each species implemented in the package Ade4 (Dray and Dufour 2007).

2.3.3 Drift Versus Selection

We used Lande's model (Lande 1976) to assess the relative contributions of drift and selection to geographic phenotypic variation in the two species. Lande's model is based on quantitative theory of molecular evolution but has been adapted for use on phenotypic traits (Ackermann and Cheverud 2002, 2004; de Azevedo et al. 2015; Smith 2011). The theory uses drift as a null model for phenotypic variation. Rejection of this null model allows the inference that selection has influenced phenotypic variation (Smith 2011).

Lande's model assesses the contribution of drift to phenotypic variation through patterns of variance/covariance between versus within groups (in this case, populations of a species). If a species has diversified through neutral evolutionary processes (mutation and drift), phenotypic variation between populations (B) should be directly proportional (i.e. the log log relationship should have a slope of 1) to the variation within-populations (W), that is, $B \propto W$ (Ackermann and Cheverud 2002). Significant deviations from such proportionality imply natural selection is responsible for the divergence of populations. Slopes that are significantly > 1 (W < B) imply higher phenotypic variation in the highly variable PC than expected under drift. This could be the result of diversifying selection on the variable PC or stabilizing selection on the less variable PC. Slopes < 1 occur when populations are highly divergent along relatively minor PCs. This could result from diversifying selection on traits in these minor PCs or stabilizing selection on the other PCs (Marroig and Cheverud 2004). Given our small sample sizes we do not draw any conclusions in this regard.

The within (W) and between (B) variances were obtained as follows for each species. The variance/covariance matrix for each species was extracted using a Manova on the phenotypic traits for each species with locality and sex as categorical predictors. A PCA was done on the variance/covariance matrix for each species to obtain the eigenvalues for each factor. Eigenvalues measure the amount of variation in the total sample accounted for by each factor and is therefore a measure of the within-population variance, W.

The between population variance, B is obtained by multiplying the matrix of eigenvectors, extracted from this PCA, by the matrix of population phenotypic means. This results in a set of new principal components. The variance of each of the new PCs is calculated to represent B. The log (B) is regressed against log (W). If the slope of this regression is 1, the null model of drift is confirmed. Significant deviation from a slope of 1 rejects drift, and selection is inferred.

We implemented an iterative process within Lande's model in which the first analyses used all populations and all PCs (the new set of PCs on which the calculation of B is based) and then repeated with W and B being re-calculated after removing one population or one PC at a time, or removing a combination of population and PC. This was done to investigate if results of Lande's model were sensitive to small samples, outliers and low numbers of variables or dependent on which phenotypic variables or populations were included. Only the PC with the lowest eigenvalue was excluded to simplify the analyses (Tables 4 and 5).

3 Results

3.1 Sexual Dimorphism

In *R. Clivosus,* there were phenotypic differences between localities (Manova: $F_{119,166}$ = 11.01, P < 0.0001) but not sexes ($F_{17,24}$ = 0.97, P > 0.51). However, in *R. Damarensis,* there were phenotypic differences between both localities and sexes (Manova: locality $F_{75,306}$ = 4.87, p < 0.0001; sex $F_{15,63}$ = 6.66, P < 0.0001). Thus, in the Lande's model for this species, we incorporated sex as a categorical predictor together with locality. Variation due to sex differences in *R. damarensis* was therefore accounted for in the within-populations V/CV matrix used in the modelling (Mutumi et al. 2017).

3.2 Geographic Variation

There was phenotypic geographic variation (Tables 2 and 3) in both species as illustrated in the two-dimensional plot of the canonical roots 1 and 2 from the DFA (Fig. 2). These roots collectively explained 94 and 61% of the variation in *R. damarensis* (Fig. 2a) and *R. clivosus* (Fig. 2b), respectively. Total classification success for *R. damarensis* was 74% (Wilks' Lambda 0.0231, $F_{42,439}$ = 12.9, P < 0.0001) and for *R. clivosus* 100% (Wilk's Lambda 0.00001, $F_{119,179}$ = 8.28, P < 0.0001).

In *R. Damarensis,* root 1 explained 80% of the variation and was dominated by FMdur, arm-wing area, wing area, wing loading and aspect ratio. These are variables associated with accurate ranging (FMdur) and flight manoeuvrability. Root 2 explained 14% and was dominated by hand-wing area and RF. These are variables associated with flight manoeuvrability and detection. In *R. clivosus* root 1 explained 39% of the variation and was dominated by arm-wing length, wing area, wingspan, hand-wing length, arm-wing area and aspect ratio. Root 2 explained 22% of the variation and was dominated by bandwidth and sweep rate of the terminal FM component, duration of the whole call and RF frequency (Fig. 2b). Thus, in both *R. damarensis* and *R. Clivosus,* the variables associated with the two dominant roots were associated with flight and detection and suggest inter-population differences in manoeuvrability and orientation.

Most of the *R. damarensis* populations clustered together in two groups, one comprised of GH, RM and SF (in the middle of the range of this species) and the other of AC, MKC and WC (all in the northern end of the range). The OR and UF populations were the most divergent (Fig. 2c). In *R. clivosus* although four populations (LOB, DH, SD and KN) clustered together, they were from very different latitudes, so too were KGB and KK. The most divergent *R. clivosus* populations were MN (eastern Zimbabwe) and Zomba plateau (southern Malawi).

D. S. Jacobs and G. L. Mutumi

Table 2 Phenotypic variable measurements (mean ± SE) for *R. damarensis* at each locality. Localities are in order of increasing latitude. Abbreviations for locality names (columns) and variable names rows are the same as in Fig. 1 and Table 1, respectively. Units are given in Table 1

	WC (15)	AC (17)	MKC (6)	SF (7)	RM (10)	OR (17)	GH (12)	UF (22)
Dur	33.6±3.0	36.2±3.2	33.2±3.3	33.9±4.4	31.0±3.9	33.2±2.8	34.8±6.0	24.9±2.4
RF	84.5±0.2	85.0±0.3	84.7±0.4	85.6±0.4	87.4±0.3	85.3±0.2	85.9±0.5	84.7±0.2
IPI	8.7±1.7	10.7±1.9	8.6±2.8	15.1±2.6	8.4±2.3	11.9±1.6	12.2±3.5	10.4±1.4
FMmin	67.7±0.8	68.6±0.9	68.6±1.2	69.9±1.2	71.1±1.1	71.0±0.7	74.4±1.6	68.5±0.6
FMdur	1.1±0.1	1.1±0.1	1.1±0.1	1.1±0.1	1.2±0.1	1.7±0.1	1.1±0.1	1.8±0.05
BW	16.8±0.8	16.4±0.8	16.0±1.2	15.8±1.1	16.3±1.0	14.2±0.7	11.5±1.5	16.2±0.6
FA	5.1±0.04	5.1±0.04	5.0±0.1	4.9±0.6	4.9±0.5	48.5±0.4	49.3±0.8	52.4±0.3
NLW	0.9±0.01	0.9±0.02	0.9±0.02	0.8±0.02	8.0±0.02	0.9±0.01	0.8±0.03	0.9±0.01
AWL	6.5±0.2	6.3±0.2	6.1±0.3	6.2±0.3	6.0±0.3	6.0±0.2	6.0±0.3	6.1±0.1
HWL	6.9±0.2	7.1±0.2	7.1±0.3	7.0±0.3	6.7±0.3	6.9±0.2	7.1±0.4	7.2±0.2
AWA	42.2±1.2	44.5±1.3	43.6±1.8	45.6±1.7	40.9±1.5	37.0±1.1	39.0±2.3	35.2±0.9
HWA	28.0±0.6	27.7±0.6	28.8±0.9	29.3±0.9	26.6±0.7	25.4±0.6	27.3±1.2	32.5±0.5
WA	173.8±3.7	170.6±4.1	176.2±5.9	175.0±5.5	158.8±5.0	150.3±3.5	157.4±7.7	156.53.1
WS	29.1±0.5	30.6±0.6	30.1±0.8	30.2±0.7	29.3±0.7	29.6±0.5	30.7±1.1	30.9±0.4
WL	5.5±0.2	6.3±0.2	6.6±0.3	6.2±0.3	6.5±0.3	6.2±0.2	7.3±0.4	10.1±0.2
A	5.0±0.2	5.5±0.2	5.2±0.3	5.2±0.3	5.4±0.2	5.9±0.2	6.0±0.3	6.1±0.1

Table 3 Phenotypic variable measurements (mean ± SE) for *R. clivosus* at each locality. Sample sizes in parentheses. Localities are in order of increasing latitude. Abbreviations for locality names (columns) and variable names (rows) are the same as in Fig. 1 and Table 1, respectively. Units are given in Table 1

	ZP (3)	MN (2)	LOB (6)	SD (15)	KGB (5)	KK (4)	KN (7)	DH (8)
Dur	31.2±3.9	23.0±4.5	29.2±2.6	39.7±1.7	30.8±3.5	33.9±3.1	32.2±2.6	36.4±2.2
RF	81.1±0.5	88.3±0.6	92.2±0.3	91.3±0.2	89.9±0.4	91.5±0.4	91.8±0.3	92.1±0.3
IPI	44.2±24.8	31.0±28.6	50.7±16.5	92.1±10.7	49.0±22.6	138.4±20.3	56.8±16.9	57.2±14.3
Disto	18.7±2.4	13.7±2.8	17.0±1.6	26.9±1.0	19.9±2.2	24.2±2.0	18.8±1.7	24.7±1.4
FMmin	64.0±1.6	79.1±1.8	79.7±1.0	70.1±0.7	70.8±1.4	68.5±1.3	69.4±1.1	68.9±0.9
BW	1.5±0.6	1.1±0.7	2.4±0.4	4.3±0.3	2.2±0.6	3.2±0.5	3.6±0.4	4.0±0.4
SR	10.4±1.0	8.9±1.2	4.0±0.7	5.9±0.4	9.2±0.9	8.3±0.8	6.0±0.7	5.1±0.6
HL	2.2±0.1	2.1±0.1	2.2±0.1	2.3±0.05	2.4±0.1	2.3±0.1	2.4±0.1	2.4±0.1
FA	5.3±0.1	5.4±0.1	5.5±0.1	5.4±0.05	5.7±0.1	5.4±0.1	5.3±0.1	5.4±0.1
NLW	0.9±0.1	0.8±0.2	0.7±0.1	0.7±0.1	2.1±0.1	1.5±0.1	0.8±0.1	0.8±0.1
AWL	7.0±0.3	6.5±0.3	7.6±0.2	6.9±0.1	7.8±0.3	7.2±0.2	6.4±0.2	7.3±0.2
HWL	8.2±0.4	6.8±0.4	9.0±0.3	8.5±0.2	9.1±0.4	8.7±0.3	8.3±0.3	8.8±0.2
AWA	49.0±2.9	37.5±3.3	53.1±1.9	47.3±1.2	50.8±2.6	49.1±2.4	46.4±2.0	48.0±1.7
WA	145.6±9.6	183.6±11.1	197.9±6.4	164.0±4.1	194.3±8.8	171.5±7.8	158.0±6.6	171.1±5.5
WS	34.4±1.3	29.7±1.6	37.6±0.9	34.6±0.6	37.5±1.2	36.3±1.1	34.0±0.9	36.2±0.8
WL	8.1±0.7	10.3±0.8	7.6±0.5	9.8±0.3	8.2±0.6	8.9±0.6	9.0±0.5	9.8±0.4
A	6.5±0.2	5.8±0.3	6.6±0.1	6.3±0.1	6.7±0.2	6.7±0.2	6.2±0.2	6.6±0.1

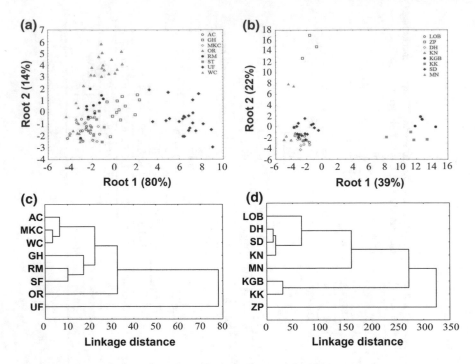

Fig. 2 Multidimensional scaling plots and cluster diagrams for *Rhinolophus damarensis* (**a** and **c**, respectively) and *R. clivosus* (**b** and **d**, respectively) from DFA. Locality abbreviations are the same as in Fig. 1

3.3 Drift Versus Selection

The relationship between within- and between-locality variances indicative of drift was not evident in either species even after removing populations, PCs or a combination of both. The slope of this relationship was always less than that predicted for drift (Tables 4 and 5, Fig. 3). Even the removal of the most distinct sites from the analyses, UF for *R. damarensis* and ZP for *R. clivosus* (Fig. 2c, d), that were also at the extreme ends of our sampling range (Fig. 1), did not yield slopes of 1.

PC scores for both species (Table 6) suggest that PCs associated with manoeuvrability, size and echolocation contributed to most of the variance in our data. The remainder of the PCs comprised of a combination of variables associated mainly with manoeuvrability and echolocation. Finally, there was an indication of co-selection between some PC pairs (Tables 4 and 5), and correlated pairs were highly variable across the different cases analysed. Such correlations are not expected under drift (Ackermann and Cheverud 2002).

Geographic variation was not correlated with geographic distance in *R. damarensis* (Mantel test, observation 0.277, simulated P value 0.139, based on 10000 repli-

Table 4 Results of Lande's model analyses for *R. damarensis*. Abbreviations for locality names are the same as in Fig. 1. The minus sign before the locality abbreviations and PC number indicates that they have been removed from that analyses

Locality	PCs used	Slope *b*	S.E.	$p(b \neq 1)$	Correlated PCs	Consistent with drift?
All	All	0.356	0.074	<0.05	2–4; 2–6; 2–7; 2–11; 2–14; 4–6; 4–11; 6–7; 6–11; 6–14; 7–14; 9–11	No
	−14	0.356	0.074	<0.05	2–4; 2–6; 2–7; 2–11; 4–6; 4–11; 6–7; 6–11; 9–11	No
−AC	All	0.368	0.067	<0.05		No
	−14	0.323	0.083	<0.05		No
−GH	All	0.314	0.073	<0.05		No
	−14	0.284	0.096	<0.05		No
MKC	All	0.360	0.077	<0.05		No
	−14	0.310	0.094	<0.05		No
−OR	All	0.346	0.067	<0.05		No
	−14	0.308	0.082	<0.05		No
−RM	All	0.351	0.074	<0.05		No
	14	0.309	0.091	<0.05		No
−ST	All	0.327	0.084	<0.05		No
	−14	0.262	0.100	<0.05		No
−UF	All	0.323	0.073	<0.05		No
	−14	0.258	0.081	<0.05		No
−WC	All	0.338	0.052	<0.05		No
	−14	0.321	0.069	<0.05		No

cates), but variation and distance were correlated in *R. clivosus* (observation 0.446, simulated P value 0.0328, based on 10000 replicates).

4 Discussion

Lande's model did not support drift in either *R. damarensis* or *R. clivosus*. There was also strong evidence for co-selection among the different traits analysed which suggests that trait divergence is not random. Phenotypic differences across populations was also not a function of geographic distance in *R. damarensis* but was a function

Table 5 Results of Lande's model analyses for *R. clivosus*. Abbreviations for locality names are the same as in Fig. 1. The minus sign before the locality abbreviations and PC number indicates that they have been removed from that analyses

Locality	PCs used	Slope b	S.E.	$p(b \neq 1)$	Correlated PCs	Consistent with drift?
All	All	0.111	0.092	<0.05	7–8; 7–11; 7–12; 8–12; 9–11; 9–13; 9–14; 9–15; 11–12; 11–13; 11–14; 11–15; 12–13; 12–14; 12–15; 13–14; 13–15; 14–15	No
	−15	0.111	0.092	<0.05	7–8; 7–11; 7–12; 8–12; 9–11; 9–13; 9–14; 11–12; 11–13; 11–14; 12–13; 12–14; 13–14	No
−LOB	All	0.032	0.088	<0.05		No
	−15	0.057	0.101	<0.05		No
−ZP	All	0.114	0.082	<0.05		No
	−15	0.046	0.091	<0.05		No
−DH	All	0.045	0.062	<0.05		No
	−15	0.058	0.071	<0.05		No
−KN	All	0.078	0.077	<0.05		No
	−15	0.092	0.091	<0.05		No
−KGB	All	0.145	0.086	<0.05		No
	−15	0.186	0.096	<0.05		No
−KK	All	0.095	0.083	<0.05		No
	−15	0.094	0.097	<0.05		No
−SD	All	0.081	0.067	<0.05		No
	−15	0.103	0.076	<0.05		No
−MN	All	0.125	0.090	<0.05		No
	−15	0.168	0.100	<0.05		No

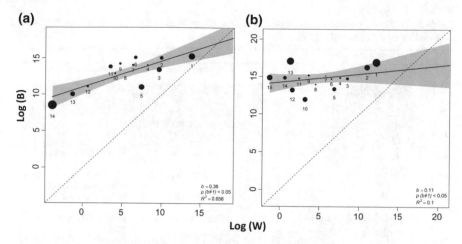

Fig. 3 Regression of B (between-group) and W (within-group variance) for *Rhinolophus damarensis* (**a**) and *R. clivosus* (**b**). PCs generated using all variables. Dot sizes indicate the PC's relative influence on the regression slope (calculated as the difference between the slope values with and without that particular PC point). The regression line (red line) is compared to the null hypothesis of drift b = 1 (dashed line)

of geographic distance in *R. clivosus*. Selection did not appear to be differentially exerted across populations because there was no significant change in the results when localities were excluded one at a time. Population divergence appeared to be mediated by selection on traits associated with manoeuvrability, detection and size in both species despite their disparate distributions.

Our results support those from studies on other rhinolophid bat species which have taken a purely genetic approach, e.g. on *R. ferrumequinum* (Sun et al. 2013) or a purely phenotypic approach like ours, e.g. Mutumi et al. (2017), and found that selection was the dominant process associated with geographic variation. The correlation between variation and geographic distance in *R. clivosus* but not *R. damarensis* is probably associated with the strong regional influence on population divergence reported for *R. damarensis*. The regional influences have apparently resulted in isolation by environment with environmental disjunctions restricting gene flow even between relatively close populations (Jacobs et al. 2013; Maluleke et al. 2017). This is supported by the strong regional groupings we found among *R. damarensis* populations (Fig. 2c). The correlation between distance and variation among populations of *R. clivosus* suggests that environmental disjunctions may not be as restrictive to dispersal as distance and gene flow between populations with shorter distances between them may counteract selection to some degree. However, confirmation of this requires an investigation of the levels of gene flow using suitable genetic markers such as microsatellites. Any natural population is likely to be subjected to several evolutionary processes over its history either at the same time or sequentially (Wright 1943). The consequences of one or more of these processes may be readily observed

Table 6 Dominant variables in each principal component used in the analysis for Lande's model grouped by their functional significance

	Rhinolophus clivosus		Rhinolophus damarensis	
	With echolocation	Without echolocation	With echolocation	Without echolocation
PC1	Dur; **IPI**; BW	AWL; HWL; AWA; WA; WL	Dur; IPI	AWL; HWL; AWA; HWA; WA; WS; WL; A
PC2	IPI; BW; Sweep	AWA; WA; WS; **WL**	Dur; FMdur; BW_Dur	AWL; HWL; WS; WL; A
PC3	Dur; **Disto**	HL; AWA; WA; A	Dur; AWA; HWA; WA	**HWL**; WL; A
PC4	Dur; AWL; AWA; WA; WL	HL; NLW; HWL; WS; A	Dur; FMdur; AWA; WL; A	NLW; **AWL**; HWL; A
PC5	WA; WS; WL	HL; AWA; WA; A; WL	Dur; AWL; HWL; WS; WL; A	NLW; AWL; HWL; AWA; HWA; WA; WS; WL
PC6	**Dur**; Disto; WL	AWL; HWL; AWA	FMdur; BW_Dur; HWL; WL; A	NLW; AWL; AWA; HWA
PC7	BW; Sweep	HWL; WA; WS; WL	FMdur; BW_Dur; HWL; A	NLW; AWL; AWA; HWA; WL; A
PC8	HL; AWA; A; WL	FA; HL; NLW	NLW; AWL; HWL; HWA; A	FA; AWA; HWA; WA; WS
PC9	Dur; HL; NLW; WS; A; WL	HWL; WA; WS; A	AWA; WA; WS; WL	FA; WA; WS; A
PC10	HL; AWA; WA; WL	NLW	**NLW**; AWL	FA; WS
PC11	**AWL**		AWA; HWA	
PC12	FMmin; HWL; WA; WS; WL		**FA**	
PC13	**NLW**; WS		AWA; HWA; WA; WS	
PC14	FMmin; FA; WS; A		WA; **WS**; A	
PC15	FA; HWL; A		PF; FMmin	
PC16	FMmin; HWL		PF; FMmin; WS	
PC17	**RF**			

Maneuvrability	Size
Mixed	Echolocation

Bold face represents a contribution ~30% or more. The rest made a contribution of less than 25% but close to 10%

or obscured depending on their relative strengths and the degree of gene flow among populations.

Many of the traits we used have direct fitness benefits associated with locomotion (wing variables) and orientation (echolocation variables) and are implicated in successful food acquisition (flight and detection of prey, Norberg and Rayner 1987) and reproduction. Echolocation may have a secondary function in communication and mate choice (Siemers et al. 2005; Knöornschild et al. 2012; Puechmaille et al. 2014; Bastian and Jacobs 2015). Several of these traits form adaptive complexes (Jacobs et al. 2007) as evidenced by correlations between them, e.g. body size and echolocation (Jones 1996), wing loading and echolocation (Norberg and Rayner 1987; Jacobs et al. 2007) as well as between several skull features and echolocation components (Jacobs et al. 2014). There are also tight associations between the phenotypic variables we have considered and environmental factors, e.g. between habitat and each of wing loading (Kalcounis and Brigham 1995) and echolocation (Schnitzler and

Kalko 2001) and between echolocation and climatic factors (Mutumi et al. 2016; Maluleke et al. 2017). The lack of support for drift among these variables is therefore not surprising and their divergence is probably the result of habitat-mediated selection.

4.1 What Are Bats Adapting To?

It is likely that the phenotypic divergence in *R. damarensis* and *R. clivosus* is the result of adaptation to local habitats reinforced by limited gene flow among populations which allows adaptive differences to accumulate (e.g. Odendaal et al. 2014). Adaptive complexes have evolved in traits associated with echolocation and flight, and several studies have been done on the adaptive divergence, in other rhinolophids, of both echolocation (Schnitzler and Kalko 2001; Odendaal et al. 2014; Jiang et al. 2015; Mutumi et al. 2016) and flight (Norberg and Rayner 1987; Jacobs et al. 2016) but also in *R. damarensis* (Maluleke et al. 2017) and *R. clivosus* (Jacobs et al. 2017).

The latter two studies suggest that climate might exert selection pressure on different populations of bats resulting in adaptive divergence in echolocation. This is not surprising because echolocation pulses are propagated through the atmosphere and atmospheric conditions should have a marked influence on such propagation to the extent that bats of the same species experiencing different climatic conditions should vary their echolocation parameters to optimize the operational range of their echolocation.

Sound travels through the atmosphere by the generation of pressure waves. Each pressure wave is formed by energy being transferred from the vibrating vocal chords of the bat to the air molecules surrounding them and then to other adjacent air molecules. As each pressure wave is formed, the initial energy in the vibrating chords is diminished by absorption and spreading losses as the energy is spread over an increasingly larger area of the expanding wavefront. This loss of energy is known as atmospheric attenuation (AA), and the degree of AA is dependent on several factors; the most important of which are: the relative humidity (RH) of the air, the temperature of the air and the frequency of the sound being propagated. The effects of these three factors on AA are not linear and are interactive (Luo et al. 2014). However, AA increases as the frequency of the sound increases. All else being equal, low-frequency sound will travel longer distances than high-frequency sound which means that bats using low-frequency echolocation can detect objects at longer distances than bats using high-frequency echolocation, provided climatic conditions are similar. It seems therefore that one of the ways that bats can optimize their detection distance is to alter the frequencies of the sound. This is in fact what was found. In an analysis in which the operational range of the echolocation of *R. damarensis* and *R. clivosus* was modelled over the same localities sampled in our study, with temperature and RH as climatic predictors, it was found that the geographic variation in RF was partly explained by these climatic variables after correction for body size (Jacobs et al. 2017; Maluleke et al. 2017). The best climatic predictor in both *R. damarensis* which has an arid distribution, and *R. clivosus* which has a distribution which extends over both arid and mesic biomes was temperature (Jacobs et al. 2017; Maluleke et al. 2017). RH had little influence. However, in similar analyses on two other rhinolophids, *R. simulator* and *R. swinnyi*, both of which have mesic distributions, the best climatic predictors were RH and temperature but temperature had a relatively lower influence than RH (Mutumi et al. 2016). Thus, in hot arid conditions, bats are adapting to temperature, and in mesic conditions, they adapt to RH, and a lesser extent temperature. Note also that in all of these studies, FA as a measure of body size was not a main predictor of RF which means that climatic effects on body size were not responsible for the variation in RF via the correlation between frequency and body size (Mutumi et al. 2016; Jacobs et al. 2017; Maluleke et al. 2017). Of course, these studies could not exclude phenotypic plasticity as the cause of the divergence in RF, albeit mediated by responses to climatic conditions.

The specialized echolocation system of HDC bats restricts them to foraging in dense vegetation where distances to background vegetation are short and atmospheric attenuation (Lawrence and Simmons 1982) is minimized allowing detectable acoustic glints off the flapping wings of insects (Schnitzler and Denzinger 2011). Flying in dense vegetation means that these bats have to fly slowly and manoeuvrably which requires short, broad wings of low wing loading (and therefore low body mass) that generate lift at low flight speeds (Norberg and Rayner 1987). This could result in particularly pronounced convergent morphology and echolocation in the family Rhinolophidae. Yet here we report differentiation in wing morphology between populations of the same species in both *R. damarensis* and *R. clivosus*. Such divergence suggests difference in flight patterns and may be the result of adaptation to local habitats, particularly to differences in the density of vegetation because the flight and echolocation variables responsible for most of the variation in both species (Fig. 2) are associated with manoeuvrability (wing components; Norberg and Rayner 1987), detection (RF; Schnitzler and Denzinger 2011) and ranging (variables associated with frequency-modulated components; Schuller et al. 1971; Schnitzler and Denzinger 2011). Unfortunately, very little is known about the way that these bats use their habitat at the local scale and an explanation for this divergence in wing morphology and body size awaits more detailed study at the local level.

Differences in wing morphology and echolocation could be an allometric response to selection on body size (Stoffberg et al. 2011; Taylor et al. 2012). Both wing morphology and echolocation form tight adaptive complexes with body size in rhinolophid bats (Jones 1996; Norberg and Rayner 1987; Jacobs et al. 2007), and selection on body size in environments that differ in climate, for example, may result in differences in body sizes with allometric responses in traits associated with flight, echolocation and feeding. Climatic factors at different altitudes may have been responsible for divergence in body size with allometric responses in echolocation frequencies in southern African rhinolophids (Taylor et al. 2012). However, evidence for size climate relationships in bats is not convincing (Findley and Wilson 1982), and investigations of the influence of body size, in the form of a test of James' rule (James 1970), on echolocation frequencies have found no evidence for such a relationship in several rhinolophid species (Yoshino et al. 2006; Chen et al. 2009; Jiang et al. 2010; Sun et al. 2013; Mutumi et al. 2016) including *R. damarensis* (Maluleke et al. 2017) and *R. clivosus* (Jacobs et al. 2017).

4.2 Limitations

The Lande's model results rely on whether the sample size and the number variables adequately represent the patterns of variation in the study system. Even though attempts were made to address these potential limitations, the following must be noted. Firstly, sample sizes for R. clivosus were below the recommended optimum for regression-based analyses (Kleinbaum et al. 1988) and therefore R. clivosus results should be treated with caution. Secondly, there is evidence from human skulls (de Azevedo et al. 2015; von Cramon-Taubadel 2013) of variation in the extent to which different trait complexes fit a null model of drift. Our study might have not adequately accounted for this because we could take only a limited number of reliable measurements from live specimens without compromising their wellbeing. Relatively fewer variables may under-estimate the regression between B and W possibly giving a result biased by outliers. However, we addressed the two challenges with the sensitivity test (iterating analyses after excluding a site and/or a PC at a time). These iterative results did not change significantly showing that variable numbers and perhaps even our sample sizes allowed optimum statistical power. Future studies can maximise sample size and variable numbers by incorporating museum specimens (these were not yet available for our species).

5 Conclusions

In phenotypic traits that have severe fitness consequences (e.g. those involved with survival and reproduction) separately and in combination with other variables within adaptive complexes, selection rather than drift is still likely to be the dominant influence on the evolution of phenotypic variation and ultimately lineage divergence. The agents of such selection is very likely to vary across the distribution of a species. Here temperature was more important in *R. damarensis* and *R. clivosus* which, respectively, occur wholly and partially, in arid regions. In the more mesic areas, both temperature and RH impact on frequency variation, the interaction between the two being important. Thus, phenotypic divergence must be studied at the level of the population, which are after all the units of evolution, if the evolution of biodiversity is to be correctly understood.

Acknowledgements We thank Pierre Pontarotti for inviting us to contribute to this volume. We are grateful to many people who helped with sample collection and logistics in the field particularly, Prof. Peter Mundy of the National University of Science and Technology, Dr. H. Madzikanda of the Zimbabwe Parks and Wildlife Authority, Tinyiko Maluleke, Blessing Buleya, Reason Nyengera,

Orsila Smit, Megan Cunnama, Ellenor Salo, Lizelle Odendaal, Serena Doole and Nyasha Gondo. Special thanks to Eugene Marais who provided invaluable information on Namibian bat roosts and the Namibian Ministry of Environment and Tourism and the Northern Cape for permission to do this research. For guidance in statistical analysis and editing the R script, we thank Rebecca Ackermann and Andrew Conith. This research was supported by grants to DSJ from the University of Cape Town and the South African Research Chair Initiative of the Department of Science and Technology, administered by the National Research Foundation (GUN 64798).

References

Ackermann RR, Cheverud JM (2002) Discerning evolutionary processes in patterns of tamarin (genus *Saguinus*) craniofacial variation. Am J Phys Anthropol 117:260–217

Ackermann RR, Cheverud JM (2004) Detecting genetic drift versus selection in human evolution. Proc Natl Acad Sci 101:17946–17951

Anthony ELP (1988) Age determination in bats. In: Kunz TH (ed) Ecological and behavioural methods for the study of bats. Smithsonian Institution Press, Washington D.C., pp 47–58

Bastian A, Jacobs DS (2015) Listening carefully: increased perceptual acuity for species discrimination in multispecies signalling assemblages. Anim Behav 101:141–154

Betti L, Balloux F, Hanihara T, Manica A (2010) The relative role of drift and selection in shaping the human skull. Am J Phys Anthropol 141:76–82

Brandon RN (2005) The difference between selection and drift: a reply to Millstein. Biol Philos 20:153–170

Brandon RN, Carson S (1996) The indeterministic character of evolutionary theory: no "no hidden variables proof" but no room for determinism either. Philos Sci 63:315–337

Campbell P, Pasch B, Pino JL, Crino OL, Phillips M, Phelps SM (2010) Geographic variation in the songs of neotropical singing mice: testing the relative importance of drift and local adaptation. Evolution 64:1955–1972

Chen S-F, Jones G, Rossiter SJ (2009) Determinants of echolocation call frequency variation in the Formosan lesser horseshoe bat (*Rhinolophus monoceros*). Proc R Soc B 276:3901–3909

Coyne JA, Orr HA (2004) Speciation. Sinauer Associates Sunderland, Sunderland, MA

Csorba G, Ujhelyi P, Thomas N (2003) Horseshoe bats of the world: (Chiroptera: Rhinolophidae). Alana Books, Berkshire

de Azevedo S, Quinto-Sánchez M, Paschetta C, González-José R (2015) The first human settlement of the New World: a closer look at craniofacial variation and evolution of early and late Holocene Native American groups. Quat Int 431:152–167

Dool SE, Puechmaille SJ, Foley NM, Allegrini B, Bastian A, Mutumi GL, Maluleke T, Odendaal LJ, Teeling EC, Jacobs DS (2016) Nuclear introns outperform mtDNA in phylogenetic reconstruction: lessons from horseshoe bats (Rhinolophidae: Chiroptera). Mol Phylogenet Evol 97:196–212

Dray S, Dufour AB (2007) The ade4 package: implementing the duality diagram for ecologists. J Stat Softw 22(4):1–20

Fenton MB (1999) Describing the echolocation calls and behaviour of bats. Acta Chiropterol 1(2):127–136

Findley JS, Wilson DE (1982) Ecological significance of chiropteran morphology. In: Kunz TH (ed) Ecology of bats. Boston University Press, Boston, pp 243–260

Finger NM, Bastian A, Jacobs DS (2017) To seek or speak? Dual function of an acoustic signal limits its versatility in communication. Anim Behav 127:135e152

Freeman PW, Lemen CA (2010) Simple predictors of bite force in bats: the good, the better and the better still. J Zool 282:284–290

Griffin DR (1958) Listening in the dark. Yale University Press, New Haven

Irwin DE, Thimgan MP, Irwin JH (2008) Call divergence is correlated with geographic and genetic distance in greenish warblers (Phylloscopus trochiloides): a strong role for stochasticity in signal evolution? J Evol Biol 21:435–448

Jacobs DS, Barclay RMR, Walker MH (2007) The allometry of echolocation call frequencies of insectivorous bats: why do some species deviate from the pattern? Oecologia 152:583–594

Jacobs DS, Babiker H, Bastian A, Kearney T, van Eeden R, Bishop JM (2013) Phenotypic convergence in genetically distinct lineages of a Rhinolophus species complex (Mammalia, Chiroptera). Plos ONE 8(12):e82614. https://doi.org/10.1371/journal.pone.0082614

Jacobs DS, Bastian A, Bam L (2014) The influence of feeding on the evolution of sensory signals: a comparative test of an evolutionary trade-off between masticatory and sensory functions of skulls in southern African Horseshoe bats (Rhinolophidae). J Evolut Biol 27:2829–2840

Jacobs DS, Mutumi GL, Maluleke T, Webala PW (2016) Convergence as an evolutionary trade-off in the evolution of acoustic signals: echolocation in horseshoe bats as a case study. In: Pontarotti P (ed) Evolutionary biology: convergent evolution, evolution of complex traits, concepts and methods. Springer Interntaional Publishing, Switzerland, pp 89–103

Jacobs DS, Catto S, Mutumi GL, Finger N, Webala PW (2017) Testing the sensory drive hypothesis: geographic variation in echolocation frequencies of Geoffroy's horseshoe bat (Rhinolophidae: *Rhinolophus clivosus*). Plos ONE 12(11):e0187769. https://doi.org/10.1371/journal.pone.01877 69

James FC (1970) Geographic size variation in birds and its relationship to climate. Ecology (3):365–390

Jiang T, Metzner W, You Y, Liu S, Lu G, Li S, Wang L, Feng J (2010) Variation in the resting frequency of *Rhinolophus pusillus* in mainland China, effect of climate and implications for conservation. J Acoust Soc Am 128:2204–2221

Jiang T, Wu H, Feng J (2015) Patterns and causes of geographic variation in bat echolocation pulses. Integr Zool 10:241–256

Jones G (1996) Does echolocation constrain the evolution of body size in bats? In: Symposia of the Zoological Society of London, The Society, London, 1960–1999, pp 111–128

Kalcounis MC, Brigham RM (1995) Intraspecific variation in wing loading affects habitat use by little brown bats (*Myotis lucifugus*). Can J Zool 73:89–95

Kirschel ANG, Slabberkoorn H, Blumstein DT, Cohen RE, de Kort SR, Buermann W, Smith TB (2011) Testing alternative hypotheses for evolutionary diversification in an African songbird: rainforest refugia versus ecological gradients. Evolution 65:3162–3174

Kleinbaum DG, Kupper LL, Muller KE (1988) Applied regression analysis and other multivariable methods, 2nd ed. PWS-KENT Publishing Company, Boston

Knöornschild M, Jung K, Nagy M, Metz M, Kalko E (2012) Bat echolocation calls facilitate social communication. Proc R Soc B: Biol Sci 279(1748):4827e4835

Kunz TH, Parsons S (2009) Ecological and behavioral methods for the study of bats. Johns Hopkins University Press, Baltimore

Lande R (1976) Natural selection and random genetic drift in phenotypic evolution. Evolution 30:314–334

Lawrence BD, Simmons JA (1982) Measurements of atmospheric attenuation at ultrasonic frequencies and the significance for echolocation by bats. J Acoust Soc Am 71(3):585–590

Leinonen T, O'Hara, R, Cano J, Merilä J (2008) Comparative studies of quantitative trait and neutral marker divergence: a meta-analysis. J Evolut Biol 21:1–17

Luo J, Kosel K, Zsebok S, Siemers BM, Goerlitz HR (2014) Global warming alters sound transmission: differential impact on the prey detection ability of echolocating bats. J R Soc Interface/R Soc 11:20130961. https://doi.org/10.1098/rsif.2013.0961PMID:24335559

Malhotra A, Thorpe RS (2000) The dynamics of natural selection and vicariance in the dominican anole: patterns of within-island molecular and morphological divergence. Evolution 54:245–258

Maluleke T, Jacobs DS, Winker H (2017) Environmental correlates of geographic divergence in a phenotypic trait: a case study using bat echolocation. Ecol Evolut 2017,00:1–15. https://doi.org/10.1002/ece3.3251

Marroig G, Cheverud JM (2004) Did natural selection or genetic drift produce the cranial diversification of neotropical monkeys? Am Nat 163:417–428

Millstein RL (2002) Are random drift and natural selection conceptually distinct? Biol Philos 17:33–53

Monadjem A, Taylo, PJ, Cotterill FPD, Schoeman CM (2010) Bats of southern and central Africa: a biogeographic and taxonomic synthesis. Wits University Press, Johannesburg

Morrone JJ (2009) Evolutionary biology: an intergrative approach with case studies Columbia University Press, Columbia

Moussy C, Hosken D, Mathews F, Smith G, Aegerter J, Bearhop S (2013) Migration and dispersal patterns of bats and their influence on genetic structure. Mamm Rev 43:183–195

Mutumi GL, Jacobs DS, Winker H (2016) Sensory drive mediated by climatic gradients partially explains divergence in acoustic signals in two horseshoe bat species, *Rhinolophus swinnyi* and *Rhinolophus simulator*. Plos ONE 11:e0148053. https://doi.org/10.1371/journal.pone.0148053

Mutumi GL, Jacobs DS, Winker H (2017) The relative contribution of drift and selection to phenotypic divergence: a test case using the horseshoe bats *Rhinolophus simulator* and *Rhinolophus swinnyi*. Ecol Evolut 2017:1–14. https://doi.org/10.1002/ece3.2966

Neuweiler G (1984) Foraging, echolocation and audition in bats. Naturwissenschaften 71:446–455

Neuweiler G (1989) Foraging ecology and audition in echolocating bats. Trends Ecol Evol 4(6):160–166

Norberg UM, Rayner JMV (1987) Ecological morphology and flight in bats (Mammalia: Chiroptera): wing adaptations, flight performance, foraging strategy and echolocation. Philos Trans R Soc Lond B Biol Sci 316:335–427

Odendaal LJ, Jacobs DS, Bishop JM (2014) Sensory trait variation in an echolocating bat suggests roles for both selection and plasticity. BMC Evolut Biol 14:60. https://doi.org/10.1186/1471-2148-14-60

Ohmer ME, Robertson JM, Zamudio KR (2009) Discordance in body size, colour pattern, and advertisement call across genetically distinct populations in a Neotropical anuran (*Dendropsophus ebraccatus*). Biol J Linnean Soc 97:298–313

Orr M R, Smith TB (1998) Ecology and speciation. Trends Ecol Evolut 13:502–506

Puechmaille SJ, Borissov IM, Zsebok S, Allegrini B, Hizem M, Kuenzel S, Schuchmann, M, Teeling, EC, Siemers BM (2014) Female mate choice can drive the evolution of high frequency echolocation in bats: a case study with *Rhinolophus mehelyi*. Plos ONE 9(7):e103452. https://doi.org/10.1371/journal.pone.0103452

R Development Core Team (2013) R: a language and environment for statistical computing, 3.1 edn. R Foundation for Statistical Computing, Vienna, Austria

Racey PA (1988) Reproductive assessment in bats. In: Kunz TH (ed) Ecological and behavioural methods for the study of bats. Smithsonian Institution Press, Washington DC, pp 31–45

Rogell B, Eklund M, Thörngren H, Laurila A, Höglund J (2010) The effects of selection, drift and genetic variation on life-history trait divergence among insular populations of natterjack toad, *Bufo calamita*. Mol Ecol 19:2229–2240

Rughetti M, Toffoli R (2014) Sex-specific seasonal change in body mass in two species of Vespertilionid bats. Acta Chiropterologica 16(1):149–155

Schnitzler H-U, Denzinger A (2011) Auditory fovea and Doppler shift compensation: adaptations for flutter detection in echolocating bats using CF-FM signals. J Comp Physiol A: Neuroethol Sens Neural Behav Physiol 197(5):541–559

Schnitzler H-U, Kalko EK (2001) Echolocation by insect-eating bats: we define four distinct functional groups of bats and find differences in signal structure that correlate with the typical echolocation tasks faced by each group. BioScience 51:557–569

Schuchmann M, Siemers B (2010) Behavioural evidence for community-wide species discrimination from echolocation calls in bats. Am Nat 176:72–82

Schuller G, Neuweiler G, Schnitzler H-U (1971) Collicular responses to the frequency modulated final part of echolocation sounds in *Rhinolophus ferrrumequinum*. Zeitschrift für vergleichende Physiologie 74:153–155

Siemers BM, Beedholm K, Dietz C, Dietz I, Ivanova T (2005) Is species identity, sex, age or individual quality conveyed by echolocation call frequency in European horseshoe bats? Acta Chiropterologica 7:259–274

Smith HF (2011) The role of genetic drift in shaping modern human cranial evolution: a test using microevolutionary modeling. Int J Evolut Biol 2011 Article ID 145262, 11 pp. https://doi.org/1 0.4061/2011/145262

Stoffberg S, Jacobs DS, Matthee CA (2011) The divergence of echolocation frequency in horseshoe bats: moth hearing, body size or habitat? J Mamm Evolut 18:117–129. https://doi.org/10.1007/s 10914-011-9158-x

Sun KP, Luo L, Kimball R, Weiet X, jin L, Jiang T, Li G, Feng J (2013) Geographic variation in the acoustic traits of greater horseshoe bats, testing the importance of drift and ecological selection in evolutionary processes. Plos ONE 8:e70368. https://doi.org/10.1371/journal.pone.0070368

Taylor PJ, Stoffberg S, Monadjem A, Schoeman MC, Bayliss J, Cotterill FPD (2012). Four new bat species (*Rhinolophus hildebrandtii* Complex) reflect Plio-Pleistocene divergence of dwarfs and giants across an Afromontane Archipelago. Plos ONE 7(9):e41744. https://doi.org/10.1371/jour nal.pone.0041744

Turelli M (1988) Phenotypic evolution, constant covariances, and the maintenance of additive variance. Evolution 42:1342–1347

von Cramon-Taubadel N (2013) Evolutionary insights into global patterns of human cranial diversity: population history, climatic and dietary effects. J Anthropol Sci 91:1e36

Weaver TD, Roseman CC, Stringer CB (2007) Were neandertal and modern human cranial differences produced by natural selection or genetic drift? J Hum Evolut 53:135–145

Whitlock MC (2000) Fixation of new alleles and the extinction of small populations: drift load, beneficial alleles, and sexual selection. Evolution 54:1855–1861

Wright S (1943) Isolation by distance. Genetics 28:114–138

Yoshino H, Matsumura S, Kinjo K, Tamura H, Ota H, Izawa M (2006) Geographical variation in echolocation call and body size of the Okinawan least horseshoe bat, *Rhinolophus pumilus* (Mammalia, Rhinolophidae), on Okinawa-jima Island, Ryukyu Archipelago, Japan. Zool Sci 23:661–667

Metagenomic Approaches Highlight the Organization and Dynamics of Plankton at the Species Level

Thomas Vannier

Abstract Plankton is composed of all organisms that drift along currents. The planktonic ecosystem is a complex system that groups together organisms interacting with each other and with their environment. This ecosystem plays an important role in the elementary processes necessary to maintain life on Earth such as photosynthesis; however, the evolutionary and organizational mechanisms that govern it remain poorly known. With the progress of high-throughput sequencing in the last decade, it is possible to obtain almost all the genomic sequences of the microorganisms present in an environmental sample, including uncultured organisms. Today, using the reads resulting from this metagenomic sequencing combined with contextual environmental conditions, we have information access to the genes composition, the diversity, and the biogeography of planktonic organisms right down to the species level. This review highlights the ecological and evolutionary insights gained from metagenomic sequencing projects applied to large-scale environmental surveys. The environmental genomics data produced by the *Tara* Oceans expedition (2009–2012) is by far the largest such dataset available today. The careful choice of oceanographic situations in each sampled region, the study of different size fractions of plankton and the significant sequencing depth compared to other studies, makes this project uniquely adequate to carry out studies of the plankton ecosystem in the surface oceans on a global planetary scale. This review focuses on analyses using the *Tara* Oceans data at different scales; individual organisms, populations, and communities. Their findings improve our understanding of the evolutionary and organizational mechanisms that govern the planktonic ecosystem.

T. Vannier (✉)
CNRS, IRD, MIO UM 110, Aix Marseille Université, Université de Toulon,
13288 Marseille, France
e-mail: thomas.vannier@mio.osupytheas.fr

© Springer International Publishing AG, part of Springer Nature 2018
P. Pontarotti (ed.), *Origin and Evolution of Biodiversity*,
https://doi.org/10.1007/978-3-319-95954-2_14

1   Introduction

1.1   Plankton: A Multitude of Organisms Evolving and Interacting Within a Complex Ecosystem

The diversity of the planktonic ecosystem plays a key role in the balances necessary for life on Earth. This diversity spans genes, organisms, and planktonic communities over space and time. Plankton includes a wide range of free-living organisms that drift along ocean currents (Fig. 1). They consist of organisms of different sizes that range from several meters, such as certain jellyfish, to less than a picometer as is the case for some viruses. Several types of plankton coexist and interact in the oceans from

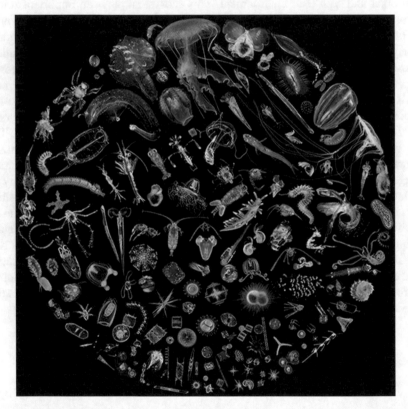

Fig. 1   Mandala of plankton. Picture of the complex and wide variety of plankton. At the upper part, are represented the larger organisms (jellyfish, siphonophorae, ctenophora, salps) corresponding to the zooplankton. In the middle, zooplankton organisms whose sizes vary from a few millimeters to several centimeters (chaetognatha, annelids, pteropods, copepods). In the lower part, we find the microscopic organisms measuring less than one millimeter (protists, radiolaria, foraminifera, diatoms, dinoflagellates). The size scale is not respected, and viruses and bacteria are not represented. *Source* Plancton, Aux origins du vivant (2013). © C. SARDET

multicellular eukaryotes such as zooplankton, as well as unicellular protists such as microalgae, all the way to bacteria, archaea, and viruses. Zooplankton are animal organisms, which feed on phytoplankton. The latter includes unicellular microalgae as well as photosynthetic bacteria. Photosynthesis is a central bioenergetic process that synthesizes organic matter, together with oxygen, in the sunlit ocean that covers 71% of the Earth surface. Ninety-eight percent of the ocean's biomass is composed of plankton which earns it its name of the invisible multitude. There are several billion phytoplanktonic organisms in a single liter of seawater. Hundreds of thousands of species floating in the surface ocean capture solar energy and produce as much oxygen as all forests and terrestrial plants combined, thus contributing about half of global primary production (Field et al. 1998). The organic matter they produce is food for zooplankton, which itself is eaten by larger species all the way to whales. Hence, plankton is at the basis of the ocean food web. Plankton is also key players in planetary-scale biogeochemical cycles and drive the biological carbon pump that sequesters carbon in the deep ocean (Guidi et al. 2016). Due to their large distribution and their substantial biogeochemical roles, they are considered major actors in the global climate regulation (Falkowski 2012) and also constitute a potential source of innovations for emerging blue biotechnology (Kennedy et al. 2008).

The planktonic ecosystem is a complex network where planktonic organisms interact with each other via parasitism, predation, symbiosis, or competition relationships (Cornell and Lawton 1992; Lima-Mendez et al. 2015), as well as with their environment (Moore et al. 2013). In comparison to terrestrial ecosystems, planktonic ecosystems are always on the move, carried by ocean currents. Indeed, the circulation of seawater in the global ocean is governed by the combined effects of wind, differences in temperature, density, salinity, but also interactions within planetary rotation-driven currents. Plankton, continuously transported by these movements of water, are intimately associated to these different physical factors. Combined with the different modes of biotic interactions, the variations in these abiotic environmental parameters have an impact on the malleability of genomes, on the species variations in diversity and abundance, and on the organization of planktonic communities. All these microscopic to planetary-scale interactions together form the planktonic ecosystem as we observe it today. Surprisingly, the ocean ecosystem is the least known ecosystem on our planet. Many planktonic species are yet to be discovered; the majority of the genes that compose them are absent from public reference databases. In addition, the impact of environmental conditions on the composition, diversity, and biogeography of communities, species, and planktonic genes at the global scale is still poorly understood. As the climate of the planet continues to evolve, better knowledge of this ecosystem, its dynamics, and its ability to adapt to the physicochemical changes of the ocean water masses is essential to understand and predict the impact of human activities on the ocean and the life on Earth equilibrium that depend on it.

1.2 Strategies to Studying the Planktonic Ecosystem

Many oceanographic expeditions have sailed out since the nineteenth century to study the planktonic ecosystem. The invention of the pelagic net as early as 1845 by Müller made it possible to carry out the first sampling of these microorganisms. Thus, the naturalist Charles Darwin was able to describe in his book *the Voyage of the Beagle* the diversity of these marine microorganisms. Modern oceanography began when HMS Challenger carried out the first ocean floor exploration from 1872 to 1876. A sampling protocol was designed to harvest planktonic microorganism while including physical, chemical, and biological parameters. During this expedition, several planktonic organisms were described and drawn in detail by the German naturalist Ernst Haeckel. Concomitant physicochemical measurements revealed the diversity of ocean environment, showing that it was possible to study ocean life in its natural habitat. However, the paucity of materials and the rudimentary technological observation methods as well as the gaps in the knowledge of ocean currents and certain oceanic regions did not allow scientists to appreciate the impact of the environment on the organization of marine life.

During the twentieth century, technological advances made it possible to fill these gaps. Two areas have particularly contributed to improving our knowledge of the marine ecosystem: imaging and genomics. Two imaging methods are generally used to study marine biology. The first one uses remote satellite imagery to track the ocean circulation in real time across the world's oceans. Simulation models have been developed to predict the amount of time for a particle to be transported to different parts of the globe. Furthermore, specific satellite fluorescence sensors tell us about the amount of chlorophyll *a* as a proxy of the abundance of phytoplankton present in the water. These pictures make it possible to visualize the phytoplankton blooms from space when the environmental conditions are met for its proliferation. The advent of planetary-scale observation greatly facilitates biogeography, the study of the geographical distribution of species or communities of species. Biogeography helps identify ecosystems by studying the spatial distribution of organisms in relation to environmental conditions. Thus, satellite imagery played a central role in the biogeochemical classification of the oceans into provinces initiated by Longhurst in the 2000s (Longhurst 2007; Oliver and Irwin 2008; Reygondeau et al. 2013). However, a limitation of this classification is that its biology input is mostly restricted to the spatial distribution of phytoplankton organisms since there is the only plankton class readily measured from space through chlorophyll *a*. The second imaging method emerged via the evolution of high-resolution and high-throughput microscopic imaging technologies. These methods have not only contributed to the discovery of many planktonic species, but also allow the measurement of the absolute abundance and taxonomic composition of plankton present in a marine environment. However, taxonomic classification has a limited resolution because of the close morphologic resemblance of many species and deterioration during the sample preparation process. Furthermore, there is a tremendous amount of work needed for manualy analyze these images, even though actual progress in computational methods for matterns recognition in images allows high-throughput automated analysis and classification.

The development of sequencing technologies has greatly contributed to deepen our knowledge of the marine ecosystem. DNA sequencing involves determining the sequence of nucleotides of a DNA fragment. It makes it possible to obtain, from the genome of an individual, all the information stored in the DNA sequences that compose his genetic material. This method, invented in the second half of the 1970s, has become an essential step in understanding biological phenomena at the molecular and cellular levels. The genomes of many microbial species have been sequenced since, including the first bacterial genome, *Haemophilius influenza* in 1995 (Fleischmann et al. 1995), and the first eukaryotic genome, *Saccharomyces cerevisiae* in 1996 (Goffeau et al. 1996). The sequencing of these organisms nevertheless required their culturing. Driven by the realization that many organisms are uncultivable, the beginning of the twenty-first century marked by the advent of new generation of sequencing technologies that are now able to sequence tens of billions of nucleotide bases a day has seen the emergence of a group of environmental genomics culture free approaches. The first of these metagenomics consists in the study of all the DNA sequences belonging to the microorganisms taken directly from their environment without having to go through a culture step (Handelsman et al. 1998). More recently, thanks to ultra high-throughput sequencers and big data management, several very large-scale metagenomics projects have been carried out. Sequencing of the human intestinal microbiota with the MetaHIT project (Qin et al. 2010) was one of the first metagenomics projects applied to humans. This project revealed an ecosystem including previously unknown microorganisms that consists of nearly 100,000 billion microorganisms, which is ten times more than the host's own cells. Several studies of this community of microorganisms have shown the existence of mutualist interactions between the intestinal microbiota and the host that have important impacts on health and disease. The ocean ecosystem is far more complex and more difficult to sample, with a far wider range of environmental conditions and with important impacts on climate. Thus, many fundamental research projects in oceanic metagenomics have been initiated to identify species and genes that escape observation by culture-dependent techniques, in order to improve our understanding of the organization of microbial communities and their ecological role.

1.3 Large-Scale Oceanographic Survey

Craig Venter, who participated in the sequencing of the human genome in the 2000s, also studied the microbial ecosystem in the oceans. He first led a pilot Sargasso Sea metagenomics project off the coast of Bermuda (Venter et al. 2004). The goal was to make an inventory of microbial genes and genomes present in the ocean surface. More than two million sequences were obtained and annotated revealing more than one million sequences not listed in the reference databases and greatly increasing the diversity of certain gene families. The Venter Institute pursued with a sampling

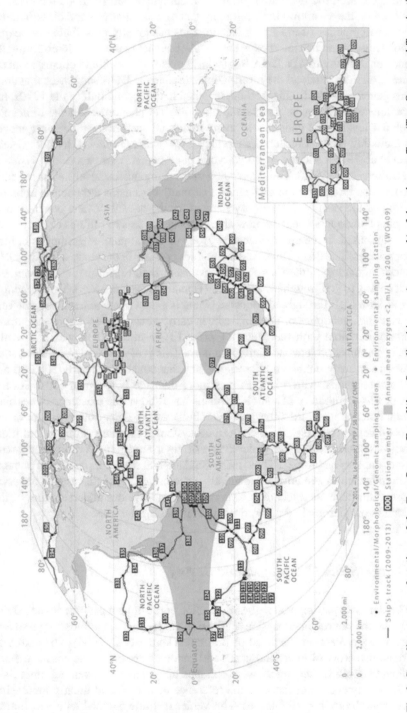

Fig. 2 Sampling route and stations of the *Tara* Oceans Expedition. Sampling itinerary (green track) of the schooner *Tara* (*Tara* Oceans and *Tara* Arctic campaigns, 2009–2013). The 210 stations where samples were taken are shown as red dots. *Source* Pesant et al. (2015). © N. LE BESCOT

campaign at a global scale. The purpose of the *Global Ocean Sampling* expedition (GOS) on board the *Sorcerer II* (Rusch et al. 2007) was to compare genomic information and microbial diversity of oceans in a variety of biogeographic contexts. Samples were collected from the water surface in about 40 different sites separated by more than 2 miles from each other to obtain different environmental samples. During this cruise, 6.1 million bacterial and viral proteins were annotated. Taken together, the GOS was the first project to demonstrate that high-throughput sequencing could be successfully used for the large-scale study of bacteria present in the oceans and thus reveal the genetic complexity of these microorganisms. However, by not taking into account the more complex plankton organisms that are eukaryotic protists and zooplankton, this project only explored a fraction of the marine plankton ecosystem. Furthermore, the physicochemical parameters measured are poor and limit the study of these microorganisms into their environmental context. Technologies have significantly evolved since the GOS campaign increasing the depth of sequencing and make it possible to sequence a larger portion of the organisms present in a sample and to observe a greater variety of plankton. One hundred and forty years after the expedition of the HMS Challenger, the *Tara* Oceans expedition (2009–2012) aimed to carry out a study of the plankton ecosystem on all oceans across all plankton classes (virus to fish larvae) while recording exhaustive environmental contextual parameters (Karsenti et al. 2011). The sampling sites were chosen for their oceanographic complementarities (Fig. 2). Sampling was carried out at different depths, with priority given to surface and deep chlorophyll maximum (DCM) where phytoplankton is abundant. Different filtration processes were used to separate the organisms according to their sizes. Thus, viruses, prokaryote, and eukaryotes could be separated and sequenced separately. Among eukaryotes, four size fractions were obtained to study plankton ranging from unicellular protists (0.8–5 μm) to multicellular zooplankton (5–20 μm, 20–180 μm and 180–2000 μm). In order to be able to carry out comparative genomics studies, i.e., to compare the genomes contained in different samples, the same standardized sampling protocols were used for all 35,000 samples. Via real-time satellite observations, the schooner collected samples in specific areas with particular physicochemical phenomena. This is the case of upwellings, where wind brings up deep, cold water, rich in nutrients where plankton frequently bloom at the surface. Samples were also collected from the heart of oceanic eddies that are known to trap and transport plankton over long distances, sometimes between two oceans (Villar et al. 2015). The schooner followed, where ever possible, the travel of water masses along ocean gyres, in order to study the impact of fluid movement on the organization and evolution of plankton communities. Comprehensive physicochemical parameters were measured in each sampling area (Pesant et al. 2015). Thanks to recent advances in the field of genomics, extensive sequencing was performed at unprecedented depth for this unique collection of samples (Alberti et al. 2017). A multipronged strategy based on different complementary approaches such as metabarcoding, metagenomics, metatranscriptomics, and single-cell genomics was chosen for the analysis of size-fractionated plankton communities. Two major gene sets have been released so far from the *Tara* Oceans sequencing effort: The Ocean Microbial Reference Gene Catalog (OM-RGC) and The Marine Atlas of *Tara*

Oceans Unigenes (MATOU). The OM-RGC is a comprehensive collection of 40 million genes from viruses, prokaryotes, and picoeukaryotes with a size up to 3 μm (Sunagawa et al. 2015) built by combining *Tara* Oceans sequences with other public marine plankton metagenomes as well as reference genomes. The MATOU is a catalog of 116 million unigenes obtained from poly-A + cDNA sequencing of different organismal size fractions ranging from 0.8 to 2000 μm (Carradec et al. 2018). About half of the unigenes have a predicted taxonomic assignation representing genes from more than 8000, mostly eukaryotic organisms. These global ocean reference gene catalogs have been recently made available through a Web service (Ocean Gene Atlas: http://tara-oceans.mio.osupytheas.fr/ocean-gene-atlas/) that allows biologists to easily explore the complex heterogeneous datasets generated by such large-scale campaigns, to produce one click plankton gene biogeographies (Villar et al.). Furthermore, a DNA metabarcoding survey based on *Tara* Oceans global plankton samples has revealed the existence of an unexpectedly high number of heterotrophic protist taxa in eukaryotic communities (de Vargas et al. 2015). To obtain a precise genome-level description of these abundant marine heterotrophic protists, a single-cell sequencing was also performed (Seeleuthner et al. 2018; Vannier et al. 2016). Because the higher latitudes could not be sampled during the initial *Tara* Oceans campaign for logistics reasons, an additional *Tara* Oceans Polar Circle expedition (2012–2013) was launched during which 58 stations were sampled in a 6 months circumnavigation of the Arctic Ocean. These new data are currently being integrated with the previous expedition and are undergoing analysis. Thus, a larger portion of the genomic sequences of the microorganisms has now been sampled from the global sunlit oceans. This is the first time that such a massive amount of genomic data can be combined with rich physicochemical measurements and physical oceanographic models for a global study of the plankton ecosystem. It is therefore through multidisciplinary approaches that the *Tara* Oceans dataset can hope to contribute to solve the questions we ask ourselves about the ocean ecosystem. However, such an interdisciplinary study is a real challenge for scientists who need to develop methods and tools to analyze heterogeneous data.

In addition, other large-scale campaigns have set out to explore the diversity of the planktonic ecosystem. For instance, the Malaspina expedition (Duarte 2015) aims to achieve a more comprehensive knowledge of microbial diversity the deep ocean. Between December 2010 and July 2011, a total of 91 samples were collected in 13 stations spread over the global ocean at different depths: surface, DCM and 2–3 depths in mesopelagic (200–1000 m) and bathypelagic waters (1000–4000 m). Sequencing of the prokaryotic size fraction is underway. An alternative project, the Ocean Sampling Day (Kopf et al. 2015), consists in a simultaneous sampling campaign of the world's oceans (typically located in coastal regions) during the summer solstice (21st June) to obtain a yearly snapshot of the marine microbial taxonomic and functional diversity in the world's oceans. During OSD 2014, scientists around the world collected 155 metabarcoding samples (16S/18S rRNA amplicons), 150 metagenome samples, and a rich set of environmental metadata, generating the largest standardized dataset on marine microbes taken on a single day. OSD sampling was repeated in 2015 and will be organized again in June 2018 after new funding was secured via the

EMBRC ASSEMBLE Plus project (http://www.assembleplus.eu/research/ocean-sa mpling-day-2018). A future project *Pourquoi pas les abysses?* will aim to study the distribution of the marine biodiversity in deep-sea marine sediments. The increasing availability of environmental metagenomes in a complementary and a wide range of environments provide opportunities to investigate the evolutionary and organizational mechanisms that govern the planktonic ecosystem at the scale of individual organisms, population, and planktonic communities.

2 Evolutionary Dynamics of an Organism Within Its Environment

2.1 Different Metagenomic Approaches to Study Plankton Diversity

Genomic analyzes can be performed to understand how the genomic structure and the geographic distribution of an individual organism are influenced within its environment. Thus, in the last decade, several families of environmental genomics tools have emerged to help answer this question from the scale of individual organism to global microplanktonic communities. A first approach consists in obtaining the catalog of genes present in a metagenome and estimating their abundance. This quantitative metagenomic analysis allows abundance estimate of gene families/functions as well as taxa present in the samples using single marker genes (Lindner and Renard 2013). This makes it possible to compare the different samples according to the taxa and genes shared. Because they have a well-defined taxonomic signature, the hypervariable regions of the rRNA genes are generally used to reveal the taxonomic origin of the organisms present in a sample (Weisburg et al. 1991; Woese et al. 1990). The use of DNA barcoding in molecular phylogeny has made it possible to identify noncultivable organisms from gene marker sequences (Hebert et al. 2003). This method consists in extracting the DNA from an organism and amplifying by PCR the marker gene target fragment (called barcodes) using a previously defined universal pair of primers. The variety of sequences obtained can be clustered to define operational taxonomic units (OTUs). Quantitative environmental genomics makes it possible, for example, to observe fluctuations in diversity of a class of organisms in distinct samples (Le Bescot et al. 2016; de Vargas et al. 2015). Using such methods, the photosynthetic protists, *Micromonas, Ostreococcus* and *Bathycoccus,* have been identified as particularly important ecologically because they are found in a wide variety of oceanic ecosystems, from the poles to the tropics (Demir-Hilton et al. 2011; Foulon et al. 2008; Lovejoy et al. 2007; Simmons et al. 2016; Šlapeta et al. 2006). The cosmopolitan distribution of these genera raises the questions of their diversity and their adaptation to local environmental conditions. Monier and colleagues carried out a study on the diversity and biogeography of this lineage using barcodes sequenced in *Tara* Oceans samples from the 2009–2012 expedition (Monier et al. 2016). However, this approach relies on the availability of known reference sequences. The absence of these references may be an obstacle by creating bias artifacts in genetic diversity estimates. To avoid this limitation, targeted investigations aim to measure the abundance and relative distribution of microbial genomes across environments through read recruitment analyses (process of aligning sequencing reads to reference genomes). It is hence possible from a reference genome to know if this organism is present or not in a sample, represented by all or just a sub-portion of its full gene complement. It is possible by this approach to study organisms or genes poorly represented in a metagenome (Suenaga 2012). By observing the presence or absence of genes in environments with varied physicochemical conditions, it is possible to study the genomic adaptations of non-cultivable organisms (Cuvelier et al. 2010). Since it is necessary to use a reference genome or marker gene that has already been sequenced, assembled, and annotated, the targeted metagenomics approaches focus on known taxa. The following sections present the application of targeted metagenomic approaches to the study of the biogeography of two green algae genomes and their gain or loss of genes in the metagenomic samples of the *Tara* Oceans expedition (2009–2012).

2.2 Biogeography of the Green Picoalga Bathycoccus Genomes in the Global Ocean

Bathycoccus prasinos is a cosmopolitan green microalgae belonging to the Mamiellophyceae, a class of picophytoplankton that contains important contributors to oceanic primary production. *Bathycoccus* was initially collected 100 m deep at the DCM in the Mediterranean Sea (Eikrem and Throndsen 1990). Cells with polysaccharide scales and having the same morphology were also observed in the Atlantic Ocean (Johnson and Sieburth 1982). *Bathycoccus* was subsequently found in diverse oceanic environments including coastal waters (Collado-Fabbri et al. 2011; Not et al. 2004), leading to its classification as a cosmopolitan organism. The genome of the coastal strain *Bathycoccus prasinos* RCC1105 was sequenced (Moreau et al. 2012) and shown to be composed of 19 chromosomes for a total of about 15 Mb. Metagenomic data suggested the existence of two *Bathycoccus* ecotypes (Monier et al. 2013; Vaulot et al. 2012), recently named BI and BII (Simmons et al. 2016). These two ecotypes have identical 18S ribosomal RNA sequences and therefore cannot be differentiated using barcode markers such as V4 or V9 regions of the 18S rRNA genes (Monier et al. 2013). Currently, because only one *Bathycoccus* genome is available in public databases, it is not possible to observe the geographical distribution and ecological preferences of these two ecotypes or to compare their genomic structure in order to study the evolutionary mechanisms that govern these unicellular green algae.

Recently, a second *Bathycoccus* genome was generated by the single-cell amplified genome (SAG) as part of the *Tara* Oceans expedition (2009–2012). Indeed, it is now possible to sequence the genome of an organism from a single cell taken directly from the environment. This makes it possible to identify and assemble non-cultivable genomes (Marcy et al. 2007). In addition, using the single-cell amplified genome method, cell functions can be finely dissected (Eberwine et al. 2014) since, in the case of multicellular organism, cells are addressed individually rather than on the whole body, as was done previously (Amberger et al. 2009; Tringe et al. 2005). This technique involves the isolation of a single cell (often through microflu-idics cell sorting) and then whole genome amplification (WGA) before sequencing and assembly. The amplification step is therefore crucial since it is impossible to sequence a genome from essentially a single DNA molecule contained in a single cell. Amplification techniques have the disadvantage of introducing amplification biases where certain regions of the genome are preferentially copied, to the detri-ment of other regions which can end up undetectable during sequencing. In addition, WGA can introduce mutations or even chimeric sequences (Gawad et al. 2016). These approaches cover an average of 40% of the genome (Baker 2012). Alterna-tives to improve sequencing coverage as well as detection of nucleic variations exist. For instance, it is possible to sequence several cells belonging to the same organism. Co-assembly is then carried out with the sequences resulting from the sequencing of each single cell. Each cell participates in the assembly of scaffolds which generally increases the size of the sequenced genomes. However, care must be taken to use cells belonging to the same species without excessive intrapopulation variation which could fragment the assembly or even generate chimeric sequences. This method was used for cells harvested in the wild during the *Tara* Oceans expedition (2009–2012) (Seeleuthner et al. 2018; Vannier et al. 2016). The *Bathycoccus* SAG had a length of 10.3 Mb, and the assembly coverage was estimated at 64% of the reference genome. Four cells with the same 18S RNA sequence as strain RCC1105 were harvested from the Arabian Sea at station 39 of the *Tara* Oceans expedition (2009–2012), hence its denomination TOSAG39-1 (first SAG of station 39 of *Tara* Oceans).

The cosmopolitan distribution of *Bathycoccus* in the oceans raises questions about its diversity and adaptations to environmental conditions. This is why the spatial diversity of *Bathycoccus* was explored using its two available genomes: the Mediter-ranean coastal strain (RCC1105) and the SAG of *Bathycoccus* (Vannier et al. 2016). The comparative genomic analysis of these two genomes sharing the same 18S rRNA sequence revealed that they are genetically distant since these two genomes share an average protein identity of only 78% and have incomplete synteny. In addition, both genomes have different internal splicing marker (ITS). Thus, these genomes correspond to two distinct but cryptic species of *Bathycoccus*. The analysis of their geographical distribution in 122 metagenomic samples shows that both ecotypes are found in nutrient-rich environment, but have distinct ecological niches (Fig. 3a). TOSAG39-1 is found in warmer waters and at the DCM, while RCC1105 is detected in colder surface and oxygen-rich waters. Finally, both *Bathycoccus* are generally absent from stations with low iron concentrations (Fig. 3b).

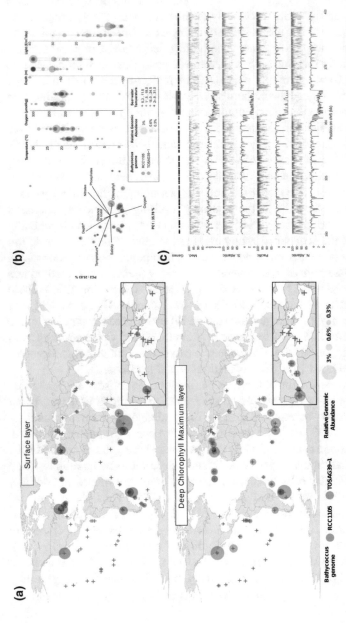

Fig. 3 **Biogeography and variability of gene structure in** *Bathycoccus prasinos*. **a** Geographical distribution of two *Bathycoccus* genomes, RCC1105 and TOSAG39-1, along *Tara* Oceans expedition stations (2009–2012) from recruitments of metagenomic reads. Top and bottom maps correspond to the surface and deep chlorophyll maximum (DCM) samples, respectively. Gray crosses indicate *Tara* oceans sampling stations and the sizes of the red or blue circles indicate the relative genomic abundances of the two *Bathycoccus* types. The two *Bathycoccus* are parameters and *Bathycoccus* genome abundance. Left: Principal component analysis. Crosses indicate stations, with a color scale corresponding to the water temperature. The two *Bathycoccus* are distributed along temperature and oxygen axes. Stars indicate parameters that statistically discriminate the two *Bathycoccus*. Right: Range of values of temperature, oxygen, and sampling depth for parameters where a significant difference was detected between RCC1105 and TOSAG39-1. **c** Evidence for cassettes of dispensable genes in *B. prasinos* RCC1105. Fragment recruitments plots are displayed by marine zones. Each dot corresponds to a given number of mapped reads at a given identity percent (indicated on the Y-axis). The density of mapped read is displayed as the black line plotted below each fragment recruitment plot. Gene position is represented by black boxes on the top of the first fragment recruitment plot, and dispensable genes are highlighted in red. Figure reproduced, with modification, from Vannier et al. (2016) with permission

2.3 Genomic Plasticity in Bathycoccus prasinos: An Evolutionary Dynamics for Such Cosmopolitan Organisms?

The genome of the coastal strain *Bathycoccus prasinos* RCC1105 is composed of two chromosomes 14 and 19 which are considered "outliers" because of their structural and functional differences compared to other chromosomes. Indeed, these two chromosomes are small, their guanine-cytosine content is low, and their proportion of horizontal genes transfers is high compared to other chromosomes. Chromosome 19, called SOC for "small outlier chromosome," is the smaller of the two. It has no collinearity with the genomes of other Mamiellales and contains many genes without homologs in other species. Chromosome 14, known as BOC for "big outlier chromosome," has higher intron content and expression levels than other chromosomes.

A study of the genomic variation of the Mediterranean strain RCC1105 in different samples was carried out using the targeted metagenomics approach to detect the gain or loss of genes in the metagenomic samples (Vannier et al. 2016). Dispensable genes, which correspond to genes present or absent in some samples, were detected. The metatranscriptomic approach allows to distinguish the potentialities due to the gene content from the effective functional activities of the organisms in situ. Thereby, from the metatranscriptomic data, the validation of the existence of these genes and their level of expression was quantified. A total of 108 dispensable genes (representing approximately 1% of the genome) were thus observed. The study of their structures and their positions on the chromosomes showed that half of these genes are located on chromosome 19, previously described as an outlier chromosome in *Bathycoccus*. The other half is randomly positioned on the genome, often in successive gene cassettes (Fig. 3c). Using metagenomic assemblages, the continuity of the genomic regions around these cassettes was verified. A synteny study between the genomes of RCC1105 and TOSAG39-1 was also performed to compare regions containing dispensable genes positioned in cassettes. This analysis showed the absence of these cassettes in the genome of TOSAG39-1. In addition, the read coverage that aligns with these gene cassettes differs across ocean basins. This suggests the existence of several genomic types within the same species that vary according to the ecological environment. In that respect, the environment has an impact on the gene structure and the biogeography of *Bathycoccus*. Previous analyses have already shown the existence of gene repertoire variations in bacteria (Acuña et al. 2013; Coleman et al. 2006; Fernández-Gómez et al. 2012; Gonzaga et al. 2012; Kashtan et al. 2014). In particular, it has been shown by comparing two ecotypes of *Prochlorococcus* that variation in gene content is related to adaptation to different light intensities (Rocap et al. 2003). Similar adaptation hypotheses have been proposed for *Ostreococcus* (Rodríguez et al. 2005) but are still a matter of debate (Demir-Hilton et al. 2011). The two *Bathycoccus* ecotypes were shown to be present at different depth levels and light intensities (Fig. 3b). However, these genomes can sometimes share the same location. In fact, TOSAG39-1 which is mostly found in DCM is also observed in surface samples where vertical currents occur leading to mixing of the two species. This is the case in the Agulhas currents as well as in the Gulf Stream. In the samples present along the currents from the Indian Ocean to the South Atlantic Ocean, there is a gradual replacement from east to west of TOSAG39-1 by the genome of RCC1105. The same distribution of the two ecotypes from west to east is observed along the North Atlantic Gyre. The evolution along this gyre of the temperature gradient and nutrient concentration likely influences the distribution of these two ecotypes (Fig. 3a). Thus, the vertical flux might transport TOSAG39-1 from the DCM depth to the surface at certain sites, before being carried away by the horizontal currents. Along these horizontal currents, the *Bathycoccus* distribution evolves with changing environmental parameters. In eukaryotes like MAST-4 (Lin et al. 2012; Rodríguez-Martínez et al. 2013) as the Arctic ecotype of *Micromonas* (Lovejoy et al. 2007), temperature also appears to influence the geographical distribution of these microorganisms. Future analysis of additional genomes and transcriptomes of wild and cultured *Bathycoccus* will improve the accuracy with which the environmental niches of the two-type *Bathycoccus* can be characterized. However, in this analysis, biotic interactions are not taken into account. Biotic factors represent all the interactions between organisms in an ecosystem. For instance, an experimental evolution experiment of *Ostreococcus tauri* inoculated with a large quantity of virus, Otv5, provided evidence that genes on outlier chromosome 19 are up-regulated in viral-resistant cell lines and that the size of this chromosome varies in resistant lines (Yau et al. 2016). Furthermore, a recent study of the Mamiellophyceae class showed that the resistance of isolates to large double-stranded DNA viruses, which abound in their natural environment, is positively correlated with the size of a single hypervariable chromosome. This finding highlights the role of viruses in shaping genome diversity in marine picoeukaryotes (Blanc-Mathieu et al. 2017). The results on gene content plasticity in chromosome 19 of *Bathycoccus* are consistent with the immunity chromosome hypothesis: frequent events of gene birth and gene loss may thus be the genomic traces of a microalgal—virus evolutionary arms race. Dispensable genes possess features of so-called de novo genes, genes emerging from previously noncoding regions. These genes are an important class of unknown genes and challenge evolutionary sciences (Carvunis et al. 2012; Schlötterer 2015). It has been hypothesized that cosmopolitan bacteria might hold specific genes or gene variants due to their ecological properties (Ramette and Tiedje 2007). Cosmopolitan marine lineages are exposed to a range of contrasted environmental constraints, raising the question of their genomic plasticity. The high turnover of certain classes of genes restricted to some environmental conditions might be an evolutionary advantage for rapid acclimatation of cosmopolitan organisms.

2.4   Biogeography and Genomic Diversity of Heterotrophic Stramenopiles

Another study has exploited single-cell genomics and metagenomics data from the *Tara* Oceans expedition to analyze the genome content and apparent oceanic distribution of uncultured heterotrophic stramenopiles (Seeleuthner et al. 2018). These marine protists have been described as key players in the biogeochemical cycles in the oceans (Worden et al. 2015). Marine stramenopile group 4 (MAST-4) representatives are small, flagellated, bacterivorous cells that are abundant in temperate and tropical oceans (Massana et al. 2004, 2014). Seeleuthner and colleagues selected 40 single-cell representatives of different lineages of MAST: 23 from three MAST-4 lineages (MAST-4A, MAST-C, and MAST-E), six from two lineages of MAST-3 (MAST-3A and MAST-F), and 11 from two lineages of chrysophytes (chrysophytes H1 and H2). A partial genome of a MAST-4 clade D, previously characterized using single-cell sequencing (Lasken 2012), was integrated into a geographical distribution analysis of these lineages (Fig. 4). For this, metagenomic fragment recruitment was performed from the 0.8–5 μm size fraction of the *Tara* Oceans metagenomics dataset. Similarity to the *Bathycoccus* study, each of the distributions was compared to the environmental parameters recorded at each sampling site. MAST-4A appears

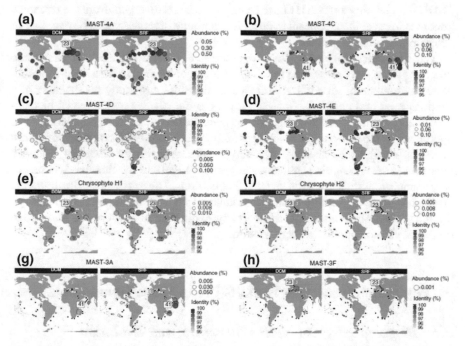

Fig. 4 Biogeographic distribution of the SAG lineages based on metagenome read recruitment with separation between deep chlorophyll maximum and subsurface. Global maps showing the presence of the SAG lineages based on metagenomics read mapping at each *Tara* Oceans (2009–2012) station either as a black dot (no signal detected) or as a circle whose diameter indicates the species relative abundance. Abundance in samples from deep chlorophyll maximum (DCM, left panel) often differs from surface samples (SRF, right panel): only MAST-4A shows the same pattern in DCM and SRF samples (**a**). The color inside each circle provides the median percentage similarity of the reads to the reference. The station from where the SAG originates is indicated by its number. **a** MAST-4A; **b** MAST-4C; **c** MAST-4D; **d** MAST-4E; **e** chrysophyte H1; **f** chrysophyte H2; **g** MAST-3A; and **h** MAST-3F. *Source* Seeleuthner et al. (2018)

to be cosmopolitan, with the exception of the Southern Ocean. MAST-4C has a geography range restricted mostly to tropical and sub-tropical waters, except in the sub-tropical Atlantic Ocean. Finally, chrysophyte H2 and MAST-3F are low abundance species encountered in different regions. The authors observed that each of the examined taxa may have a specific spatial distribution that correlates with environmental parameters, principally ocean provinces, temperature, and depth. Water temperature was the most significant parameter that discriminated the distributions, suggesting that some of these species likely have preferential temperature ranges in which they are maximally abundant. Interestingly, Seeleuthner and colleagues also used the metagenomics data to obtain indications of genetic diversification by using the similarity of nucleotide sequences to each reference genome as a measure of divergence. The previously sequenced MAST-4 D genome is encountered in only one coastal sample from the South Atlantic Ocean, indicating that open ocean populations of MASTs can differ from coastal populations. Additionally, MAST-4E, MAST-3F, and chrysophyte H1 have genotypes subsets divergent from the reference genomes. Hence, there appears to be a high genomic diversity yet to be explored in heterotrophic marine protists.

The characterization of the biogeography and genetic plasticity of planktonic microorganisms has shown the capacity of metagenomic data generated by large-scale environmental survey like *Tara* Oceans expedition to study the diversity of a single species and its genes. However, it is possible, always by a targeted metagenomic approach, to identify genomic variants inside a plankton population. The hypothesis of the existence of several genomic types of *Bathycoccus prasinos* across ocean basins and the existence of genotype subsets divergent from the reference genomes of several marine stramenopile group 4 highlight the fact that it is necessary to analyze further plankton genomic populations. The next section illustrates the efforts developed to identify genomic variants inside a plankton population through the integration of metagenomic and oceanographic data to advance our knowledge on the planktonic population organization.

3 Population Genomics Variants Using Metagenomic Data

3.1 Biogeography, Population Structure, and Evolution of the Copepod Oithona

Other approaches propose to identify genomic variants inside a plankton population at the whole genome level using metagenomic data. A recent study performed this type of analysis for the first time using copepod *Oithona nana* (Madoui et al. 2017). The study of pelagic copepod populations at the molecular level contributed to identify environmental factors that drive the appearance and fixation of adaptive traits. *Oithona* is considered very abundant and widespread in the ocean's surface (Gallienne and Robins 2001). This genus plays an important ecological role as grazers and secondary producers in the marine trophic food chain (Turner 2004). Within this genus, three species were described as particularly widespread based on morphological identifications: *Oithona similis* Claus, 1866, *Oithona atlantica* Farran, 1908, and *Oithona nana* Giesbrecht, 1892. The present knowledge on the biogeography of *Oithona* species has been mainly conducted through morphological identification of specimens collected by independent and geographically restricted studies. Madoui and colleagues performed a global biogeography of *Oithona* species using the metagenomic data sequenced from the *Tara* Oceans expedition (2009–2012). To this end, the metagenomic reads from the sequencing of the DNA extracted from the 20 to 180 μm size fractions at the surface and DCM waters were aligned on a manually curated 28S reference sequence database. To allow the identification of *O. similis* and *O. nana* without ambiguity, the authors selected the reads with stringent alignments (reads with perfect identity over 98% of their length). From this read mapping-based biogeography, the authors observed a predominance of *O. similis* in temperate and polar waters which illustrates its ecological relevance as a small-sized zooplankton in these oceanic areas. Due of the general lack of references for most of the 44 *Oithona* species, several samples were identified as containing undefined *Oithona* spp.; this point highlights the fact that more efforts are required to generate new references for under-represented species in the public databases. Finally, the species *O. nana* was described as limited to coastal waters, since it was identified only in a few *Tara* Oceans samples, mostly in the Mediterranean Sea (MS) (Fig. 5a). The populations of *O. nana* in the MS seem to be structured between and within basins; however, the intrapopulation genetic differentiation could not be measured. From this biogeography, the authors performed a population genomics analysis using *O. nana* in the MS. The genomic variation landscape of *O. nana* was investigated in five stations (TARA_10, 11, 12, 24 and 26) located in the northern part of the MS (Fig. 2). The alignment of metagenomic reads from these stations against the *O. nana* genome detected genomic variants including 221,018 biallelic variable loci. The B-allele frequency (BAF) was calculated and the pairwise F_{st} was used to estimate the genetic distance between the five populations (Table 1). The pairwise F_{st} between stations showed that the populations from the Adriatic Sea (TARA_24 and 26) were structured and presented moderate differentiation compared to two populations from the MS western basin (TARA_10 and 11). The authors observed a moderate genetic structure within the western basin population group and a low value between TARA_10 and 11 ($F_{st} = 0.048$) due to the short distance between these stations. Furthermore, a PCA based on the BAF clustered the Adriatic populations with those from TARA_12 and separated the populations from TARA_10 and 11 (Fig. 5b). These results suggested that the *O. nana* populations are structured between the MS basins but also within the two basins.

Next, the variation in the BAF was used to identify genomic clines in the *Oithona* populations along the MS circulation patterns (Algerian, Lybio-Egyptian, and Northern currents) (Fig. 5c). Madoui and colleagues selected five stations (TARA_7, 8, 14, 17 and 18) for their localization in the southern part of the MS and integrated them with the previous Mediterranean stations. The BAFs of these ten MS stations were calculated from metagenomic read alignments which identified 754,669 bial-

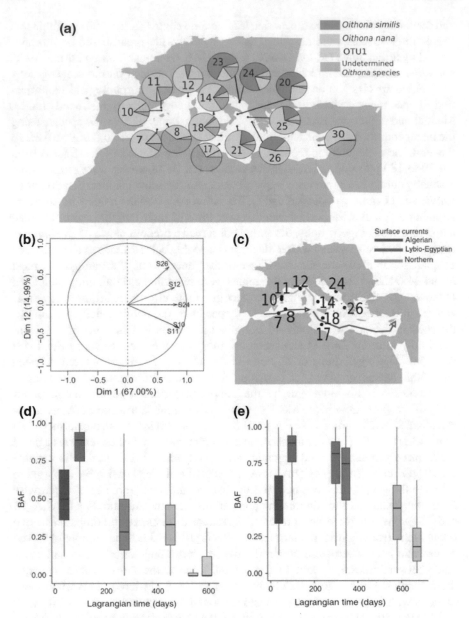

Fig. 5 Biogeography of *Oithona* species and genomic variants in *Oithona nana* populations on the Mediterranean Sea (MS). a Biogeography of the *Oithona* species on the MS at the surface layer using the *Tara* Oceans (2009–2012) metagenomic data and 28S sequences. The proportion of each species is represented by a pie chart. **b** Principal component analysis results of the populations based on the B-allele frequency. **c** Dominant Mediterranean surface currents. **d** BAF variation along the surface currents of the first way and **e** on the second way. X-axis is the Lagrangian distance in days. Y-axis is the BAF. Figure reproduced, with modification, from Madoui et al. (2017) with permission

Table 1 Median pairwise F_{st} distances between *Oithona nana* populations sampled in five stations of the Mediterranean Sea. *Source* Madoui et al. (2017)

	Western MS basin			Eastern MS basin	
	TARA_10	TARA_11	TARA_12	TARA_24	TARA_26
TARA_10	–	0.05	0.13	0.1	0.22
TARA_11		–	0.11	0.1	0.22
TARA_12			–	0.1	0.09
TARA_24				–	0.14
TARA_26					–

lelic loci. The comparison of BAF between stations allowed the authors to confirm the presence of another *Oithona* species, closely related but distinct to *O. nana*. Then, the variation in the BAF distribution was analyzed along the Mediterranean currents as a function of the Lagrangian distance (Berline et al. 2014) from station TARA_7. This approach allows to study the genomic variation along gyres. This variation was analyzed along two trajectories following the main circulation patterns (Fig. 5c). Along these two trajectories, they observed a genomic cline between *O. nana* and the other species with an admixture in the populations from different stations (Fig. 5d, e). Thus, two genomic variations along marine gyres were identified. This study provided novel findings in zooplankton population genomics through the original integration of metagenomic and oceanographic data.

3.2 Metapangenome *Approach to Analyze the Fitness of Microbial Populations*

A recent analysis and visualization strategy developed to tackle genes content dynamics in closely related microbial genomes was performed using a combination of pangenomes and metagenomes approaches (Delmont and Eren 2018). Pangenomes offer detailed characterizations of core and accessory genes found in a set of microbial genomes, generally by clustering genes based on sequence homology. This approach has been used to characterize the gene content of microbial populations in environmental samples through metagenomic read recruitment (Delmont and Eren 2016; Nayfach et al. 2016; Scholz et al. 2016). Delmont and Eren proposed a novel *metapangenome* approach which combines the identification of gene clusters and the relationships between genomes based on shared genes (pangenomics) with the ability to track populations precisely across environments through genome-wide read recruitment (metagenomics). Such an approach can be used to identify the ecological role of gene clusters that may be linked to niche partitioning and microbial population fitness. With the specifically developed Anvi'o workflow (Eren et al. 2015), the authors characterized the metapangenome of *Prochlorococcus* isolates and single-cell genomes in surface marine waters using metagenomic reads from the

Tara Oceans expedition (2009–2012). *Prochlorococcus*, the smallest known photo-synthetic bacterium (Chisholm et al. 1988), was described as abundant in the ocean's surface layer despite large variation in environmental conditions (Bouman et al. 2006). Two groups of *Prochlorococcus* were observed: One group adapted to high-light and the other to low-light (Biller et al. 2014). Several genetically divergent lineages exist within *Prochlorococcus*, upon which phylogenetic diversity is super-imposed extensive gene gain and loss (Kettler et al. 2007). The genomic presence of many highly variable genes may be a driving force in the evolution of *Prochloro-coccus*. The use of large-scale metagenomic data can provide a solution to under-stand the distribution of phylogenetic and genomic diversity across environments. A first study using the metagenomes from the Global Ocean Sampling expedition showed a strong association between the *Prochlorococcus* accessory gene functions and the community composition of 12 *Prochlorococcus* isolates on a large geo-graphical scale (Kent et al. 2016). More recently, Delmont and Eren investigated the gene cluster distribution from 31 *Prochlorococcus* isolates using their occur-rence estimated from metagenomic reads counts from the *Tara* Oceans expedition (2009–2012) surface layer samples. The authors did not detect the low-light group in the metagenomic dataset, as previously described in the literature. They revealed that closely related *Prochlorococcus* populations sharing the same high-light niche exhibit differences in their relative abundance. Furthermore, the integration of 74 *Prochlorococcus* SAGs in their analyses revealed intriguing patterns of *Prochloro-coccus* hypervariable genomic islands by quantifying the correlation between indi-vidual gene clusters and the environment. Their results, consistent with the literature, show the potential of the metapangenomics approach to facilitate the recovery of key insights from novel and less studied microbial populations, especially those with no cultured representatives. Interestingly, the authors point out that a vast majority of isolates and single-amplified genomes contain only a subset of the complete set of genes microbial populations. Thus, metagenomic short read recruitment does not provide access to genes that are lacking in available genomes.

Another study by the same research team used the metapangenome approach to analyze the genome-wide diversity and the evolutionary dynamics of a single marine microbial population at the level of the amino acid variants across distant geographies (Delmont et al. 2017). For this finer analysis, the authors used the SAR11 family *Pelagibacteraceae* (Thrash et al. 2011) which represent the most ubiquitous free-living lineage of heterotrophic bacteria in the world's oceans (Morris et al. 2002). This dominance in surface seawater makes them a good candidate to study their diversity and their evolution in marine habitats. Delmont and colleagues performed a metagenomic reads alignment on 21 SAR11 genomes using 93 samples from *Tara* Oceans expedition and 10 samples from the Ocean Sampling Day project to cover high-latitude areas of the Northern hemisphere (Kopf et al. 2015). A pangenomic analysis in conjunction with read recruitment produced a metapangenome, revealing distinct distribution patterns for each clade within SAR11. A low-latitude subclade alone recruited more than one percent of total surface ocean metagenomic reads worldwide. The authors named this population "SAR11 Low-Latitude Population A" (S-LLPA). A total of 799 genes were selected as core genes in S-LLPA. These core genes recruited on average 1.25% of reads in the 74 metagenomes. The authors proposed that this high abundance offered the opportunity to study the genome-wide diversity and evolutionary dynamics of a single marine microbial population across distant geographies. To this end, they investigated the genomic diversity of S-LLPA within the boundaries of its niche. In a first step, they characterized all single-nucleotide variants (SNV) in these 799 core genes, revealing a total of 10,807,040 SNVs across the 74 metagenomes. From these SNVs, the genomic variations that impact amino acid sequence were identified in metagenomic data. The frequency of single-amino acid variants (SAAVs) in translated protein sequence was determined using the metagenomic short reads covering all three nucleotides in a given codon. After a filtering criterion of the preliminary selected SAAVs designed to improve downstream beta-diversity analyses, a total of 738,324 SAAVs within the core genes were identified. The analysis of the patterns of amino acid variability suggests that within S-LLPA, solvent accessibility may explain the differential involvement of amino acids in SAAVs. This reveals a strong correspondence between functional classes of amino acids and their stability in the environment, and a systematic influence of purifying selection against deleterious variants in the hydrophobic core of S-LLPA proteins. Finally, Delmont and colleagues performed a hierarchical clustering of samples based on deep learning-estimated distances to estimate relationships between metagenomes based on the collection of SAAVs. The distribution of the samples in two main groups which emerged from the clustering reflected large-scale ocean current temperatures. A first group was associated with warm currents and a second group was associated with cold currents. Two main ecological niches were suggested for this prevalent population. The analysis of the six sub-clusters (S-LLPA proteotype) that emerged from the dendrogram showed that they linked samples from distant geographical regions. The authors suggest that similar influence from environmental pressures like temperature may have resulted in increased protein sweep events in genes in S-LLPA. In this, observation appears to be in conflict with a simulation based on a neutral evolution model (Hellweger et al. 2014). Indeed, neutral theory holds that most variation at the molecular level does not affect fitness, and, therefore, the evolutionary fate of genetic variation is best explained by stochastic processes. To explore this concept further, analyses the evolutionary dynamics of plankton will have to take into account the different components constituting the planktonic ecosystem, including ocean currents. Furthermore, to mitigate the absence of many planktonic species in public databases and the fact that some genes are lacking in available genomes, e.g., SAGs, another family of metagenomics analysis emerged: comparative metagenomics. In the paper *Metagenomics and the Units of Biological Organization* published in 2010, W. Ford Doolittle and Olga Zhaxybayeva wrote (Doolittle and Zhaxybayeva 2010):

> First, we assert that metagenomics as a set of research techniques came along just in time to rescue microbiology from one of its more intractable practical concerns, culturability, and also one of its more onerous concepts, the species. Second, we claim that data bearing on the key concerns of environmental microbiologists, which we consider to be the diversity, dispersal, and niche differentiation of microbial cells, can be collected and interpreted perfectly well without reference to the concept of species. And third, we note that as a research

field, metagenomics is in the process of replacing the species as its fundamental unit with the community (or ecosystem).

Eight years later, this vision of the impact of metagenomics is proving to be true especially concerning comparative metagenomics which now allows us to reach observation at the scale of global microplanktonic communities. Applications using this method to study the organization of the planktonic ecosystem are illustrated in the next section.

4   Biogeography and Dynamics of Planktonic Communities

Biogeography is the study of the geographical distribution of species or communities of species in relation to environmental conditions. A first classification of the oceans was proposed in the 2000s using biogeochemistry data (Longhurst 2007; Oliver and Irwin 2008). However, this division mainly reflects the spatial distribution of phytoplanktonic organisms since it relies on the measurement of chlorophyll a. In previous sections, plankton biogeography was investigated at different scales (individuals or populations of microorganisms), for specific classes of plankton at specific geographical localization. With the progress of high-throughput sequencing in the last few years, it is now possible to survey the biogeography of a whole community of plankton at the global scale using comparative metagenomics approaches.

4.1   Geographical Organization of Planktonic Communities

Comparative metagenomics aims to detect similarities and differences between distinct metagenomes. A first approach is to use reference sequences to separate out known data (homologous to the references) in the metagenomes and to compare the samples based on these known subsets. However, the large amount of unknown sequences in complex environment means that these methods often only end up using a minor fraction of the total genetic information present in the metagenomes. Thus, an alternative comparative metagenomics approach does not require the use of known sequences. This approach, called de novo metagenomics, allows the comparison of all sequences contained in different metagenomes regardless of their known or unknown status. To perform this comparison, a local alignment algorithm like BLAST (Altschul et al. 1990) can be used for small datasets. Dinsdale and colleagues performed such a metagenomic comparison of almost 15 million random reads obtained by pyrosequencing from 45 distinct microbiomes and 42 distinct viromes from 9 environment types (Dinsdale et al. 2008). They showed that there exists a strong discriminatory metabolic profile across distinct environments. Furthermore, while most of the functional diversity was maintained in all of the communities, the relative occurrence of metabolisms varied, and the differences between metagenomes predicted the biogeochemical conditions of each environment. Interestingly, in their review, Doolittle and Zhaxybayeva highlight the fact that the variance based on the differential metagenomic reads between environments (79.9% of the combined microbiome and 69.9% of the virome variance) differs from a barcode analysis of the 16S rRNA genes from multiple environments witch explained only about 10% of the variance, suggesting that different ecosystems cannot be distinguished by their taxa alone. In fact, this last decade, several papers report the low resolution of the 16S or 18S rRNA variable regions for estimating the diversity of communities (Sawabe et al. 2007; Piganeau et al. 2011; Tang et al. 2012; Bendif et al. 2014; Wu et al. 2015; Machado and Gram 2015; Mohrbeck et al. 2015; Vannier et al. 2016; Machado and Gram 2017; Seeleuthner et al. 2018; Rodriguez-R et al. 2018). Due to its finer resolution, metagenomics is often better suited than metabarcoding to study the organization of planktonic ecosystems.

Such a de novo metagenomic analysis was performed with the metagenomes obtained from the sequencing of the *Global Ocean Sampling* expedition by Rusch and colleagues (2007). A de novo metagenomic comparison aims to measure the number of similar metagenomic sequences between different samples. The degree of genomic similarity can be used to group the metagenomic samples on the basis of their whole genetic content. Thus, Rusch and colleagues calculated the genetic similarity between two samples from the comparison of their full complement of metagenomic reads, rather than from the 16S rRNA subset as is frequently the case. For this purpose, an all against all pairwise sequence alignment was computed, which in turn allowed the calculation of the genetic similarity between all samples pairs. Via a hierarchical clustering method, the grouping of genetically similar samples showed that metagenomes from geographically close locations or sharing common environmental factors clustered together. However, this method is unable to associate a similarity. A distance is only defined between two objects which would indicate the number of shared sequences within samples. This can be problematic when a sample has redundant reads, and therefore, this distance does not provide information on the complexity of these samples. Finally, the alignment method used by Rush and colleagues (BLAST) does not scale and rapidly becomes intractable with comparisons of many metagenomes containing several million reads, as is the case for instance in the context of the *Tara* Oceans expedition. Indeed, despite significant efforts and success in optimizing the alignment time of classical alignment tools (Kent 2002; Edgar 2010; Kiełbasa et al. 2011; Drezen et al. 2014), the acceleration of alignment computation time has been unable to keep up with the acceleration of sequencing throughput.

As an alternative to classical sequence comparison approaches, methods based on analysis of k-mers distributions allows significant improvements. Alignment approaches have recently been specifically developed to de novo compare large number of metagenomes containing several million of sequences. For instance, the tool *Compareads* (Maillet et al. 2012) used a time and memory-efficient method for extracting similar reads between two metagenomic datasets of arbitrary size. For this, the similarity is based on shared *k-mers*. A shared *k-mer* corresponds to a word of size k whose forward and/or reverse complement version exists in the two samples

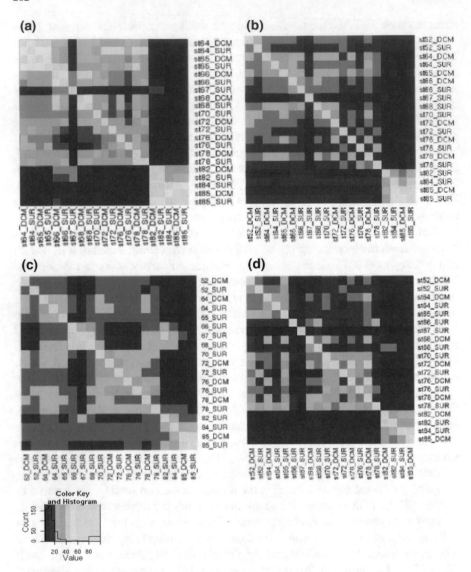

Fig. 6 Similarity matrix resulting from the comparison of the samples collected in the Indian, South Atlantic and Austral Oceans during the *Tara* Ocean expedition (2009–2012). Percentage of similarity was calculated with the *Compareads* tool on surface and DCM for filter: **a** 0.22–3 μm; **b** 0.8–5 μm; **c** 20–180 μm; **d** 180–2000 μm. Each color corresponds to a percentage of similarity. Figure reproduced, with modification, from Villar et al. (2015) with permission

to be compared. Thus, similar sequences between two samples are defined with this tool as being two sequences that share with a strict identity at least n non-overlapping k-mers (n and k being user selectable parameters, higher values of which yield more stringent alignment conditions). The derived similarity score between metagenomic samples has the advantage of being simple, and the indexing structure used does not require any sequence alignments per se to be performed. Indeed, despite of a few false positives and the simple definition of similarity, the *Compareads* tool makes it tractable to calculate the similarity of two metagenomic datasets of several hundred million reads in a few hours and deliver reliable biological conclusions compared to the classical methods. This tool was used on a subset of samples from the *Tara* Oceans expedition to study the genetic connection between the Indian Ocean and the Atlantic Ocean via Agulhas rings, anticyclonic mesoscale structures that transport water westwards across the Agulhas choke point (Villar et al. 2015). Eleven sampling stations from three oceans (Indian, South Atlantic, and Southern) in four size fractions (0.22–3 µm, 0.8–5 µm, 20–180 µm and 180–2000 µm) at surface and DCM depths were used. The Compareads-based pairwise comparisons of the genetic material of all organismal size fractions shared by these three basins were performed (Fig. 6). It was observed that a sample in the South Atlantic Ocean shared a low read simi- larity with all other samples (67_SRF) since it was sampled in a specific upwelling zone which contains a microplankton community different from other samples. Fur- thermore, Villar and colleagues observed a sample similarity within the same ocean (intra-ocean) higher than sample similarity between distinct oceans (inter-oceans). However, despite a larger potential exchange interface, samples from the Southern Ocean had fewer shared reads with samples from other oceans than shared between the South Atlantic and Indian Oceans. It is likely that the steep difference in environ- mental conditions in the Southern Ocean compared to the two other oceans means that a large proportion of the planktonic microorganisms present in this ocean are not the same as those present in the South Atlantic and Indian Oceans. Remarkably, the inter-ocean percentage of similar reads between samples from the Indian Ocean and the South Atlantic Ocean was close to the percentage of reads shared between the same intra-ocean samples. Hence, despite a limited physical connection via the Agulhas choke point, these two oceans appear to share similar plankton communi- ties. *Compareads* gave access to the difference in complexity between two samples. Indeed, by analyzing the dissymmetry of the similarity matrices of each sample size fraction, a large fraction of the sequences of the Atlantic Ocean was found to be similar to sequences of the Indian Ocean, while a lesser fraction of sequences of the Indian Ocean was similar to sequences in the South Atlantic Ocean. This could be explained by the connection of these two oceans by Agulhas rings. These rings forming in the Indian Ocean trap plankton from this ocean and transport it in the South Atlantic Ocean. During the expedition, sampling was also carried out inside a ring in the South Atlantic gyre 9 months after its formation, as well as another ring close to the Brazilian coast 3 years after its formation. Reads percentage shared by the sample collected in the "younger" ring was larger with the Indian Ocean than with the surrounding Atlantic Ocean. These sequence similarities between metagenomic samples were in line with the hypothesis of the capture and transport of planktonic microorganism by the Agulhas rings. To explore deeper the impact of ocean currents on the genomic organization of planktonic communities, it is possible to go even further by comparing plankton community genetic similarities with particle travel times between sampling stations.

4.2 Impact of Currents on the Plankton Community Biogeography

Plankton, transported by ocean currents, is constrained by multiscale processes, whether physicochemical or biological, and which together form the *seascape*. All these interactions along currents have an essential role on plankton population genetic structure (White et al. 2010). The evolutionary dynamics of plankton community organization along currents can be evaluated; however, such studies have usually focused on narrow spatio-temporal windows and/or on selected plankton groups. For instance, an overview of the ocean viral community patterns and its ecological drivers was described by Brum and colleagues (Brum et al. 2015). By modulating microbial population size, diversity, metabolic outputs, and gene flow, viruses deeply influence the planktonic ecosystem. Brum and colleagues used the datasets from 43 *Tara* Oceans expedition samples to assess viral community patterns and structure. A biogeographic pattern was revealed using metagenomic data whereby viral communities were passively transported by oceanic currents and locally structured by environmental conditions that affect host community structure. In this study, the movement of viral population throughout the oceans was visualized using calculations based on reciprocal comparison of viral population abundances between neighboring samples. To this end, for each sample pair, the average relative population abundances in one sample originating from a neighboring sample were calculated using targeted metagenomics and compared. The sign of the relative abundance difference between neighboring samples was used to estimate the movement direction, and the absolute value of the difference was interpreted as reflecting the movement magnitude. The authors observed that this metagenomic signal revealed that population exchange between viral communities was directed along major oceanic current systems such as the Agulhas current. Further, current strength was generally related to the magnitude of inter-sample population exchange. Brum and colleagues suggested that the intensity of water mass movement, in addition to environmental conditions, drives the biogeography of viral population.

The global plankton biogeography can also be interpreted in light of the spatio-temporal context of the seascape by comparing plankton community dissimilarities (β-diversity) to simulated connectivity times between sampling stations (Lagrangian distance). β-diversity is the comparison of species diversity between ecosystem along environmental gradients. This term was introduced by Whittaker (1960). β-diversity therefore measures how different local systems are. But this simple definition means β-diversity can be measured in different ways and is still subject of debate (Moreno and Rodríguez 2010). It can represent a *turnover* when quantifying the changes of communities along a spatial, temporal, or environmental gradient; or a variation when quantifying the differences in communities in a spatial or temporal extent (Anderson et al. 2011; Vellend 2001). The correlation between taxonomic similarity and geographic/temporal distance produces a distance–decay relationship which corresponds to the decline in similarity with increasing geographic/temporal distance. In their review, Hanson and colleagues discussed the processes which they claim create and maintain microbial biogeographic patterns (selection, drift, dispersal, and mutation) on ecological and evolutionary scales (Hanson et al. 2012). They consider how the interplay of these processes affects biogeographic patterns and the distance–decay relationship. This distance–decay indicates not only that composition is different between locations, but also that this variation is spatially autocorrelated (Nekola and White 1999). Distance–decay relationships have been repeatedly observed for microorganisms in a range of habitats at various taxonomic resolutions (Casteleyn et al. 2010; Green et al. 2004; Hewson et al. 2006; Horner-Devine et al. 2004; Martiny et al. 2011; Salazar et al. 2016). But few studies have identified a robust distance–decay pattern across taxa or across intrinsic factors related to organism characteristics, such as body size (Soininen et al. 2007).

A recent study focused on the role that abundance and body size play in determining spatial patterns of diversity (Villarino et al. 2018). Part of this study focused on the analysis of the estimated dispersal scales of several planktonic and nektonic organisms from prokaryotes to small mesopelagic fishes collected during the Malaspina 2010 Circumnavigation Expedition (Duarte 2015). Where possible, analysis of macroorganisms was conducted at the species level and that of mesozooplankton and heterotrophic prokaryotes and eukaryotes was conducted at the OTU level. To obtain estimated dispersal scales, a biotic similarity matrix and a surface ocean transit time matrix were used. The β-diversity matrix was calculated using pairwise "species" similarities for each group using the Jaccard similarity index. The surface ocean transit time matrix was built using estimates of minimum connection times or surface ocean transit times between sampling sites obtained from previously published global surface ocean Lagrangian particle simulations (Jönsson and Watson 2016). Where β-diversity was correlated with the oceanic transit time, they estimated rates of community dispersal and species spatial turnover using the halving-time metric and the time-decay slope. Thus, the authors observed that the estimated dispersal scales for different groups show a negative correlation with body size, where less abundant large-bodied communities have both significantly shorter dispersal scales and larger species spatial turnover rates than more abundant small-bodied plankton. This observation goes against the previously proposed neutral patterns (Hellweger et al. 2014) which suggested that microbes evolve faster than ocean currents can disperse them. With their faster growth rates, shorter generation times, higher metabolic rates, smaller organisms are indeed expected to follow this organization. Furthermore, in the global oceans, the influence of physical transport of both biotic and abiotic components and processes was recently reported to be strongest in smaller plankton, which contains most prokaryotic and eukaryotic phototrophs (Sunagawa et al. 2015; de Vargas et al. 2015). Whereas larger plankton, mostly multicellular eukaryotes with longer lifetimes, are therefore likely to be transported over long distances, resulting in a slower community distance-decay relationship. Villarino and colleagues made their observations with a distance–decay relationships analyzed among taxa based on 16S and 18S rRNA. As seen previously with *Bathycoccus*, the use of metabarcoding will not differentiate two cryptic species possessing the same barcode, and it is likely that many other microplankton species share the same genetic marker. The analysis of distances from such metabarcodes between different samples will therefore be impacted by this limited resolution bias, affecting both estimates of variations of microorganisms diversity and the *turnover* of communities. It is possible that the low microbial taxonomic resolution at the basis of the analyses of Villarino and colleagues could have led to artefactual differences between small and large-bodied plankton. Global-scale investigations on the drivers of microbial biogeographic pattern have focused on the analyses of taxa typically defined by barcodes such as 16S or 18S ribosomal RNA genes. The observed patterns are directly dependent on the units considered in examination of biogeographic patterns. The broader the examined units, the more cosmopolitan they naturally become.

Caro-Quintero and Konstantinidis review the findings emerging from metagenomic surveys of natural microbial populations (Caro-Quintero and Konstantinidis 2012). They highlight the potential of this method to study microbial communities at the species level, showing that microbial communities are predominantly organized in genetically and ecologically discernible populations, which possess the attributes expected for species. It is possible to explore with these high-resolution investigations, how populations are organized, interact, and evolve within communities. Caro-Quintero and Konstantinidis proposed that the discussion concerning the spatial scaling of microbial biodiversity be recast as follows: "Is there a spatial scale, a degree of sampling effort and a level of taxonomic resolution at which microbial biodiversity scaling relationships approach those of macroorganisms?". The sampling effort realized during the *Tara* Oceans expedition seems to be a good opportunity to answer this question, since comparative metagenomic approaches using *Tara* Oceans metagenomic reads have opened access to the biogeography of the viral, bacterial, and eukaryotic plankton communities in the surface ocean. Furthermore, this global plankton biogeography can be interpreted in light of the spatio-temporal and ecological context of the seascape by comparing the β-diversity to estimated connectivity times between sampling stations and differences in environmental conditions (Richter, Watteaux, Vannier et al., submitted). This latter analysis is found to be more fairly resolutive and to provide a better understanding of the evolutionary and organizational mechanisms that govern the planktonic ecosystem.

5 Conclusion and Perspectives

With the progress of high-throughput sequencing in the last few years and with the resulting metagenomic approaches, it is possible to study the DNA sequences of microorganisms directly sampled from the ocean. In this way, the environmental genomics data produced by the *Tara* Oceans expedition combined with a broad spectrum of analyses strategies improved our knowledge on the biogeography and on the evolutionary mechanisms that govern individual organisms all the way to populations. Furthermore, the integration of oceanographic data with the comparative metagenomic approaches at the population and community scales has enhanced our understanding of the organizational mechanisms that govern the planktonic ecosystem. In fact, the resolution of the metagenomic approach is currently unparalleled as it allows microorganisms to be described at the species level. Moreover, it provides access to almost all genomic sequences of the microorganisms present in a sample, including uncultured organisms. This approach triggers a new vision of microbial ecology that supplants the concept of "species." Comparative metagenomics is making a major contribution to the understanding of marine ecosystems by pushing observations to the scale of global microplanktonic communities. Communities will likely become the units of evolutionary and ecological study. In a community, organisms coexist and interact with each other and with the environment, and indeed, the network of biotic/abiotic interactions in the global ocean is beginning to be studied (Lima-Mendez et al. 2015). A future challenge would be to integrate these biotic interactions within the spatio-temporal and environmental context to take into account all the factors contributing to the planktonic ecosystem. Indeed, even if datasets such *Tara* Oceans are vast and represent unparalleled sequencing efforts, these data are nonetheless still fragmentary and preliminary since based essentially on one shot sampling and as such would need to be verified independently and also do not capture temporal trends such as seasonal variability or multi annual trends. Thus, the addition of metagenomic data resulting from past and future expeditions at different depths and geographical localizations should help to deepen our understanding of this ecosystem, its dynamics, and its ability to adapt to the physicochemical changes of ocean water, especially in the face of climate change.

Acknowledgements I would like to thank Pascal Hingamp (Aix-Marseille University) for his critical revision of this chapter. I would also like to thank my present and past colleagues/collaborators who have worked with me on the analyses of the *Tara* Oceans data over the past six years, particularly Patrick Wincker, Olivier Jaillon, Eric Pelletier, Jean-Marc Aury, Yoann Seeleuthner, Jade Leconte, Emilie Villar, Magali Lescot, Youri Timsit, Pierre Peterlongo, Nicolas Maillet, Daniele Iudicone, Colomban de Vargas, Daniel Richter, and Daniel Vaulot.

References

Acuña LG, Cárdenas JP, Covarrubias PC, Haristoy JJ, Flores R, Nuñez H, Riadi G, Shmaryahu A, Valdés J, Dopson M et al (2013) Architecture and gene repertoire of the flexible genome of the extreme acidophile Acidithiobacillus caldus. PLoS ONE 8:e78237

Alberti A, Poulain J, Engelen S, Labadie K, Romac S, Ferrera I, Albini G, Aury J-M, Belser C, Bertrand A et al (2017) Viral to metazoan marine plankton nucleotide sequences from the Tara Oceans expedition. Sci Data 4

Altschul SF, Gish W, Miller W, Myers EW, Lipman DJ (1990) Basic local alignment search tool. J Mol Biol 215:403–410

Amberger J, Bocchini CA, Scott AF, Hamosh A (2009) McKusick's Online Mendelian Inheritance in Man (OMIM®). Nucleic Acids Res 37:D793–D796

Anderson MJ, Crist TO, Chase JM, Vellend M, Inouye BD, Freestone AL, Sanders NJ, Cornell HV, Comita LS, Davies KF et al (2011) Navigating the multiple meanings of β diversity: a roadmap for the practicing ecologist. Ecol Lett 14:19–28

Baker M (2012) Method offers DNA blueprint of a single human cell. Nature

Bendif EM, Probert I, Carmichael M, Romac S, Hagino K, de Vargas C (2014) Genetic delineation between and within the widespread coccolithophore morpho-species Emiliania huxleyi and Gephyrocapsa oceanica (Haptophyta). J Phycol 50:140–148

Berline L, Rammou A-M, Doglioli A, Molcard Λ, Petrenko A (2014) A connectivity-based eco-regionalization method of the Mediterranean Sea. PLoS ONE 9:e111978

Biller SJ, Berube PM, Berta-Thompson JW, Kelly L, Roggensack SE, Awad L, Roache-Johnson KH, Ding H, Giovannoni SJ, Rocap G et al (2014) Genomes of diverse isolates of the marine cyanobacterium *Prochlorococcus*. Sci Data 1:140034

Blanc-Mathieu R, Krasovec M, Hebrard M, Yau S, Desgranges E, Martin J, Schackwitz W, Kuo A, Salin G, Donnadieu C et al (2017) Population genomics of picophytoplankton unveils novel chromosome hypervariability. Sci Adv 3:e1700239

Bouman HA, Ulloa O, Scanlan DJ, Zwirglmaier K, Li WKW, Platt T, Stuart V, Barlow R, Leth O, Clementson L et al (2006) Oceanographic basis of the global surface distribution of Prochlorococcus ecotypes. Science 312:918–921

Brum JR, Ignacio-Espinoza JC, Roux S, Doulcier G, Acinas SG, Alberti A, Chaffron S, Cruaud C, de Vargas C, Gasol JM et al (2015) Patterns and ecological drivers of ocean viral communities. Science 348:1261498

Caro-Quintero A, Konstantinidis KT (2012) Bacterial species may exist, metagenomics reveal. Environ Microbiol 14:347–355

Carradec Q, Pelletier E, Silva CD, Alberti A, Seeleuthner Y, Blanc-Mathieu R, Lima-Mendez G, Rocha F, Tirichine L, Labadie K et al (2018) A global ocean atlas of eukaryotic genes. Nat Commun 9:373

Carvunis A-R, Rolland T, Wapinski I, Calderwood MA, Yildirim MA, Simonis N, Charloteaux B, Hidalgo CA, Barbette J, Santhanam B et al (2012) Proto-genes and de novo gene birth. Nature 487:370–374

Casteleyn G, Leliaert F, Backeljau T, Debeer A-E, Kotaki Y, Rhodes L, Lundholm N, Sabbe K, Vyverman W (2010) Limits to gene flow in a cosmopolitan marine planktonic diatom. Proc Natl Acad Sci 107:12952–12957

Chisholm SW, Olson RJ, Zettler ER, Goericke R, Waterbury JB, Welschmeyer NA (1988) A novel free-living prochlorophyte abundant in the oceanic euphotic zone. Nature 334:340–343

Coleman ML, Sullivan MB, Martiny AC, Steglich C, Barry K, Delong EF, Chisholm SW (2006) Genomic islands and the ecology and evolution of Prochlorococcus. Science 311:1768–1770

Collado-Fabbri S, Vaulot D, Ulloa O (2011) Structure and seasonal dynamics of the eukaryotic picophytoplankton community in a wind-driven coastal upwelling ecosystem. Limnol Oceanogr 56:2334–2346

Cornell HV, Lawton JH (1992) Species interactions, local and regional processes, and limits to the richness of ecological communities: a theoretical perspective. J Anim Ecol 61:1–12

Cuvelier ML, Allen AE, Monier A, McCrow JP, Messié M, Tringe SG, Woyke T, Welsh RM, Ishoey T, Lee J-H et al (2010) Targeted metagenomics and ecology of globally important uncultured eukaryotic phytoplankton. Proc Natl Acad Sci 107:14679–14684

Delmont TO, Eren AM (2016) Linking comparative genomics and environmental distribution patterns of microbial populations through metagenomics. bioRxiv 58750

Delmont TO, Eren AM (2018) Linking pangenomes and metagenomes: the Prochlorococcus meta-pangenome. PeerJ 6:e4320

Delmont TO, Kiefl E, Kilinc O, Esen OC, Uysal I, Rappe MS, Giovannoni S, Eren AM (2017) The global biogeography of amino acid variants within a single SAR11 population is governed by natural selection. bioRxiv 170639

Demir-Hilton E, Sudek S, Cuvelier ML, Gentemann CL, Zehr JP, Worden AZ (2011) Global distribution patterns of distinct clades of the photosynthetic picoeukaryote Ostreococcus. ISME J 5:1095–1107

Dinsdale EA, Edwards RA, Hall D, Angly F, Breitbart M, Brulc JM, Furlan M, Desnues C, Haynes M, Li L et al (2008) Functional metagenomic profiling of nine biomes. Nature 452:629–632

Doolittle WF, Zhaxybayeva O (2010) Metagenomics and the units of biological organization. Bioscience 60:102–112

Drezen E, Durand P, Lavenier D (2014) KLAST, a Blast-like tool for fast sequence similarity searches. Boston

Duarte CM (2015) Seafaring in the 21st century: the Malaspina 2010 circumnavigation expedition. Limnol Oceanogr Bull 24:11–14

Eberwine J, Sul J-Y, Bartfai T, Kim J (2014) The promise of single-cell sequencing. Nat Methods 11:25–27

Edgar RC (2010) Search and clustering orders of magnitude faster than BLAST. Bioinform Oxf Engl 26:2460–2461

Eikrem W, Throndsen J (1990) The ultrastructure of Bathycoccus gen. nov. and B. prasinos sp. nov., a non-motile picoplanktonic alga (Chlorophyta, Prasinophyceae) from the Mediterranean and Atlantic. Phycologia 29:344–350

Eren AM, Esen ÖC, Quince C, Vineis JH, Morrison HG, Sogin ML, Delmont TO (2015) Anvi'o: an advanced analysis and visualization platform for omics data. PeerJ 3:e1319

Falkowski P (2012) Ocean science: the power of plankton. Nature 483:S17–20

Fernández-Gómez B, Fernàndez-Guerra A, Casamayor EO, González JM, Pedrós-Alió C, Acinas SG (2012) Patterns and architecture of genomic islands in marine bacteria. BMC Genom 13:347

Field CB, Behrenfeld MJ, Randerson JT, Falkowski P (1998) Primary production of the biosphere: integrating terrestrial and oceanic components. Science 281:237–240

Fleischmann RD, Adams MD, White O, Clayton RA, Kirkness EF, Kerlavage AR, Bult CJ, Tomb JF, Dougherty BA, Merrick JM et al (1995) Whole-genome random sequencing and assembly of Haemophilus influenzae Rd. Science 269:496–512

Foulon E, Not F, Jalabert F, Cariou T, Massana R, Simon N (2008) Ecological niche partitioning in the picoplanktonic green alga Micromonas pusilla: evidence from environmental surveys using phylogenetic probes. Environ Microbiol 10:2433–2443

Gallienne CP, Robins DB (2001) Is Oithona the most important copepod in the world's oceans? J Plankton Res 23:1421–1432

Gawad C, Koh W, Quake SR (2016) Single-cell genome sequencing: current state of the science. Nat Rev Genet 17:175–188

Goffeau A, Barrell BG, Bussey H, Davis RW, Dujon B, Feldmann H, Galibert F, Hoheisel JD, Jacq C, Johnston M et al (1996) Life with 6000 genes. Science 274(546):563–567

Gonzaga A, Martin-Cuadrado A-B, López-Pérez M, Megumi Mizuno C, García-Heredia I, Kimes NE, Lopez-García P, Moreira D, Ussery D, Zaballos M et al (2012) Polyclonality of concurrent natural populations of Alteromonas macleodii. Genome Biol Evol 4:1360–1374

Green JL, Holmes AJ, Westoby M, Oliver I, Briscoe D, Dangerfield M, Gillings M, Beattie AJ (2004) Spatial scaling of microbial eukaryote diversity. Nature 432:747–750

Guidi L, Chaffron S, Bittner L, Eveillard D, Larhlimi A, Roux S, Darzi Y, Audic S, Berline L, Brum JR et al (2016) Plankton networks driving carbon export in the oligotrophic ocean. Nature 532:465–470

Handelsman J, Rondon MR, Brady SF, Clardy J, Goodman RM (1998) Molecular biological access to the chemistry of unknown soil microbes: a new frontier for natural products. Chem Biol 5:R245–249

Hanson CA, Fuhrman JA, Horner-Devine MC, Martiny JBH (2012) Beyond biogeographic patterns: processes shaping the microbial landscape. Nat Rev Microbiol 10:497–506

Hebert PDN, Cywinska A, Ball SL, deWaard JR (2003) Biological identifications through DNA barcodes. Proc R Soc B Biol Sci 270:313–321

Hellweger FL, van Sebille E, Fredrick ND (2014) Biogeographic patterns in ocean microbes emerge in a neutral agent-based model. Science 345:1346–1349

Hewson I, Steele JA, Capone DG, Fuhrman JA (2006) Temporal and spatial scales of variation in bacterioplankton assemblages of oligotrophic surface waters. Mar Ecol Prog Ser 311:67–77

Horner-Devine MC, Lage M, Hughes JB, Bohannan BJM (2004) A taxa—area relationship for bacteria. Nature 432:750–753

Johnson PW, Sieburth JM (1982) In-situ morphology and occurrence of eucaryotic phototrophs of bacterial size in the picoplankton of estuarine and oceanic waters 1. J Phycol 18:318–327

Jönsson BF, Watson JR (2016) The timescales of global surface-ocean connectivity. Nat Commun 7:11239

Karsenti E, Acinas SG, Bork P, Bowler C, De Vargas C, Raes J, Sullivan M, Arendt D, Benzoni F, Claverie J-M et al (2011) A holistic approach to marine eco-systems biology. PLoS Biol 9:e1001177

Kashtan N, Roggensack SE, Rodrigue S, Thompson JW, Biller SJ, Coe A, Ding H, Marttinen P, Malmstrom RR, Stocker R et al (2014) Single-cell genomics reveals hundreds of coexisting subpopulations in wild Prochlorococcus. Science 344:416–420

Kennedy J, Marchesi JR, Dobson AD (2008) Marine metagenomics: strategies for the discovery of novel enzymes with biotechnological applications from marine environments. Microb Cell Fact 7:27

Kent WJ (2002) BLAT—the BLAST-like alignment tool. Genome Res 12:656–664

Kent AG, Dupont CL, Yooseph S, Martiny AC (2016) Global biogeography of *Prochlorococcus* genome diversity in the surface ocean. ISME J 10:1856–1865

Kettler GC, Martiny AC, Huang K, Zucker J, Coleman ML, Rodrigue S, Chen F, Lapidus A, Ferriera S, Johnson J et al (2007) Patterns and implications of gene gain and loss in the evolution of Prochlorococcus. PLoS Genet 3:e231

Kiełbasa SM, Wan R, Sato K, Horton P, Frith MC (2011) Adaptive seeds tame genomic sequence comparison. Genome Res 21:487–493

Kopf A, Bicak M, Kottmann R, Schnetzer J, Kostadinov I, Lehmann K, Fernandez-Guerra A, Jeanthon C, Rahav E, Ullrich M et al (2015) The ocean sampling day consortium. GigaScience 4:1–5

Lasken RS (2012) Genomic sequencing of uncultured microorganisms from single cells. Nat Rev Microbiol 10:631–640

Le Bescot N, Mahé F, Audic S, Dimier C, Garet M-J, Poulain J, Wincker P, de Vargas C, Siano R (2016) Global patterns of pelagic dinoflagellate diversity across protist size classes unveiled by metabarcoding. Environ Microbiol 18:609–626

Lima-Mendez G, Faust K, Henry N, Decelle J, Colin S, Carcillo F, Chaffron S, Ignacio-Espinosa JC, Roux S, Vincent F et al (2015) Determinants of community structure in the global plankton interactome. Science 348:1262073

Lin Y-C, Campbell T, Chung C-C, Gong G-C, Chiang K-P, Worden AZ (2012) Distribution patterns and phylogeny of marine stramenopiles in the north pacific ocean. Appl Environ Microbiol 78:3387–3399

Lindner MS, Renard BY (2013) Metagenomic abundance estimation and diagnostic testing on species level. Nucleic Acids Res 41:e10

Longhurst A (2007) Ecological geography of the sea. Academic Press, London

Lovejoy C, Vincent WF, Bonilla S, Roy S, Martineau M-J, Terrado R, Potvin M, Massana R, Pedrós-Alió C (2007) Distribution, phylogeny, and growth of cold-adapted picoprasinophytes in Arctic Seas 1. J Phycol 43:78–89

Machado H, Gram L (2015) The fur gene as a new phylogenetic marker for Vibrionaceae species identification. Appl Environ Microbiol 81:2745–2752

Machado H, Gram L (2017) Comparative genomics reveals high genomic diversity in the genus Photobacterium. Front Microbiol 8:1204

Madoui M-A, Poulain J, Sugier K, Wessner M, Noel B, Berline L, Labadie K, Cornils A, Blanco-Bercial L, Stemmann L et al (2017) New insights into global biogeography, population structure and natural selection from the genome of the epipelagic copepod Oithona. Mol Ecol 26:4467–4482

Maillet N, Lemaitre C, Chikhi R, Lavenier D, Peterlongo P (2012) Compareads: comparing huge metagenomic experiments. BMC Bioinform 13:S10

Marcy Y, Ouverney C, Bik EM, Lösekann T, Ivanova N, Martin HG, Szeto E, Platt D, Hugenholtz P, Relman DA et al (2007) Dissecting biological "dark matter" with single-cell genetic analysis of rare and uncultivated TM7 microbes from the human mouth. Proc Natl Acad Sci USA 104:11889–11894

Martiny JBH, Eisen JA, Penn K, Allison SD, Horner-Devine MC (2011) Drivers of bacterial β-diversity depend on spatial scale. Proc Natl Acad Sci 108:7850–7854

Massana R, Castresana J, Balagué V, Guillou L, Romari K, Groisillier A, Valentin K, Pedrós-Alió C (2004) Phylogenetic and ecological analysis of novel marine stramenopiles. Appl Environ Microbiol 70:3528

Massana R, del Campo J, Sieracki ME, Audic S, Logares R (2014) Exploring the uncultured microeukaryote majority in the oceans: reevaluation of ribogroups within stramenopiles. ISME J 8:854–866

Mohrbeck I, Raupach MJ, Arbizu PM, Knebelsberger T, Laakmann S (2015) High-throughput sequencing—the key to rapid biodiversity assessment of marine metazoa? PLoS ONE 10:e0140342

Monier A, Sudek S, Fast NM, Worden AZ (2013) Gene invasion in distant eukaryotic lineages: discovery of mutually exclusive genetic elements reveals marine biodiversity. ISME J 7:1764–1774

Monier A, Worden AZ, Richards TA (2016) Phylogenetic diversity and biogeography of the Mamiellophyceae lineage of eukaryotic phytoplankton across the oceans. Environ Microbiol Rep

Moore CM, Mills MM, Arrigo KR, Berman-Frank I, Bopp L, Boyd PW, Galbraith ED, Geider RJ, Guieu C, Jaccard SL et al (2013) Processes and patterns of oceanic nutrient limitation. Nat Geosci 6:701–710

Moreau H, Verhelst B, Couloux A, Derelle E, Rombauts S, Grimsley N, Bel MV, Poulain J, Katinka M, Hohmann-Marriott MF et al (2012) Gene functionalities and genome structure in Bathycoccus prasinos reflect cellular specializations at the base of the green lineage. Genome Biol 13:R74

Moreno CE, Rodríguez P (2010) A consistent terminology for quantifying species diversity? Occologia 163:279–282

Morris RM, Rappé MS, Connon SA, Vergin KL, Siebold WA, Carlson CA, Giovannoni SJ (2002) SAR11 clade dominates ocean surface bacterioplankton communities. Nature 420:806–810

Nayfach S, Rodriguez-Mueller B, Garud N, Pollard KS (2016) An integrated metagenomics pipeline for strain profiling reveals novel patterns of bacterial transmission and biogeography. Genome Res 26:1612–1625

Nekola JC, White PS (1999) The distance decay of similarity in biogeography and ecology. J Biogeogr 26:867–878

Not F, Latasa M, Marie D, Cariou T, Vaulot D, Simon N (2004) A single species, Micromonas pusilla (Prasinophyceae), dominates the eukaryotic picoplankton in the Western English Channel. Appl Environ Microbiol 70:4064–4072

Oliver MJ, Irwin AJ (2008) Objective global ocean biogeographic provinces. Geophys Res Lett 35:L15601

Pesant S, Not F, Picheral M, Kandels-Lewis S, Le Bescot N, Gorsky G, Iudicone D, Karsenti E, Speich S, Troublé R et al (2015) Open science resources for the discovery and analysis of *Tara* Oceans data. Sci Data 2:150023

Piganeau G, Eyre-Walker A, Jancek S, Grimsley N, Moreau H (2011) How and why DNA barcodes underestimate the diversity of microbial eukaryotes. PLoS ONE 6:e16342

Qin J, Li R, Raes J, Arumugam M, Burgdorf KS, Manichanh C, Nielsen T, Pons N, Levenez F, Yamada T et al (2010) A human gut microbial gene catalogue established by metagenomic sequencing. Nature 464:59–65

Ramette A, Tiedje JM (2007) Biogeography: an emerging cornerstone for understanding prokaryotic diversity, ecology, and evolution. Microb Ecol 53:197–207

Reygondeau G, Longhurst A, Martinez E, Beaugrand G, Antoine D, Maury O (2013) Dynamic biogeochemical provinces in the global ocean. Glob Biogeochem Cycles 27:1046–1058

Rocap G, Larimer FW, Lamerdin J, Malfatti S, Chain P, Ahlgren NA, Arellano A, Coleman M, Hauser L, Hess WR et al (2003) Genome divergence in two Prochlorococcus ecotypes reflects oceanic niche differentiation. Nature 424:1042–1047

Rodríguez F, Derelle E, Guillou L, Le Gall F, Vaulot D, Moreau H (2005) Ecotype diversity in the marine picoeukaryote Ostreococcus (Chlorophyta, Prasinophyceae). Environ Microbiol 7:853–859

Rodríguez-Martínez R, Rocap G, Salazar G, Massana R (2013) Biogeography of the uncultured marine picoeukaryote MAST-4: temperature-driven distribution patterns. ISME J 7:1531–1543

Rodriguez-R LM, Castro JC, Kyrpides NC, Cole JR, Tiedje JM, Konstantinidis KT (2018) How much do rRNA gene surveys underestimate extant bacterial diversity?. Appl Environ, Microbiol

Rusch DB, Halpern AL, Sutton G, Heidelberg KB, Williamson S, Yooseph S, Wu D, Eisen JA, Hoffman JM, Remington K et al (2007) The Sorcerer II Global Ocean Sampling Expedition: Northwest Atlantic through Eastern Tropical Pacific. PLoS Biol 5:e77

Salazar G, Cornejo-Castillo FM, Benítez-Barrios V, Fraile-Nuez E, Álvarez-Salgado XA, Duarte CM, Gasol JM, Acinas SG (2016) Global diversity and biogeography of deep-sea pelagic prokaryotes. ISME J 10:596–608

Sawabe T, Kita-Tsukamoto K, Thompson FL (2007) Inferring the evolutionary history of vibrios by means of multilocus sequence analysis. J Bacteriol 189:7932–7936

Schlötterer C (2015) Genes from scratch—the evolutionary fate of de novo genes. Trends Genet 31:215–219

Scholz M, Ward DV, Pasolli E, Tolio T, Zolfo M, Asnicar F, Truong DT, Tett A, Morrow AL, Segata N (2016) Strain-level microbial epidemiology and population genomics from shotgun metagenomics. Nat Methods 13:435–438

Seeleuthner Y, Mondy S, Lombard V, Carradec Q, Pelletier E, Wessner M, Leconte J, Mangot J-F, Poulain J, Labadie K et al (2018) Single-cell genomics of multiple uncultured stramenopiles reveals underestimated functional diversity across oceans. Nat Commun 9:310

Simmons MP, Sudek S, Monier A, Limardo AJ, Jimenez V, Perle CR, Elrod VA, Pennington JT, Worden AZ (2016) Abundance and biogeography of picoprasinophyte ecotypes and other phytoplankton in the Eastern North Pacific Ocean. Appl Environ Microbiol AEM 02730-15

Šlapeta J, López-García P, Moreira D (2006) Global dispersal and ancient cryptic species in the smallest marine eukaryotes. Mol Biol Evol 23:23–29

Soininen J, Lennon JJ, Hillebrand H (2007) A multivariate analysis of beta diversity across organisms and environments. Ecology 88:2830–2838

Suenaga H (2012) Targeted metagenomics: a high-resolution metagenomics approach for specific gene clusters in complex microbial communities. Environ Microbiol 14:13–22

Sunagawa S, Coelho LP, Chaffron S, Kultima JR, Labadie K, Salazar G, Djahanschiri B, Zeller G, Mende DR, Alberti A et al (2015) Structure and function of the global ocean microbiome. Science 348:1261359

Tang CQ, Leasi F, Obertegger U, Kieneke A, Barraclough TG, Fontaneto D (2012) The widely used small subunit 18S rDNA molecule greatly underestimates true diversity in biodiversity surveys of the meiofauna. Proc Natl Acad Sci 109:16208–16212

Thrash JC, Boyd A, Huggett MJ, Grote J, Carini P, Yoder RJ, Robbertse B, Spatafora JW, Rappé MS, Giovannoni SJ (2011) Phylogenomic evidence for a common ancestor of mitochondria and the SAR11 clade. Sci Rep 1:13

Tringe SG, von Mering C, Kobayashi A, Salamov AA, Chen K, Chang HW, Podar M, Short JM, Mathur EJ, Detter JC et al (2005) Comparative metagenomics of microbial communities. Science 308:554–557

Turner JT (2004) The importance of small planktonic copepods and their roles in pelagic marine food webs. Zool Stud 255–266

Vannier T, Leconte J, Seeleuthner Y, Mondy S, Pelletier E, Aury J-M, de Vargas C, Sieracki M, Iudicone D, Vaulot D et al (2016) Survey of the green picoalga Bathycoccus genomes in the global ocean. Sci Rep 6:37900

de Vargas C, Audic S, Henry N, Decelle J, Mahé F, Logares R, Lara E, Berney C, Le Bescot N, Probert I et al (2015) Ocean plankton. Eukaryotic plankton diversity in the sunlit ocean. Science 348:1261605

Vaulot D, Lepère C, Toulza E, De la Iglesia R, Poulain J, Gaboyer F, Moreau H, Vandepoele K, Ulloa O, Gavory F et al (2012) Metagenomes of the Picoalga Bathycoccus from the Chile Coastal Upwelling. PLoS ONE 7:e39648

Vellend M (2001) Do commonly used indices of β-diversity measure species turnover? J Veg Sci 12:545–552

Venter JC, Remington K, Heidelberg JF, Halpern AL, Rusch D, Eisen JA, Wu D, Paulsen I, Nelson KE, Nelson W et al (2004) Environmental genome shotgun sequencing of the Sargasso Sea. Science 304:66–74

Villar E, Farrant GK, Follows M, Garczarek L, Speich S, Audic S, Bittner L, Blanke B, Brum JR, Brunet C et al (2015) Environmental characteristics of Agulhas rings affect interocean plankton transport. Science 348:1261447

Villar E, Vannier T, Vernette C, Lescot M, Cuenca M, Alexandre A, Bachelerie P, Rosnet T, Pelletier E, Sunagawa S et al. The Ocean Gene Atlas: exploring the biogeography of plankton genes online. Nucleic Acids Res

Villarino E, Watson JR, Jönsson B, Gasol JM, Salazar G, Acinas SG, Estrada M, Massana R, Logares R, Giner CR et al (2018) Large-scale ocean connectivity and planktonic body size. Nat Commun 9:142

Weisburg WG, Barns SM, Pelletier DA, Lane DJ (1991) 16S ribosomal DNA amplification for phylogenetic study. J Bacteriol 173:697–703

White C, Selkoe KA, Watson J, Siegel DA, Zacherl DC, Toonen RJ (2010) Ocean currents help explain population genetic structure. Proc R Soc Lond B Biol Sci. https://doi.org/10.1098/rspb2 0092214

Whittaker RH (1960) Vegetation of the Siskiyou Mountains, Oregon and California. Ecol Monogr 30:279–338

Woese CR, Kandler O, Wheelis ML (1990) Towards a natural system of organisms: proposal for the domains Archaea, Bacteria, and Eucarya. Proc Natl Acad Sci USA 87:4576–4579

Worden AZ, Follows MJ, Giovannoni SJ, Wilken S, Zimmerman AE, Keeling PJ (2015) Rethinking the marine carbon cycle: factoring in the multifarious lifestyles of microbes. Science 347:1257594

Wu S, Xiong J, Yu Y (2015) Taxonomic resolutions based on 18S rRNA genes: a case study of subclass Copepoda. PLoS ONE 10:e0131498

Yau S, Hemon C, Derelle E, Moreau H, Piganeau G, Grimsley N (2016) A viral immunity chromosome in the marine picoeukaryote, Ostreococcus tauri. PLOS Pathog 12:e1005965

Part IV
Origin of Life

Ion–Molecule Reactions as a Possible Synthetic Route for the Formation of Prebiotic Molecules in Space

Riccardo Spezia, Yannick Jeanvoine and Debora Scuderi

Abstract Thanks to many astrophysical observations, the number of prebiotic molecules observed in space is growing daily. Organic molecules, which can be the first building blocks for appearance of life, were found in both interstellar medium and comets. As an example, several molecules with the peptide bond moiety were reported, like formamide and urea. The glycine detection has a long and controversial history, and it was recently reported on the comet 67P/Churyumov-Gerasimenko. A general question concerns how these molecules could be formed given the extreme conditions of space. Theoretical chemistry, combined in some cases with laboratory experiments, can help in quantifying the physical chemistry conditions which can allow their synthesis. Here, we summarize some studies on the particular case of ion–molecule collisions.

1 Introduction

Thanks to progresses in astronomical observations, the number of molecules observed in space is growing daily. In particular, radioastronomy was able to detect (or suggest) the presence of several molecules in the interstellar medium (ISM) and the recent

R. Spezia (✉)
Laboratoire de Chimie Théorique, LCT, CNRS, Sorbonne Université,
4, Place Jussieu, 75252 Paris Cedex 05, France
e-mail: riccardo.spezia@sorbonne-universite.fr

R. Spezia · Y. Jeanvoine
LAMBE, CNRS, CEA, Université Evry, Université Paris-Saclay, 91025 Evry, France
e-mail: yannick.jeanvoine@univ-evry.fr

D. Scuderi
Laboratoire de Chimie Physique, UMR8000, CNRS, University of Paris-Sud,
Orsay 91405, France
e-mail: debora.scuderi@u-psud.fr

© Springer International Publishing AG, part of Springer Nature 2018
P. Pontarotti (ed.), *Origin and Evolution of Biodiversity*,
https://doi.org/10.1007/978-3-319-95954-2_15

spatial mission on the comet 67P/Churyumov-Gerasimenko has obtained many data on its chemical composition which are becoming nowadays available. In particular, several complex organic molecules (COMs) with prebiotic roles have been reported. In the ISM, one of the first COMs detected was formamide, NH_2CHO, in 1971 (Rubin et al. 1971) followed by isocyanic acid, HNCO, in 1972 (Snyder and Buhl 1972). More recently, the cyanic acid, HOCN, was reported (Brünken et al. 2010), as well as urea (Remijan et al. 2014), propylene oxide, CH_3CHCH_2O, the first chiral molecule (McGuire et al. 2016) and benzonitrile (McGuire et al. 2018). Among the different classes of molecules, some alcohols were reported [like methanol (Ball et al. 1970), ethanol (Zuckerman et al. 1975), vinyl alcohol (Turner and Apponi 2001), phenol (Kolesniková et al. 2013)], and other organic compounds, like esters or carboxylic acids. Interestingly, the molecule with brute formula $H_4C_2O_2$ can correspond to three different structures, glycolaldehyde, the first sugar detected in the ISM (Hollis et al. 2000), methyl formate, [also detected Brown et al. (1975); Neill et al. (2012)], and acetic acid [detected in Mehringer et al. (1997)].

The recent mission on the 67P/Churyumov-Gerasimenko is providing many information on the chemical variety of comets. Previously, organic molecules like ethylene glycol (Crovisier et al. 2004) or formamide (Bockelée-Morvan et al. 2000) have been reported in the Hale-Bopp comet. The recently detected COMs on the 67P/Churyumov-Gerasimenko comet include methyl isocyanate, acetone, acetamide, propionaldehyde, chloromethane (Goesmann et al. 2015; Fayolle et al. 2017) and possibly glycine (Altwegg et al. 2016). The presence of glycine, the simplest amino-acid, in space is at the center of a longstanding debate. In particular, its detection in the ISM was proposed in Kuan et al. (2003) but, after deeper studies, it was finally disconfirmed (Snyder et al. 2005; Jones et al. 2007). Amino-acids, however, were detected long time ago in the Murchison meteorite (Cronin and Moore 1971), and this was an important discovery which supports the hypothesis of their extraterrestrial origin (Engel et al. 1990; Engel and Macko 1997).

An exhaustive and daily updated list of molecules observed (or suggested to be present) in space can be found in two very useful Websites.[1]

One important aspect that concerns astrochemistry is the question: How these COMs can be formed? This is particularly critical, since the ISM conditions are very extreme: low density (ultra-vacuum) and very low temperature. Thermal reactions are then very unlikely, and reactivity would occur with very low activation energies or, even better, through barrier-less pathways. Different hypothesis are proposed to circumvent these obstacles for the synthesis of COMs under ISM conditions. One possibility is that the reactions responsible to COMs synthesis occur in the interstellar grains or on the cometary ice. The solid or solid/vacuum interface would thus act as a catalyzer. In laboratory studies were reported showing that prebiotic molecules like aldehydes and sugars can be formed in cometary ice analogs (Marcellusa et al. 2015), or aminoethanol (Duvernay et al. 2010), acetonitrile and amino-acetonitrile in interstellar ice analogs (Danger et al. 2011) also assisted by VUV radiation. However, quantum chemistry calculations, while in some cases show a tendency of lowering

---

[1] See http://www.astrochymist.org/ and https://www.astro.uni-koeln.de/cdms/molecules.

the reaction barrier (Chen and Woon 2011), reported in other cases that the final barriers are relatively high (Koch et al. 2008; Rimola et al. 2014).

Another possibility is direct bimolecular gas-phase reactions. To decrease the activation energy, two cases are generally considered: radical reactions and ion–molecule reactions. Theoretical chemistry was used to explore both possibilities. In the case of radical reactions, Barone and co-workers have recently proposed the formation of formamide (Barone et al. 2015) and a genealogical tree connecting ethanol with glycolaldehyde, acetic acid, and formic acid (Skouteris et al. 2018). Experimentally, neutral reactions were studied by Kaiser and co-workers in particular for the formation of carbonaceous species (Kaiser 2002). Few years ago, experiments at very low temperature (between 63 and 82 K) have shown that thanks to quantum mechanical tunneling OH radical can react with methanol (Shannon et al. 2013), showing that this can be a relevant effect to build COMs under astrophysical conditions.

Ion–molecule reactions were considered as possible source of different simple molecules (Larsson et al. 2012; Petrie and Bohme 2007) and recently also for COMs (Blagojevic et al. 2003). Experimentally, Bohme and co-workers have reported that glycine and alanine, for example, can be obtained by ion–neutral reaction, in both protonated and ionized forms (Snow et al. 2007). Recently, we have studied and discussed, by using different theoretical chemistry approaches, the formation of formamide (Spezia et al. 2016), urea (Siro Brigiano et al. 2018) and glycine (Jeanvoine et al. 2018). Here, we report and summarize these studies which were performed by coupling static and dynamical calculations. The static calculations are aimed to provide information on the activation energies on the minimum energy pathway, while dynamics can provide the real-time information on how a reaction evolves. Moreover, as we will show, one can consider a further source of activation, which is the translation energy that the species can locally have, such that a part of this energy can be used to pass the energy barriers.

In the following, we will first provide some general information on how these calculations are done and which useful information can be obtained from them. Gas-phase ion–molecule reactions can be probed by in laboratory experiments, and some of them are reported in Sect. 3. Then, we will show how they can be applied in order to suggest possible synthetic routes to the formation of some prototypical COMs. The article ends with general conclusions and outlooks.

2 Theoretical Chemistry Approaches

Theoretical chemistry is a powerful tool for astrochemistry since it allows to study the structure and reactivity of systems under conditions very different from what easily accessible in common laboratories, and also of systems which are difficult to stabilize under usual experimental conditions, even in the gas phase. As we will briefly discuss in Sect. 3, experimental works were done in order to mimic possible space conditions, but this is always very difficult and expensive. Theoretical and computational chemistry offers a possibility of studying such systems often applying the same technical implementations used in other fields. However, as we will discuss in the following, some particular cares must be considered.

One large field of theoretical chemistry is electronic structure calculation. Given an arrangement of atoms forming a molecule, it is possible to calculate the internal energy and the wave function, from which many properties can be obtained. This corresponds to the so-called static calculations, which were used to investigate the possibility of formation of several molecules. Some basic ideas and most used methods are recalled in Sect. 2.1. However, molecules, even at low temperature conditions, can move and this can be important in particular to address the problem of their synthesis. The two approaches are often complementary, and the use of both can give a fully accomplished picture of reaction properties which can be pertinent to the formation of prebiotic molecules in space. Here, we will detail the methods used for gas-phase reactions. As already remarked, this is not the only possibility; in particular, solid–vacuum or solid–liquid interfaces can be valuable models for astrochemical conditions. They will need a theoretical approach which has some differences related to the modeling of the condensed phase, and they will not be discussed here.

2.1 Static Quantum Chemistry

Quantum chemistry is based on the resolution of the time-independent Schrödinger equation, which reads for a molecular system:

$$\mathscr{H}(\mathbf{r}; \mathbf{R})\Psi(\mathbf{r}; \mathbf{R}) = E(\mathbf{R})\Psi(\mathbf{r}; \mathbf{R}) \qquad (1)$$

where $\mathscr{H}$ is the molecular Hamiltonian operator, Ψ the wave function, and E the energy of the system. What is commonly done is the Born–Oppenheimer approximation, such that $\mathscr{H}$, Ψ, and E will depend parametrically on the nuclear coordinates $\mathbf{R}$ ($\mathbf{r}$ are the electronic coordinates). Equation 1 cannot be solved analytically for molecular systems, and thus, a number of approximations were developed in last decades to solve it. In particular, two key aspects should be considered: the level of theory used and the basis set. In quantum chemistry, two families of theories are used: (i) the wave function-based methods and (ii) the density functional theory (DFT) methods. The basis of wave function methods is the Hartree–Fock theory, but to have accurate results the so-called post-Hartree Fock methods must be employed. In particular, second-order Møller–Plesset perturbation theory (MP2) gives good results, but to have the high precision (often needed in astrochemistry) the highly correlated coupled cluster theory with single, double, and implicitly triple excitations (CCSD(T)) must be employed. Without going into the details [which can be found in books devoted on this topic (Cramer 2006)], these two methods are very expensive and can be used only for relatively small systems. DFT is another possibility which introduces dynamical correlation at a relatively low computational cost. The basic idea is grounded on the Hohenberg–Kohn theorems, and in the field of quantum chemistry, the use of Kohn–Sham formalism allows to use an approach analogous to the wave function-based methods to solve the problem. The main point of DFT methods is that the form of the functional is not known, and thus, a huge number of formulations and functionals were developed in the last decades. However, these methods do not achieve the accuracy of CCSD(T) methods and are used either for systems that are too large to be studied with highly correlated methods, either to have a first quick screening.

The second aspect is the so-called basis set, i.e., the way the wave function is represented. In the field of quantum chemistry, the linear combination of atomic orbitals (LCAO) approach is used, from which a number of basis sets were defined. Without going into details [see for example (Szabo and Ostlund 1996)], larger is the number of basis functions better is the description one can obtain. Of course, the cost of the calculation will increase with increasing basis set dimension. What is important is that highly correlated methods (like the aforementioned MP2 and CCSD(T)) need also a sufficiently large basis set to have a good quality of the results.

Quantum chemistry provides the wave function and the energy given a nuclear configuration. This is generally coupled with the search of stationary points: Minima correspond to geometries in which the first derivatives of the potential are zero and the second derivatives are positive.

The second derivatives are related to the definition of the normal modes, which can be obtained by the diagonalization of the Hessian matrix. Given a system of N atoms, the $3N$ Hessian matrix will provide, once diagonalized, $3N - 6$ normal modes, the six left being the three translational and rotational degrees of freedom. What is relevant for astrochemistry is that from this it is possible to add the zero-point vibrational energy, ZPVE. In fact, assuming the harmonic approximation, a molecular system can be represented as an ensemble of $3N - 6$ independent harmonic oscillators. From the well-known solution of the Schrödinger equation of one harmonic oscillator, the ZPVE is thus the sum of the vibrational energies of the different normal modes given that the vibrational quantum number is zero (as in the limit T $\rightarrow$ 0 K):

$$ZPVE = \frac{1}{2}h \sum_i^{3N-6} \nu_i \tag{2}$$

where ν_i is the vibrational frequency associated with the i normal mode and h is the Planck constant. Algorithms like steepest descent are often used to locate a minimum. For complex molecular systems, the initial guess is very important, since the resulting potential energy surface (PES) can have several local minima (i.e., structures which have the topology of a minimum but whose energy is not the lowest one). Thus, one has to carefully check that several possible conformers are considered. For small systems, the "chemical intuition" (and patience) is always enough, while for larger ones more developed algorithms can be used.

Other than minima, a reaction pathway is characterized (in a simple a schematic picture, see Fig. 1) also by transition states. Topologically, a (tight) transition state is a saddle point in the PES, corresponding to a structure which is a local minimum in all directions (normal modes) but one (the reaction coordinate).

Fig. 1 Schematic picture illustrating a model profile of a bimolecular reaction proceeding in one arbitrary reaction coordinate and passing through minima (the complex A:B) and transition states (TS, in the present example we have set arbitrarily a barrier higher in energy than reactants)

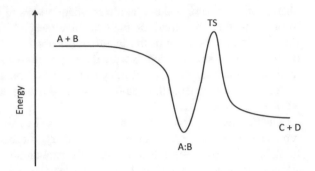

Locating minima and TSs needs the calculation of first and second derivatives of the energy, which is a relevant increase in computational cost. This can be done normally only for DFT and MP2 calculations, while it is impossible for highly correlated CCSD(T) methods. What is often done is that the geometry is optimized with the highest DFT and/or MP2 method possible, the ZPVE is calculated, and the stationary point is used to re-calculate the electronic energy (i.e., E of Eq. 1) at the highest theory level (e.g., CCSD(T)).

Once the PES is obtained, it is possible to estimate the rate constants by employing a kinetic theory. In the gas phase, the Rice–Ramsperger–Kassel–Marcus (RRKM) theory is generally employed (Baer and Hase 1996). This theory allows to obtain the microcanonical rate constant, $k(E)$, from vibrational modes and inertia moments of reactants and transition states. In the case of bimolecular reactions, capture theories are used to estimate the probability that the two reactant molecules make an intermediate complex (Su and Chesnavich 1982).

2.2 Chemical Dynamics

Chemical dynamics is a simulation approach in which the system moves following the Newton equations of motion

$$\mathbf{F} = M\mathbf{a} = -\nabla U \tag{3}$$

where the energy and the gradient are obtained from quantum chemistry calculations. The general framework, in particular for systems evolving only in the electronic ground state, is the Born–Oppenheimer approximation, leading to the so-called Born–Oppenheimer dynamics which for each atom I of mass M_I is:

$$M_I \ddot{R}_I(t) = -\nabla_I V_k^{BO}(R_I(t)) \tag{4}$$

$$M_I \ddot{R}_I(t) = -\nabla \min_{\phi_0^{FN}} \{ \langle \phi_0^{FN} | \mathscr{H}_{el} | \phi_0^{FN} \rangle \} \tag{5}$$

where $V_k^{BO}(R_I(t))$ is the potential energy of the electronic state k, which here corresponds to the ground state, $\mathcal{H}_{el}$ is the electrostatic Hamiltonian operator which depends parametrically on the nuclei positions, and ϕ_0^{FN} is the so-called fixed nuclei wave function. These equations are solved numerically, using one of the different algorithms, e.g., velocity (Verlet 1967).

This general framework can be specifically adapted for gas-phase reactions, and here, in particular we discuss the specific case of ion–molecule reactions. An ion–molecule reaction is, in fact, a reactive scattering event, in which the ion, A^+, and the neutral, B, initially at a given rovibrational state, collide giving rise to different possible products:

$$A^+ + B \rightarrow A^+(v_A = v'; j_A = j') + B(v_B = v''; j_B = j'') \tag{6}$$

$$A^+ + B \rightarrow C^+ + D \tag{7}$$

$$A^+ + B \rightarrow E^+ \tag{8}$$

Reaction (6) corresponds to inelastic scattering in which at the end the two reactants are chemically unchanged but their rovibrational states are modified. They can lately react, and this further event can be described as a simple unimolecular dissociation (Baer and Hase 1996). The collision is aimed to give energy to the reactant molecules, but it does not induce any chemical reaction between them. They cannot synthesize new molecules with more atoms than the reactants, and thus, they will not be discussed anymore.

Reaction (7) corresponds to the reactive scattering and can be at the origin of the synthesis of new molecules since the new species, C^+ and D, arise from a chemical reaction between the two products. Examples will be given in the next sections.

Reaction (8) corresponds to the formation of a new and larger molecule constituted by all the atoms of the reactants. This is a synthetic reaction which can be problematic in the gas phase. In fact, the new species formed, E^+, will have the total energy of the reactants plus the collision energy and if there is no energy dissipation it will break apart (forming at least the same reactants). In the astrochemistry literature, it is well known that three processes should be considered, each one characterized by a rate constant (Hergst 1985; Vasyunin and Herbst 2013): (i) the collisional deactivation (k_{coll}) in which the excess energy is dissipated by collision with an environment; (ii) the spontaneous emission or radiative decay (k_{rad}), in which one (or more) infrared photon(s) are emitted thus decreasing the vibrational energy of the molecule and stabilizing it; (iii) the unimolecular dissociation (k_{uni}) which provides the reactants back.

Chemical dynamics simulations in the gas phase are done by preparing the reactants with a given internal vibrational and rotation energy, generally corresponding to a given target temperature. For vibrational energy, the normal modes are considered and the energy is given by a Boltzmann sampling of the normal mode quantum numbers. This gives rise to the so-called quasi-classical trajectories (QCT) (Chapman and Bunker 1975; Peslherbe et al. 1999). Then, the collisional system is set: A^+ and B are placed at a given distance and, setting one at the origin of the cartesian coordinates system, the other is randomly rotated around the Euler's angles. Then, the impact parameter is set, which is a measure of the distance between the head-to-head collision. Then, the collision is done by fixing the collision energy, E_{coll}, in the center-of-mass framework.

Since different initial conditions must be sampled, it will result in an ensemble of trajectories for each value of E_{coll}. This can lead to a huge computational effort. Thus, a compromise between accuracy and computational power is needed. At this end, we have recently used highly correlated MP2 methods for relatively small systems (Spezia et al. 2016) and then used these results to find the best semiempirical Hamiltonian for the largest ones (Jeanvoine et al. 2018).

3   Ion Chemistry Experiments in a Laboratory

Gas-phase ion chemistry is a longstanding topic in physical chemistry. The use of ultra-vacuum techniques allows to have detailed information on chemical reactions. At this end, the species must be put in the gas phase and controlled. Often this is done on ionic species, which can be controlled thanks to their charge. It is in fact possible to select a given ion among the set of species formed by its mass-over-charge (m/z) ratio, to selectively transmit the ions from one set to the other, to give a well-controlled kinetic energy and to trap these ions. Quadrupoles, ion traps, and other mass spectrometers are used at this aim. Information on ion stability and reactivity can be obtained, generally by measuring a resulting mass spectrum. In the field of astrochemistry, many works were done by the group of Bohme, who studied ion–molecule reactions by means of a selected ion flow tube (SIFT) made of two quadrupoles separated by a flow tube. Precursor ions have been obtained by chemical ionization (CI). Characterization of ions formed by ion–molecule reactions have been done by collision-induced dissociation (CID). In particular, they have shown how ion–molecule reactions can be at the origin of the synthesis of glycine and β-alanine (Snow et al. 2007) both protonated and ionized, through the reactions:

$$(NH_2OH)H^+ + CH_3COOH \rightarrow NH_3CH_2COOH^+ + H_2O \qquad (9)$$

$$NH_2OH^+ + CH_3COOH \rightarrow NH_2CH_2COOH^+ + H_2O \qquad (10)$$

$$(NH_2OH)H^+ + CH_3CH_2COOH \rightarrow NH_3CH_2CH_2COOH^+ + H_2O \qquad (11)$$

$$NH_2OH^+ + CH_3CH_2COOH \rightarrow NH_2CH_2CH_2COOH^+ + H_2O \qquad (12)$$

The resulting molecular ions are formed in the gas phase with the m/z values corresponding to those of the respective species. Unfortunately, they did not have direct characterization to probe the structure of the formed ions. The characterization was done from the CID spectrum of the products compared with a commercial reference. In the same SIFT, Bohme and co-workers have produced also the carboxylic acid (Blagojevic et al. 2003).

Ion–molecule reactions can be done in other ultra-vacuum apparatus, like the Atomic and Molecular Physics Laboratory at the University of Trento (AMPL), where it is possible to obtain the abundance of reaction products as a function of collision energy. These reactions were done, for example, between $C_2H_2N^+$ and C_2H_6 (Fathi et al. 2016), and recently to propose the formation of prebiotic molecules from cyclopropyl cyanide in Titan's atmosphere (López et al. 2018).

4 Synthesis of Some Prebiotic Molecules

4.1 Formamide

As already mentioned, formamide was one of the first prebiotic molecules observed in space (Rubin et al. 1971). It is a simple prebiotic molecule (Saladino et al. 2012; Saitta and Saija 2014; Niether et al. 2016) containing one N–C–O bond which is chemically equivalent to a peptide bond. Recently, Largo and co-workers have studied by means of static highly correlated quantum chemistry calculations the possibility of forming it via ion–molecule reactions (Redondo et al. 2014a, b). Several reactions were tested (see Table 1 for a summary), and only one was found to have a small activation barrier:

$$NH_2OH_2^+ + H_2CO \rightarrow HCOHNH_2^+ + H_2O \tag{13}$$

Based on these results, we have studied some of these reactions by explicit ion–molecule collisions (Spezia et al. 2016). In this case, the Born–Oppenheimer trajecto-

Table 1 Summary of bimolecular gas-phase reactions studied by highly correlated quantum chemistry calculation possibly leading to formation of formamide in the gas phase

Reaction	ΔE^a	$\Delta E_a{}^b$	References
$NH_2OH + H_2CO \rightarrow$ $NH_2CHO + H_2O$	−66.41	+59.90	Redondo et al. (2014a)
$NH_3OH^+ + H_2CO \rightarrow$ $NH_2CHOH^+ + H_2O$	−70.66	+35.11	Redondo et al. (2014a)
$NH_2OH_2^+ + H_2CO \rightarrow$ $NH_2CHOH^+ + H_2O$	−96.30	+2.77	Redondo et al. (2014a)
$NH_2OH^+ + H_2CO \rightarrow$ $NH_2CHOH^+ + OH$	−46.37	+34.29	Redondo et al. (2014a)
$NH_4^+ + H_2CO \rightarrow$ $NH_2CHOH^+ + H_2$	−3.12	+60.38	Redondo et al. (2014b)
$NH_2 + H_2CO \rightarrow$ $NH_2CHO + H$	−11.60	+0.05	Barone et al. (2015)

[a]Energy difference between products and reactants in kcal/mol
[b]Activation energy (in kcal/mol), corresponding to the highest barrier reported

ries were done on the MP2/6-31G** potential energy surface (calculated on-the-fly) in which the initial rovibrational energy of the reactants was kept low (15 K) but the collision energy was tuned in the 0.04–4.3 eV range. Using such relatively high level of theory, it was possible to run only 60 trajectories for each system and each collision energy values, and to increase the statistics we run also trajectories with the semiempirical (and faster) MSINDO Hamiltonian (Ahlswede and Jug 1999), obtaining comparable results between the two methods.

Among all the reactions investigated, reaction (13) was found to provide a product composed by water and a molecular ion with the same stoichiometry of protonated formamide, H_4NCO^+. However, the structure was not the one of protonated formamide but an isomer. Based on this result, two reactions were considered: (i) isomerization; (ii) dissociative recombination (i.e., the reaction with one electron and the subsequent loss of H). In the first case, to overcome the activation energy (42.7 kcal/mol) the pathway leading to NH_2CO^+ and H_2 was also open, thus making the product unstable when enough energy to overcome the barrier is provided: If this reaction would occur and the H_4NCO^+ would have enough internal energy to isomerize, the fragmentation products would be finally obtained. On the other hand, the dissociative recombination can form the formamide structure with an activation energy of about 20.7 kcal/mol which is lower than the isomerization pathway (and further fragmentation):

$$H_4NCO^+ + e^- \rightarrow H_4NCO \rightarrow HCONH_2 + H \tag{14}$$

This provided an hypothesis for a possible synthetic route of formamide under astrophysical conditions: ion–molecule collision, formation of an intermediate with not enough internal energy to isomerize and further fragment, dissociative recombination, and formation of neutral formamide.

From other reactions, it was not possible to obtain any pathway leading to formamide, but some useful insights on how this reaction could be responsible to the formation of other astrochemically relevant molecules. This is a kind of information which can be obtained only from on-the-fly chemical dynamics simulations where no information on products is present a priori. For example, we have found that it is possible to obtain protonated hydroxyl-amino-methanol and amino-methanol, by the association of NH_2OH with H_2COH^+ and NH_3 with H_2COH^+, respectively, but these syntheses will need stabilization by photon emission in gas phase. Another, probably more interesting, reaction is the possibility of forming methanimine [observed in the ISM in Godfrey et al. (1973)], via the reactions:

$$NH_3 + H_2COH^+ \rightarrow NH_2CH_2^+ + H_2O \tag{15}$$

$$NH_2CH_2^+ + e^- \rightarrow NH_2CH + H \tag{16}$$

Recently, Barone and co-workers have proposed that formamide could be formed by a radical reaction in the gas phase (Barone et al. 2015) (see reaction in Table 1) for which a very low barrier was reported. In the calculations, it was necessary to use a highly correlated quantum chemistry method to obtain such a reaction profile, and thus, it was not studied (yet) by direct dynamics simulations. Further improvements in the ab initio methods used for chemical dynamics would make possible to study such class of radical reactions in the near future.

4.2 Urea

The presence of urea in the interstellar medium (and more specifically in the Sgr-B$_2$ cloud) was recently proposed by Remijan et al. (2014). A first theoretical chemistry study investigated the PES of different possible isomers pointing out that the most stable one should be largely dominant (Fourré et al. 2016). In a more recent work, we have studied the possibility of forming urea by several ion–molecule, radical, and neutral-neutral reactions (Siro Brigiano et al. 2018) by employing highly correlated static quantum chemistry calculations. In particular, we have investigated the thermodynamics and kinetics of such reactions. Several possibilities were considered, but only one provided a barrier-less pathway:

$$NH_2OH_2^+ + HCONH_2 \rightarrow UreaH^+ + H_2O \qquad (17)$$

which corresponds to the reaction between protonated hydroxylamine (as for formamide as its highest energy tautomer) with neutral formamide, forming protonated urea and a neutral water molecule. Once the protonated urea is formed, then the neutral species can be obtained either by a dissociative recombination reaction or via a proton exchange with one species with higher proton affinity. While the proton affinity of urea is relatively high [868.4 ± 2.5 kJ/mol experimentally (Zheng and Cooks 2002) and 868.4 kJ/mol from our calculations] in the ISM, there are some species with higher proton affinity which can potentially take the proton and neutralize this species.

However, even if the reaction pathway connecting the reactants to the products of reaction (17) is barrier-less, there are competing reaction channels open, in particular: (i) the formation of the reactants in a different protonation state (NH_3OH^+ or $HCOHNH_2^+$) which is more stable than the one forming urea; (ii) the formation of HNCO plus H_2O and NH_4^+; and (iii) the formation of neutral urea and H_3O^+. This last reaction is not problematic since it leads directly to neutral urea. The formation of HNCO can be an interesting other possible product since this molecule was also observed in the ISM (Snyder and Buhl 1972). The formation of the reactants in a different protonation state is, on the contrary, a parasitic reaction which will decrease the efficiency of the ion–molecule synthesis. By using a simple kinetic model based on RRKM theory, we estimated that if the complex between the two reactants has low internal energy, the pathway leading to the reactants tautomers is dominant, but increasing its energy the pathway leading to urea (both neutral and protonated) and HNCO increases its probability with respect to the first one. Chemical dynamics simulations will be surely useful to clarify these aspects. Unfortunately, they cannot

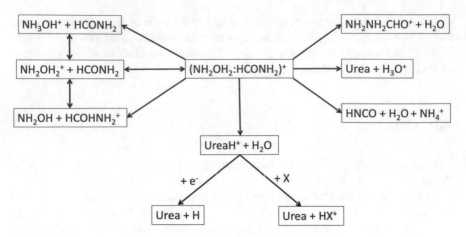

Fig. 2 Reaction network which can be responsible to the formation of urea in gas phase

be done with the same level of theory employed for the static calculations, and thus, a preliminary assessment of the methods would be useful.

A schematic representation of the different reaction pathways is reported in Fig. 2.

4.3 Glycine

About 10 years ago, Bohme and co-workers have performed the reaction between protonated hydroxylamine and acetic acid in the SIFT and they detected a reaction product with m/z 76 which corresponds to protonated glycine (Snow et al. 2007). To verify that, the product has not only the mass but also the structure of glycine; they have compared collision-induced dissociation behavior of this m/z 76 with the one of standard glycine obtained as protonated species in the gas phase. The two spectra were similar, and thus, they suggest that the reaction led to protonated glycine. There are two important aspects of this experiment: (i) the reaction is done in a flow tube where the gas buffer pressure is relatively high (0.33 Torr) such that the ions are thermalized; (ii) the ion is obtained by chemical ionization and, in particular, by collision between neutral hydroxylamine and CH_5^+, which can produce both tautomers, NH_3OH^+ and $NH_2OH_2^+$.

In the same article, DFT calculations have found that the $NH_2OH_2^+ + CH_3COOH$ reaction proceeds via a barrier-less pathway (-1.2 kcal/mol), but more recently Largo and co-workers have studied the same reaction with a highly correlated (and thus more precise) method finding a barrier of 2.3 kcal/mol (Barrientos et al. 2012), thus excluding the possibility that this reaction occurs at low temperatures without any other activation.

Recently, we have studied by collisional dynamics the two reactions which could lead, under the SIFT conditions at least, to the formation of protonated glycine (Jean-voine et al. 2018):

$$NH_3OH^+ + CH_3COOH \rightarrow GlyH^+ + H_2O \tag{18}$$

$$NH_2OH_2^+ + CH_3COOH \rightarrow GlyH^+ + H_2O \tag{19}$$

Simulations show that both reactions can produce an ion with m/z 76 and H_2O but with different spatial arrangements of the atoms (i.e., other isomers). The iso-mers formed change as a function of reactants and collision energy, but the pro-tonated glycine structure was never obtained. We should note that simulations are limited in time (few ps) such that a further isomerization can occur (in particular in the thermalized conditions of the SIFT apparatus), similarly to what observed for formamide. Unfortunately, the atomic arrangements needed for isomerization are too complex to determine the reaction pathways (and then transition states) by stan-dard techniques. Other more sophisticated methods, like transition-state search using chemical dynamics (Martínez-Núñez 2015) or temperature-dependent fragmentation dynamics (Martin-Somer et al. 2018), should be employed.

We should notice that simulations were done with reactants at both room tempera-ture (300 K as in SIFT conditions) and at very low temperature (15 K, representative of the ISM), obtaining the same results.

5 Conclusions and Outlooks

In this article, we provided a short overview of how bimolecular reactions can be involved in the formation of prebiotic molecules in space. While recently the role of grains was suggested to be at the origin of synthesis of complex organic molecules, COMs, the gas-phase reactions cannot be disregarded and they were probably over-looked. Furthermore, gas-phase reactions can be studied in a very detailed and con-trolled way both experimentally and theoretically, and they could provide information which can be lately translated to other environments.

Here, we have focused the attention in particular on ion–molecule reactions and on the synthesis of three prebiotic molecules: formamide, urea, and glycine. They were observed in different regions of space, and they can be formed if the reactions are controlled in some way. For formamide and glycine, it is necessary to add at least some translational energy (which can derive also from evaporation from a grain or from shock waves), while for urea the reaction is barrier-less but other competitive pathways can be opened. Interestingly, ion–molecule reactions suggest that the product ions can be formed as high energy isomers which can be kinetically trapped, thus suggesting to radioastronomers also to look for them. A common aspect of these reactions is that they need as reactant protonated hydroxylamine, which was not observed. Recently, the astrophysical community has questioned itself on this non-detection, suggesting that the protonated and highly reactive form is probably formed in space and either it is not yet detected either it reacts very fast being at the origin, as many studies summarized here suggest, to a series of reactions forming COMs (Pulliam et al. 2012).

Concluding, we would note that the formamide and urea molecules (containing a peptide-like bond) were observed in the interstellar medium, while glycine is not, but in comet and meteorites. One fascinating hypothesis, as suggested some years ago by Halfen et al. (2011), is that space can be a source of molecules with peptide bonds already formed, and then used as elementary bricks for the formation of more complex organic and biological molecules. Surely, this hypothesis must be further investigated.

Acknowledgements We thank ANR DynBioReact (Grant No. ANR-14-CE06-0029-01) and CNRS program INFINITI (project ASTROCOL) for support.

References

Ahlswede B, Jug K (1999) J Comput Chem 20:563–571
Altwegg K, Balsiger H, Bar-Nun A, Berthelier J-J, Bieler A, Bochsler P, Briois C, Calmonte U, Combi MR, Cottin H, De Keyser J, Dhooghe F, Fiethe B, Fuselier SA, Gasc S, Gombosi TI, Hansen KC, Haessig M, Jäckel A, Kopp E, Korth A, Le Roy L, Mall U, Marty B, Mousis O, Owen T, Rème H, Rubin M, Sémon T, Tzou C-Y, Hunter Waite J, Wurz P (2016) Sci Adv 2:e1600285
Baer T, Hase WL (1996) Unimolecular Reaction Dynamics. Oxford University Press, Oxford
Ball JA, Gottlieb CA, Lilley AE, Radford HE (1970) Astrophys J 162:L203–L210
Barone V, Latouche C, Skouteris D, Vazart F, Balucani N, Ceccarelli C, Lefloch B (2015) Mon Not R Astron Soc Lett 453:L31–L35
Barrientos C, Redondo P, Largo L, Rayón VM, Largo A (2012) Astrophys J 748:99
Blagojevic V, Petrie S, Bohme DK (2003) Mon Not R Astron Soc 339:L7–L11
Bockelée-Morvan D, Lis DC, Wink JE, Despois D, Crovisier J, Bachiller R, Benford DJ, Biver N, Colom P, Davies JK, Gérard E, Germain B, Houde M, Mehringer D, Moreno R, Paubert G, Phillips TG, Rauer H (2000) Astronom. Astrophys. 353:1101–1114
Brown RD, Crofts JG, Godfrey PD, Gardner FF, Robinson BJ, Whiteoak JB (1975) Astrophys J 197:L29–L31
Brünken S, Belloche A, Martn S, Verheyen L, Menten KM (2010) Astronom Astrophys 516:A109
Chapman S, Bunker DL (1975) J Chem Phys 62:2890–2899
Chen L, Woon DE (2011) J Phys Chem A 115:5166–5183
Cramer CJ (2006) Essential of Computational Chemistry. Wiley, Chichester, UK
Cronin JR, Moore CB (1971) Science 172:1327–1329
Crovisier J, Bockelée-Morvan D, Biver N, Colom P, Despois D, Lis DC (2004) Astronom Astrophys 418:L35–L38
Danger G, Bossa J-B, de Marcellus P, Borget F, Duvernay F, Theulé P, Chiavassa T, d'Hendecourt L (2011) Astronom Astrophys 525:A30
de Marcellusa P, Meinertb C, Myrgorodska I, Nahonc L, Buhsed T, Le Sergeant d'Hendecourt L, Meierhenrich UJ (2015) Proc Natl Acad Sci USA 112:965–970
Duvernay F, Dufauret V, Danger G, Theulé P, Borget F, Chiavassa T (2010) Astronom Astrophys 523:A79
Engel MH, Macko SA (1997) Nature 389:265–268

Engel MH, Macko SA, Silfer JA (1990) Nature 348:47–49

Fathi P, Geppert WD, Ascenzi D (2016) Int J Mass Spectrom 411:1–13

Fayolle EC, Öberg KI, Jorgensen JK, Altwegg K, Calcutt H, Müller HSP, Rubin M, van der Wiel MHD, Bjerkeli P, Bourke TL, Coutens A, van Dishoeck EF, Drozdovskaya MN, Garrod, RT, Ligterink NFW, Persson MV, Wampfler SF (2017) The ROSINA team. Nat Astron 1:702–708

Fourré I, Rosset L, Chevreau H, Ellinger Y (2016) 589:A18

Godfrey PD, Brown RD, Robinson BJ, Sinclair MW (1973) Astrophys Lett 13:119

Goesmann F, Rosenbauer H, Bredehöft, JH, Cabane, M, Ehrenfreund P, Gautier T, Giri C, Krüger, H, Le Roy L, MacDermott, AJ, McKenna-Lawlor S, Meierhenrich UJ, Munoz Caro GM, Raulin F, Roll R, Steele A, Steininger H, Sternberg R, Szopa C, Thiemann W, Ulamec S (2015) Science 349, aab0689

Halfen DT, Ilyushin V, Ziurys LM (2011) Astrophys J 743:60

Hergst E (1985) Astrophys J 291:226–229

Hollis JM, Lovas FJ, Jewell PR (2000) Astrophys J 540:L107–L110

Jeanvoine Y, Largo A, Hase WL, Spezia R (2018) J Phys Chem A 122:869–877

Jones PA, Cunningham MR, Godfrey PD, Cragg DM (2007) Mon Not R Astron Soc 374:579–589

Kaiser RI (2002) Chem Rev 102:1309–1358

Koch DM, Toubin C, Peslherbe GH, Hynes JT (2008) J Phys Chem C 112:2972–2980

Kolesniková L, Daly AM, Alonso JL, Tercero B, Cernicharo J (2013) J Mol Spectrosc 289:13–20

Kuan Y-J, Charnley SB, Huang H-C, Tseng W-L, Kisiel Z (2003) Astrophys J 593:848–867

Larsson M, Geppert WD, Nyman G (2012) Rep Prog Phys 75:066901

López E, Ascenzi D, Tosi P, Bofill JM, de Andrés J, Albert M, Lucas JM, Aguilar A (2018) Phys Chem Chem Phys 20:6198–6210

Martin-Somer A, Martens J, Grzetic J, Hase WL, Oomens J, Spezia R (2018) J Phys Chem A 122:2612–2625

Martínez-Núñez E (2015) J Comput Chem 36:222–234

McGuire BA, Carroll PB, Loomis RA, Finneran IA, Jewell PR, Remijan AJ, Blake GA (2016) Science 352:1449–1552

McGuire BA, Burkhardt AM, Kalenskii S, Shingledecker CN, Remijan AJ, Herbst E, McCarthy MC (2018) Science 359:202–205

Mehringer DM, Snyder LE, Miao Y, Lovas F (1997) Astrophys J 480:L71–L74

Neill JL, Muckle MT, Zaleski DP, Steber AL, Pate BH, Lattanzi V, Spezzano S, McCarthy MC, Remijan AJ (2012) Astrophys J 755:143

Niether D, Afanasenkau D, Dhont JKG, Wiegand S (2016) Proc Natl Acad Sci USA 113:4272–4277

Peslherbe GH, Wang H, Hase WL (1999) Adv Chem Phys 105:171–202

Petrie S, Bohme DK (2007) Mass Spectrom Rev 26:258–280

Pulliam RL, McGuire BA, Remijan AJ (2012) Astrophys J 751:1

Redondo P, Barrientos C, Largo A (2014a) Astrophys J 780:181

Redondo P, Barrientos C, Largo A (2014b) Astrophys J 793:32

Remijan AJ, Snyder LE, McGuire BA, Kuo H-L, Looney LW, Friedel DN, Golubiatnikov GY, Lovas FJ, Ilyushin VV, Alekseev EA, Dyubko SF, McCall BJ, Hollis JM (2014) Astrophys J 783:77

Rimola A, Taquet V, Ugliengo P, Balucani N (2014) C Ceccarelli Astronom Astrophys 572:A70

Rubin RH, Swenson GW Jr, Benson RC, Tigelaar HL, Flygare WH (1971) Astrophys J 169:L39–L44

Saitta AM, Saija F (2014) Proc Natl Acad Sci USA 111:13768–13773

Saladino R, Botta G, Pino S, Costanzo G, Di Mauro E (2012) Chem Soc Rev 41:5526–5565

Shannon RJ, Blitz MA, Goddard A, Heard DE (2013) Nature Chem 5:745–749

Siro Brigiano F, Jeanvoine Y, Largo A, Spezia R (2018) Astronom Astrophys 610:A26

Skouteris D, Balucani N, Ceccarelli C, Vazart F, Puzzarini C, Barone V, Codella C, Lefloch B (2018) Astrophys J 854:135

Snow JL, Orlova G, Blagojevic V, Bohme DK (2007) J Am Chem Soc 129:9910–9917

Snyder LE, Buhl D (1972) Astrophys J 177:619–623

Snyder LE, Lovas FJ, Hollis JM, Friedel DN, Jewell PR, Remijan A, Ilyushin VV, Alekseev EA, Dyubko SF (2005) Astrophys J 619:914–930

Spezia R, Jeanvoine Y, Hase WL, Song K, Largo A (2016) Astrophys J 826:107

Su T, Chesnavich WJ (1982) J Chem Phys 76:5183–5185

Szabo A, Ostlund NS (1996) Modern Quantum Chemistry. Dover, New York

Turner BE, Apponi AJ (2001) Astrophys J 561:L207–L210

Vasyunin AI, Herbst E (2013) Astrophys J 769:34

Verlet L (1967) Phys Rev 159:98

Zheng X, Cooks RG (2002) J Phys Chem A 106:9939–9946

Zuckerman B, Turner BE, Johnson DR, Clark FO, Lovas FJ, Fourikis N, Palmer P, Morris M, Lilley AE, Ball JA, Gottlieb CA, Penfield H (1975) Astrophys J 196:L99–L102

Did Gene Expression Co-evolve with Gene Replication?

Charles W. Carter Jr. and Peter R. Wills

Abstract We assemble recent experimental work on the aminoacyl-tRNA synthetase evolution into the context of theoretical studies on the nature of the problems preventing the emergence of genetic coding. What initially appeared as experimental curiosities—evidence for ancestral bidirectional coding of the two synthetase classes, the extended inversion symmetries in higher-order structure and functionality, and the strong correlations between amino acid physical chemistry and both protein folding and the tRNA identity elements used by synthetases to recognize cognate tRNAs—fit the landscape painted by the theoretical studies like a hand in a glove. We conclude that the prevailing RNA World scenario does not furnish an adequate basis for genetic coding. An important corollary is that the evolution of gene expression was very closely coupled to that of gene replication.

> The RNA World scenario.... is hardly falsifiable and is extremely difficult to verify due to a *great number of holes in the most important parts.*
>
> —Eugene Koonin

This epigram from an open peer review of a published description of the RNA World hypothesis (Bernhardt 2012) expresses, better than we can, the status of the prevailing hypothesis about the origins of life. The hypothesis itself can be summarized simply in its entirety: The earliest biological molecules were RNA molecules capable of self-replication. The "holes in the most important places" include that:

i. Inheritance of such RNA molecules required, minimally, faithfully copying many hundreds of bases, but left no molecular fossils with properties expected of such putative "founding ribozymal replicators." Their absence underscores the importance of the few legitimate molecular fossils that have been identified (e.g., Carter 2017; and references therein).

C. W. Carter Jr. (✉)
Department of Biochemistry and Biophysics, University of North Carolina at
Chapel Hill, Chapel Hill, NC 27599-7260, USA
e-mail: carter@med.unc.edu

P. R. Wills
Department of Physics, Auckland University, Auckland, New Zealand

© Springer International Publishing AG, part of Springer Nature 2018
P. Pontarotti (ed.), *Origin and Evolution of Biodiversity*,
https://doi.org/10.1007/978-3-319-95954-2_16

ii. The hypothesis is mute concerning how the impact of these ancestral RNA molecules on the earth's chemistry self-organized a supporting metabolism.

   a. The primary purpose of catalysis is to synchronize metabolic reactions whose uncatalyzed rates differ by >20 powers of ten at ambient temperatures (Wolfenden 2011).
   b. Ribozymes tend to stabilize transition states largely by substrate alignment, or favorable and *temperature-independent entropic changes* that cluster around a far smaller dynamic range (Schroeder and Wolfenden 2007).
   c. Protein catalysts form strong binding interactions to the transition state, whose strengths increase as temperature decreases, assuring that synchronization could survive large changes in the earth's temperature (Wolfenden 2011).
   d. Catalytic RNA is thus inherently far less capable than proteins of synchronizing chemical reactions with very different rates.

iii. Most strikingly, the hypothesis cannot explain how molecular parasites created and sustained by such replicators in an RNA World also created proteins—a vastly more powerful class of molecular catalysts, whose production and function require accumulation and inheritance of substantial physicochemical information about amino acid behavior (Carter and Wolfenden 2015, 2016; Wolfenden et al. 2015).

   a. The RNA World fails to account for the necessary assignment catalysts to make proteins according to a computational templating mechanism—the *sine qua non* of genetics (Wills et al. 2015; Wills 1993, 2001; Nieselt-Struwe and Wills 1997).
   b. An RNA World provides no selective advantage for taking over such RNA parasites by an inferior, binary coding system whose bidirectional ancestry is now documented experimentally and in the subsequent phylogeny of protein aaRS (Martinez et al. 2015).
   c. The RNA World furnishes no dynamic algorithm to explain the miraculously rapid discovery, from among the possible genetic codes, of an infinitesimally small fraction with optimal robustness and consistency with bidirectional genetic coding (Carter and Wills 2018b; Wills and Carter 2018).

1 Introduction: Barriers to the Self-organization of Coding

There is no disagreement that nucleic acids served as the heritable repository for information that nature accumulated as it created biology. Our thinking diverges substantively from the canonical RNA World framework by questioning how that information came to be embedded in RNA. Biological information emerges in two distinct forms. One is represented by the sequences of triplet codons, in what became messenger RNA, for those polypeptide sequences that can both fold and function as catalysts; the other by the codon table embedded in transfer RNA sequences, and which furnishes the rules used today by aminoacyl-tRNA synthetases (aaRS) to convert mRNA repeatedly and faithfully templates by ribosomes, first into correct primary sequences of amino acids, and subsequently, via folding, into functional proteins.

Protein enzymes synchronize all metabolic reactions today. The apparent difficulty of evolving genetically coded peptide catalysts from scratch suggested that the modest ability of self-splicing RNAs to accelerate phosphoryl transfer reactions might be the vestige of a substantially larger repertoire of ribozymal catalysts. Little trace remains in contemporary biology of any such ribozymes. All were presumed to have eventually been taken over by proteins somehow "invented" by those primordial RNAs. We summarize here an extensive critique of the "RNA World" hypothesis, from both experimental (Carter and Wills 2018b) and theoretical (Wills and Carter 2018) standpoints.

The most serious doubts that ribozyme catalyzed metabolism preceded protein synthesis are rooted in fundamental obstacles facing the self-organization of genetic coding articulated first by Manfred Eigen, who summarized the problems associated with copying polymeric sequence information (Eigen and Schuster 1977, 1978; Eigen 1971a, b) and elaborated subsequently by others, in relation to its translation (Wills et al. 2015; Wills 1994, 2001, 2004, 2014a, b, 2016; Bedian 1982, 2001) in pursuit of a theoretical basis for the emergence of coding. These obstacles assume three forms:

(i)   Information processing errors limit the lengths of genes that can be sustained without catastrophic informational degradation. Eigen's solution was to divide genetic statements into smaller chunks that could be replicated more accurately, and whose interdependence to form hypercycles assured their mutual survival.

(ii) Differentiation of new functions was difficult, if not impossible, as long as amino acid sequence space was sparsely populated. Large "moats" of non-functional sequences surrounded all functional "genes," so mutational trajectories eventually returned to the population centroid, irrespective of whether the original catalysts were protein or RNA. Strong attraction to centroid sequences has also been observed by Michael Levitt (Xia and Levitt 2002) in simulated evolutionary trajectories.

(iii) A tiny fraction of the possible codon tables is as robust to mutation as is the universal genetic coding table (Freeland and Hurst 1998). Of that fraction, codes consistent with bidirectional coding—i.e., expression of functional gene products from opposite strands of the same gene (Martinez et al. 2015)—probably represent a correspondingly smaller fraction. Such an unusual code could not have been discovered randomly, so the search must therefore have been accelerated by an efficient feedback algorithm, much as installing a computer system begins with a small kernel of instructions capable of bootstrapping a much more complex instruction set.

Thus, in our view the key challenge is to understand how emerging, error-ridden systems of inheritance, low-dimensional coding, and gene expression were able to populate both sequence and structure spaces with diverse inherited proteins whose ancestries can be linked persuasively to the sequences of contemporary descendants. We assemble what we feel is a convincing and multi-faceted case that bidirectional coding of ancestral aaRS (Carter 2017), combined with nanoenvironment sensing of the amino acid phase transfer equilibria (Carter and Wolfenden 2015, 2016; Wolfenden et al. 2015), provided the necessary intrinsic self-organizational feedback to engineer solutions to each obstacle (Carter and Wills 2018b).

Further, the dynamic equations for the population variables (Wills and Carter 2018) show that the selective disadvantages of hybrid systems and the thermodynamic requirement to minimize dissipative free energy losses (Katsnelson et al. 2018; Sonnino et al. 2015; Kaila and Annila 2008) probably coupled the evolution of nucleic acid replication and gene expression so tightly that they must have coevolved from intrinsically complementary structural chemistry (Carter 1975; Carter and Kraut 1974).

2 Differentiation: The aaRS Class Division Decisively Partitioned Sequence Space

Aminoacyl-tRNA synthetases translate the genetic code by activating amino acids and acylating tRNA. Our experimental deconstruction of Class I and II aminoacyl-tRNA synthetase superfamilies (Carter 2014, 2016, 2017; Martinez et al. 2015; Carter et al. 2014; Chandrasekaran et al. 2013; Li et al. 2011, 2013; Cammer and Carter 2010; Pham et al. 2007, 2010), together with the identification of separately localized recognition elements in their specific tRNAs (Carter and Wolfenden 2015, 2016; Carter and Wills 2018a; Wolfenden et al. 2015), now provides substantive inferences about the rise of genetic coding. Specifically, their respective superfamily ancestries in the bidirectional genetic coding of both ancestral Class I and II synthetases by a single gene (Martinez et al. 2015; Carter et al. 2014; Carter 2015) assured both differentiation and survival. Catalysis of amino acid activation by 46-residue peptides translated from opposite strands of the same designed bidirectional gene (Martinez et al. 2015) argues that genetic coding might have evolved from simple, low-fidelity ancestral peptides. Moreover, bidirectional coding creates inversion symmetries in aaRS primary, secondary, and tertiary structures that decisively differentiated two coding specificities from the very beginning.

The aminoacyl-tRNA synthetases are not monophyletic, but descend from two distinct lineages (Eriani et al. 1990) *with very limited structural similarity* (Carter 1993, 2017). A central challenge facing the aminoacyl-tRNA synthetase field has been to understand why nature needed two different families of aminoacyl-tRNA synthetases to activate all 20 amino acids. We recently provided evidence (Carter and Wolfenden 2015, 2016; Wolfenden et al. 2015) that this puzzling class division reflects the phase transfer equilibria that form a 2D physicochemical "basis set" for nature's 3D building blocks:

(i) Transfer free energies from vapor and water to cyclohexane are necessary and sufficient to explain the distribution of amino acid side chains in folded proteins. Because the transfer free energy from vapor to cyclohexane correlates strongly with side chain size, it is clear from Fig. 1 that amino acids activated by Class I aaRS have larger side chains and they distribute significantly into the cores of folded proteins. Those activated by Class II aaRS have smaller side chains and distribute to protein surfaces (Carter and Wolfenden 2015; Wolfenden et al. 2015).

(ii) Because of the logarithmic scale in Fig. 1, amino acid diversity is vastly greater than that of nucleic acid bases. In fact, any four-letter alphabet containing 2 aa each from Class I and II would span, on average, 50-fold greater surface area enabling exponentially greater chemical diversity.

The Class I and II aaRS have parallel structural phylogenies (Fig. 2). Like Russian Matryoshka dolls, Class I and II aaRS families reveal that successively smaller modular catalysts can be isolated and their function assayed. Bars under each construct correspond in color to the different modules we identified and the order in which we deleted them experimentally, to make catalytic domains, Urzymes, and most recently Protozymes, that retain the most highly conserved portions from the preceding construct.

Vertical histograms show that Urzyme and Protozyme transition-state stabilization free energies are 60 and 40% of those of full-length aaRS. The progression of

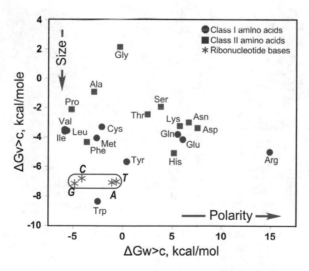

Fig. 1 Chemical diversity of monomers found in proteins and nucleic acids. Phase transfer free energies for the size and polarity of the amino acids are described in Carter et al. (2015), Wolfenden et al. (2015). Those of the nucleic acid bases are described in Cullis and Wolfenden (1981). The chemical diversity of proteins is enhanced by the significant separation in the y-axis direction, which emphasizes the significant difference between the sizes of Class I (red) and Class II (blue) amino acid side chains

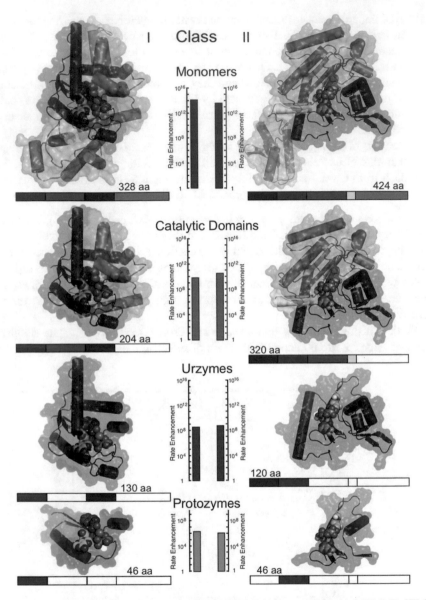

Fig. 2 Deconstruction of Class I tryptophanyl (PDB ID 1MAU)- and II histidyl (PDB ID 2EL9)-tRNA synthetases into successively smaller fragments that retain catalytic activity (Carter et al. 2014; Carter 2014; Li et al. 2011, 2013, Li and Carter 2013; Pham et al. 2010). Graphics for smaller constructs are derived from coordinates of the full-length enzymes. Colored bars underneath each structure denote the modules contained within each structure; white segments are deleted. The number of amino acids in each construct is noted. Measured catalytic rate enhancements relative to the uncatalyzed second-order rate, $(k_{cat}/K_M)/k_{non}$, are plotted on vertical scales aligned in the center of the figure and colored from blue to red (Martinez et al. 2015)

structural variation, enhanced catalytic activity, and higher fidelity imply that these constructs can reasonably be interpreted as snapshots in the parallel evolution of both synthetase classes. That conclusion is supported by linear increases in $\Delta G_{kcat/KM}$ with peptide length. That linearity exhibits the same intercepts and slopes for constructs over an eightfold range from both aaRS classes (Martinez et al. 2015).

Although Urzymes are remarkable catalysts, they are relatively non-specific (Fig. 3). Urzymes exhibit a complete set of functions observed for full-length aaRS: They accelerate amino acid activation with ATP, exhibit pre-steady-state bursts characteristic of tight binding to the activated amino acids (Li et al. 2011; Pham et al. 2010), and catalyze aminoacylation of cognate tRNAs (Li et al. 2013). Thus, they represent a rather sophisticated stage in overall synthetase evolution. Yet, experimental $\Delta G_{kcat/KM}$ values for amino acid activation by ATP show that Class I LeuRS and Class II HisRS Urzymes have similar and complementary specificities (Fig. 3). More negative histograms imply more active amino acid substrates. LeuRS Urzyme prefers Class I substrates; HisRS Urzyme prefers Class II substrates, both by ~–1 kcal/mole. That specificity differential would have allowed ancestral aaRS Urzymes to choose an amino acid from the correct class about 80% of the time.

Modern synthetases prefer their cognate amino acids by ~–5.5 kcal/mole, which enables them to choose the correct amino acid out of twenty approximately 99.99% of the time. Interestingly, the additional specificity of the contemporary full-length aaRS appears to come entirely from allosteric influences on the Urzymes from more contemporary sequences (Weinreb et al. 2014; Li and Carter 2013).

aaRS Urzymes are thus too promiscuous to support more than "statistical ensembles" of peptides, as hypothesized by Carl Woese (Woese et al. 1966; Woese 1965a, b). They thus would have had the properties of quasispecies—a collection

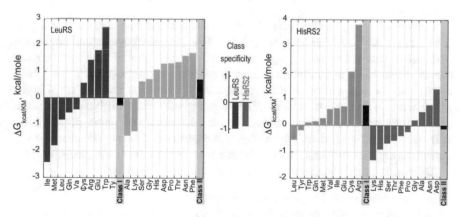

Fig. 3 Amino acid specificity spectra of Class I LeuRS and Class II HisRS2 Urzymes. The net free energy for specificity of Class I and II Urzymes for cognate (ref) versus heterologous amino acids (i), i.e., $\Delta G(k_{cat}/K_M)_{ref} - \Delta G(k_{cat}/K_M)_{amino\ acid\ (i)}$, is approximately 1 kcal/mole for both Class I and II Urzymes (center). Class I amino acids are blue; Class II amino acids are green. Bold colors denote substrates from the homologous class, and pastel colors denote heterologous substrates (adapted from Freeland and Hurst 1998)

of related sequences centered around the most active sequence that in contemporary biology is normally associated with rapidly mutating viruses (Andino and Domingo 2015), but which has considerable relevance when considering the early evolution of all biological sequences under regimes when copying was relatively unfaithful.

The Rodin–Ohno hypothesis (Rodin and Ohno 1995) *states that ancestral Class I and II aaRS resulted from bidirectional coding from a single gene.* The hypothesis makes three different kinds of testable predictions. (i) Those portions that can be aligned sense/and antisense should be the most highly conserved modules, and should retain catalytic activity. (ii) The coding sequences themselves should retain the trace of such coding in a higher frequency of middle codon-base pairing. (iii) It should be possible to construct a *bona fide* sense/antisense gene and show that both strands code for functional amino acid activating enzymes.

Our published experiments confirm each of these predictions (Carter 2015, 2016, 2017; Martinez et al. 2015; Carter et al. 2014; Chandrasekaran et al. 2013; Li et al. 2011, 2013; Cammer and Carter 2010). The first ancestral Class I and Class II aaRSs were therefore very probably coded by opposite strands of the same ancestral gene.

The most highly conserved secondary structures of Class I and II aaRS are found in the 46-residue protozymes. Both protozymes contain an α helix flanked by N- and C-terminal β strands and hence have similar secondary structures in antiparallel orientation. However, codons for all aliphatic and other nonpolar side chains that form the cores of proteins are anticodons, and in the reverse direction, for highly polar side chains. This peculiarity of the universal genetic code (Zull and Smith 1990) ensures that surfaces of both heptapeptide and binary periodicities of helices and extended strands will have opposite polarities. For this reason, their resulting tertiary structures can be considered, essentially, to be formed by inside-out assembly of secondary structures coded in two complementary ways by the same, unique, genetic information.

Conserved, catalytically essential residues in the two active sites are drawn from amino acids activated by the opposite class (Fig. 4). Although amino acids from both aaRS classes are used throughout the tertiary structures of both aaRS classes, use of amino acids as catalytic residues is distinctively segregated. Class I aaRS active sites provide electrostatic transition-state stabilization using Class II amino acid residues lysine, serine, histidine, threonine, aspartic acid, and glycine, positioned with conserved proline residues. Class II active sites coordinate active-site Mg^{2+} ions with conserved glutamate residues and ATP phosphates and adenine heterocycle with conserved arginine residues. Both glutamate and arginine are Class I amino acids.

This mutual catalytic interdependence has two consequences:

(i) It rules out sequential appearance of one class before the other.
(ii) It establishes a tight, hypercycle-like interdependence, assuring survival.

Bidirectional genetic coding of opposite synthetase classes means that codons specifying the order of amino acids in a protein by one of the two strands are complementary and hence as different as possible from those on the opposite strand. Thus, the mutational trajectory taking either strand into its complement is necessarily as long

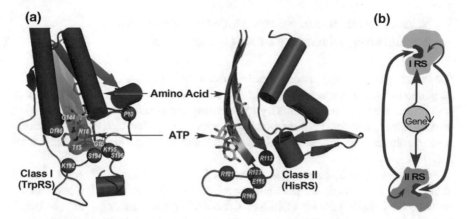

Fig. 4 Active sites of Class I (blue) and Class II (red) aminoacyl-tRNA synthetases. Binding sites for ATP and amino acid are buried in Class I, and much closer to the surface in Class II aaRS. Conserved active-site residues that motivated Rodin and Ohno to advance their sense/antisense coding hypothesis (labeled spheres) are drawn exclusively from the set of substrates activated by the other class, as indicated by respective colors. **b** Exclusive use of catalytic residues drawn from the opposite amino acid class enforces a strong interdependence with significant parallels to hypercyclic interdependence described by Eigen as a solution to the problem of sustaining meaningful sequence information at high error rates

as possible. The two complementary coding strands are maximally differentiated! Consequences of bidirectional genetic coding—similar secondary structures with opposite polarities, inside-out tertiary structures, and strict segregation in the use of catalytic residues in their active sites—ensure that this decisive differentiation propagates at all hierarchical levels of the encoded proteins. An important consequence is that bidirectional coding strongly constrains the range of possible mutations that can simultaneously improve gene products on both strands. Thus, bidirectional coding was almost certainly a transient phase that gave rise to strand specialization as soon as the gene products gained sufficiently sophisticated functionality and fidelity to sustain single-strand coding.

Bidirectional coding of the earliest genes, including the ancestral aaRS genes, thus provided a definitive mechanism for populating distinctly different regions of sequence space. Additionally, the higher-order structures, mechanisms, and substrates of the two classes of synthetases, and even the configuration of the ATP binding sites (see Fig. 2, Dutta et al. 2014) are also all differentiated by the inversion relationship inherent in bidirectional coding. As a consequence, bidirectional coding surmounted the fundamental barrier to quasispecies differentiation by establishing, de novo, two distinct and complementary quasispecies, whose subsequent bifurcation could generate stable variants with new functionality. Moreover, mutual interdependence at the mechanistic level assured survival.

3 Nanoenvironmental Sensing Created Reflexivity, Accelerating Discovery of an Optimal Code

The contemporary genetic code maps fold-determining amino acid physical chemistry onto tRNA identity elements in the acceptor stem and anticodon, creating, only in peptides, the reflexivity necessary to bootstrap the genetic code by accelerating searches through nucleic acid and protein sequence spaces for functionally matching genetic software and protein hardware (Carter and Wills 2018b; Wills and Carter 2018).

The genetic code is massively improbable. There are perhaps 10^{65} ways to build such a code. How, then, did nature discover one so optimal in such a short time? As with Levinthal's paradox (Dill and Chan 1997), the answer to this rhetorical question is that the code must have been discovered stepwise, through a sequence of most probable intermediates.

Wolfenden and Carter worked out important parts of that answer: Amino acid physical chemistry underlies both protein folding and the code (Carter and Wolfenden 2015, 2016; Wolfenden et al. 2015). Models for the ancestry of both tRNAs and aaRSs (Schimmel et al. 1993; Di Giulio 1992) suggest that the tRNA acceptor stem carries an earlier "operational RNA code" within the acceptor stem, allowing synthetase catalytic domains and the acceptor-TΨC stem-loops of tRNAs to mediate protein synthesis. Aminoacylation is efficiently catalyzed by Urzymes (Li et al. 2013), which are even smaller than the aaRS catalytic domains, and therefore cannot possibly recognize tRNA anticodons. That observation lent substantial support to the proposal of an operational RNA code, pointedly raising the question: of what did that code consist?

Amino acid phase transfer free energies correlate with tRNA identity elements. To determine how the acceptor stem selects a suitable aaRS, we examined correlations between the acceptor stem (and anticodon) identity elements (Giegé et al. 1998) and various physical metrics of the twenty amino acids. There are two experimental metrics characteristic of each amino acid side chain: its free energies of transfer, respectively, from vapor and water to cyclohexane. Cyclohexane is superior to octanol for such metrics because of the sparing solubility of water in octanol. The "basis set" afforded by phase transfer equilibria involving the vapor phase as a reference state is (at least) two-dimensional (Carter and Wolfenden 2015; Wolfenden et al. 2015), and cyclohexane differentiates most cleanly between the size and polarity of the amino acid side chains.

Potential predictors were formed by representing each acceptor stem and anticodon identity element base with a binary code; two digits per base (denoting, respectively, whether two (-1) or three (1) H-bonds are formed when it base pairs and whether the base is a pyrimidine (1) or a purine (-1). Regression models for the phase transfer equilibria of the twenty amino acid side chains as linear combinations of the resulting columns facilitated identification of predictive correlations between these predictors and the phase transfer equilibria. These models showed that acceptor-stem bases form a complete code for the free energies of transfer from vapor to cyclohexane, which correlate closely with the side chain size, but not for their polarities given for transfer from water to cyclohexane, which are coded instead by the anticodon bases (Carter and Wolfenden 2015).

Amino acid phase transfer equilibria are also strongly correlated with protein folding. (Carter and Wolfenden 2015; Wolfenden et al. 2015). Thus, the polarity and size of the twenty amino acids form a "combinatorial table" with which to build proteins, in analogy to the periodic table that organizes atoms (Carter and Wolfenden 2016). Linking amino acid phase transfer equilibria to both folding and the choice of mRNA codons convincingly unifies Crick's adaptor hypothesis (Crick 1955) with his Central Dogma of Molecular Biology (Crick 1970) (Fig. 5).

Rudimentary genetic coding may have emerged directly from a peptide•RNA collaboration involving a binary code. The moment when prebiotic chemistry transitioned into biology was thus likely characterized by the following general features: (i) aaRSs were produced from complementary strands of the same ancestral gene. (ii) They catalyzed aminoacylation of complementary groups of amino acids with significantly different physicochemical properties. Class I aaRS activated amino acids with large side chains, whereas Class II aaRS activated amino acids with small side chains. Ancestral tRNAs also fell into two groups, in which acceptor-stem sequences

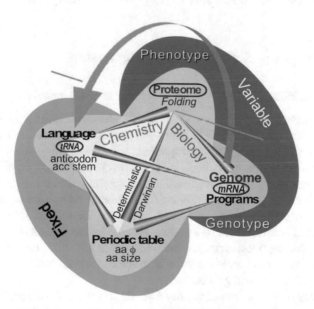

Fig. 5 Incorporating the physical chemistry of amino acid side chain phase transfer free energies into protein folding and genetic coding unifies Crick's adaptor hypothesis about the need for transfer RNA (Crick 1955) with his Central Dogma of Molecular Biology (Crick 1970), which is represented by the orange-brown triangle at the top of the tetrahedron formed by linkages of all three of its vertices with the amino acid transfer free energies at the bottom vertex. This figure clarifies biological relationships between chemistry (deterministic) and biology (Darwinian) and between fixed and variable components of living organisms (adapted from Carter and Wolfenden 2016 and Carter and Wills 2018b)

also correlated with the physical size of the amino acid side chains and could be recognized specifically by only one aaRS class. So Class I aaRs associated with and acylated ur-tRNAs coding for Class I amino acids, and similarly for Class II. This provided a binary "code" based on physicochemical principles. Additional specificity then refined that code so that initially more discriminating subgroups of amino acids and ultimately individual amino acids became associated with individual aaRS and individual tRNAs.

That unification of synthetase and tRNA phylogeny with amino acid physical chemistry illuminates a computational strange loop: Aminoacyl-tRNA synthetases are reflexive because they are made according to rules they themselves enforce (Fig. 5). An important role of such reflexivity is feedback. Only those peptides whose amino acid sequences dictate formation of persistent three-dimensional structures with specific binding pockets for amino acid and tRNA substrates will be able to aminoacylate tRNA with appropriate amino acids. That functional criterion feeds back to condition the choice, not only of codon sequences in potential genes, but also for the tRNA identity elements associated with the codons for each amino acid. Thus, reflexivity itself is, in turn, something that would have been selected much more rapidly from among peptides made as a consequence of aminoacylation by peptide aaRSs than from among ribozymes, which obey entirely different folding rules.

4   Bootstrapping Biology's Operating System

Significant consequences from the two preceding sections include:

(i) Protein aaRS evolution substantially accelerated the search for an optimal code: Koonin has acknowledged (Koonin 2011) that the probability of embedding the information necessary to implement a self-sustaining RNA World is far too small to realize in a single universe and hence would require the probabilistic resources of multi-verses. This acknowledgment is especially crippling because systems based entirely on RNA folding rules have no intrinsic self-organizational feedback circuit and would require natural selection to sense functional consequences of inserting specific amino acids into a protein polymer. Coded peptides, in contrast, naturally furnish the differentiation and reflexivity necessary to accelerate the search for codes with optimal properties like those of the universal genetic code.

(ii) Hybrid coding systems necessarily degrade specificity (Wills and Carter 2018). The differential equations governing protein production by both ribozymal and protein aaRSs show that the system with the higher error rate will adversely impact the overall error rate. Hence, purifying selection would be expected to eliminate the less-sophisticated system. As protein-based aaRS likely descended from a system with one or at most two bits of coding capacity (Martinez et al. 2015), no ribozymal system of tRNA aminoacylation could ever have operated with higher coding capacity. Similarly, it is unlikely that peptides, which lack an obvious mechanism for precise nucleic acid sequence recognition, could ever have functioned as the adaptors predicted by Crick (1955).

(iii) Minimizing dissipative losses requires that transcription/replication and translation error rates be comparable throughout the early evolution of the two processes, necessarily coupling their evolution by informational "impedance matching" (Carter and Wills 2018b).

These conclusions argue persuasively that all key elements of the Central Dogma of Molecular Biology and the universal genetic code arose nearly simultaneously, and from simple origins in a peptide•RNA partnership, in a manner analogous to previously described gene-replicase-translatase systems modeled by simulations (Füchslin and McCaskill 2001; Markowitz et al. 2006) and as suggested originally (Carter and Kraut 1974).

The translation machinery has been justly described as "biology's Operating System" (Bowman et al. 2015). Although an appropriate metaphor, the computational essence to which those authors refer is not actually the ribosome itself, but rather the universal genetic code and the machinery necessary to implement it, namely the set of tRNAs and aaRS. Thus, the evolutionary history of that machinery (aaRS and tRNA) is tantamount to installing an operating system.

Installing a computer operating system requires a minimal instruction set capable of bootstrapping a full instruction set. The minimal instruction set resides in ROM and is referred to as a kernel or "boot block." It initiates the booting process. We argue that the "boot block" for biology's OS combined a bidirectional gene encoding ancestral Class I and II aaRS with the nanoenvironmental sensing that made generating functional aaRS dependent on protein folding (Fig. 6). We recognize here an important distinction between a kernel from which a computer operating system is built, which is necessarily precise, and our putative boot block for the genetic code, which must in addition account for the emergence of specificity from only slightly non-random origins. Specificity was, in fact, discovered as the genetic code was optimized.

Details of the protozyme gene (Martinez et al. 2015) underscore its appropriateness to function as the genetic component of the boot block. The 46-residue peptides forming the ATP binding sites of the Class I and II aaRS both bind ATP tightly and accelerate amino acid activation by $\sim 10^6$-fold. Adapting the protein design program, Rosetta, to impose gene complementarity, we constructed amino acid sequences to stabilize their folds. Catalytically important residues were fixed, as they already had complementary codons. To validate experimental assays, we mutated catalytic residues H18 (Class I strand) and R113 (Class II strand) with alanine, which inactivated both Protozymes.

All four resulting distinct Protozymes—two Class I, two Class II, two with WT sequences, and two with genetically complementary coding sequences—accelerate amino acid activation by 10^6-fold (Martinez et al. 2015).

The Protozyme gene codes for amino acid activating enzymes with two different specificities, each presumably preferring amino acids from its class as we showed

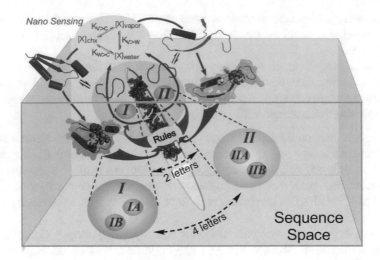

Fig. 6 Experimental characterization of bidirectional coding of early ancestral Class I and II amino acyl-tRNA synthetase evolution (represented by the bi-colored gene at the center (Carter 2017)) and coding relationships associated with tRNA identity-determining bases (Green ellipse; (Carter and Wolfenden 2016)) reveal important elements of a minimal, efficient feedback loop (arrows) that is necessary and sufficient to drive the search for an optimal genetic code. Analogy with the installation of a computer operating system suggests the term "boot block" to describe this feedback loop. Central to this notion is that gene sequences on opposite strands of the ancestral gene were able, when subjected to the physical forces in water (the green ellipse) and consequent protein folding to convert their primary amino acid sequences into three-dimensional structures capable of specific recognition of both amino acid and tRNA substrates (ribbon cartoons). These folded structures then, together, assumed responsibility for enforcing the very rules by which they were originally synthesized. This configuration has the elements of self-recognition necessary and sufficient to form a computational closed loop as described by Hofstadter (1979)

is true for the Urzymes (Carter et al. 2014; Carter 2015). Translated peptides fold as indicated by the two folding equilibria (Fig. 6), which are governed by the phase equilibria in the green ellipse. This nanoenvironmental sensing creates unique binding pockets with catalytic residues, thereby converting the translated sequences into reflexive catalysts: When folded, they can execute the rules by which their sequences were chosen.

The putative boot block uncovers a critically important difference between ribozymal and peptide translatases. RNA folding rules rely primarily on maximizing hydrogen bonding between complementary bases and ring stacking. Thus, they are distinct from protein folding rules, which arise from a more sophisticated interaction between amino acid side chain size and polarity (Carter and Wolfenden 2015, 2016; Wolfenden et al. 2015). Thus, no amino acid activating ribozymes could readily discover coding rules related to protein folding, and a population of ribozymes searching for an optimal code would require the slow, extrinsic feedback cycle of natural selection, which would have acted on system phenotype at whatever level corresponded to an organism! Moreover, differential equations that describe the behavior of these two schemes (Wills and Carter 2018) show that any hybrid system would eliminate the less sophisticated by purifying selection, preventing takeover of ribosomal translation by protein enzymes. *Asking natural selection to evolve translation from a collection of ribozymes is tantamount to asking a mechanic to assemble an internal combustion engine from raw materials without a milling machine.*

When translation is mediated by aaRS enzymes, the coding rules readily discovered by tRNA variation are intrinsically sensitive to protein folding rules. Self-organization bypasses the need for natural selection to choose the code. That difference massively accelerated the organization of the code. These conclusions are reinforced in the next section when we consider dissipative phenomena brought about by errors.

5 Minimizing Dissipative Energy and Information Losses

Polymer biosynthesis in contemporary biology is sharply differentiated: Proteins are assembled by the ribosome, which is a ribozyme—albeit with limited catalytic proficiency (Schroeder and Wolfenden 2007); note that the ribosome itself has no known "translatase" function as that is carried out by aaRS and tRNAs. DNA and RNA are invariably synthesized by polymerases, which are proteins. The emergence of what we now understand as "biology" invariably must entail the emergence of protein *replicases, posing a question* analogous in detail to that posed by the emergence of the protein *aminoacyl-tRNA synthetases*. Our treatment of the latter question answers the former. To wit, it is much more probable that protein polymerases always catalyzed RNA and DNA synthesis than that ribozymal replicases ever evolved! We elaborate below.

Living things dissipate free energy (Prigogine and Nicolis 1971). Moreover, inasmuch as no information transfer process is error-free, information flow is also subject to dissipative losses. Indeed, a significant portion of the (structural) information accumulated by both aaRS and polymerases has likely been selected because by enhancing specificity, it has reduced dissipative informational losses (Carter 2017; Weinreb et al. 2014; Li and Carter 2013).

Pathways that rapidly refined the choices of codons in the coding table and explored nucleic acid codon and amino acid sequence spaces were intimately connected to the time those processes required and, hence, they must have occurred by the fastest route. In turn, the fastest route would have been the one that dissipated the least free energy and information. It should not be surprising, therefore, to discover significant analogies, if not homologies, between such evolutionary questions and the physics of mechanical and electrical engineering processes. The central question is how nature minimizes dissipative losses (Sonnino et al. 2015) of free energy and information. We adapt recent results emerging from the generalization of impedance matching (Gladstone 2016; Martin 2005) to bring the physical principle of least action (Kaila and Annila 2008; Sharma and Annila 2007) to bear on identifying factors that influenced the efficiency of the resulting evolutionary path. What emerges is a surprising conclusion: The smallest dissipative losses in both free energy and information will occur if the error rates of replication and translation are "matched" in the sense of impedance matching in an electrical circuit. That conclusion is validated by the coupling evident in differential equations (Wills and Carter 2018) for the two processes and reinforced by an argument parallel to the improbability of crude peptide translatases taking over from a sophisticated ribozymal coding world.

The noise-to-signal ratio of information processing systems, N/S, represents the opposition to information transfer, or impedance, by analogy with the impedance of an electrical circuit. Complexity, Ψ, can be expressed as bits of information per unit monomer processed. These two variables are related by a tolerance hyperbola, $N/S = \varepsilon_{min}/\Psi$. Here, ε_{min} is a constant efficiency, which has been estimated at 70% by Schneider (Schneider 2010) for systems like repressors.

Consider genes as the source, proteins as the load; the argument implies that to minimize free energy dissipation during growth of complexity, replication and translation errors must remain coupled throughout evolution. High error rates are compatible only with low complexity, which can rise only as errors are brought under control by increased fidelity. The alphabet size of the codon table necessarily restricts the complexity and consequently the enzymatic specificity of coded proteins and is thus an impedance matcher. As for a bicycle derailleur, a two-letter alphabet is the lowest gear, as two-letter patterns will have the lowest specificity and produce low complexity objects with the least dissipation. Increasing alphabet size and bandwidth sustain informational impedance matching by facilitating the emergence of increasing complexity (Carter and Wills 2018b).

6 Conclusions

Eugene Koonin, a balanced and insightful advocate for the RNA World hypothesis (Katsnelson et al. 2018; Koonin 2011; Koonin and Novozhilov 2009), acknowledges without irony (Koonin 2011) that the RNA World is impossibly improbable!

Four billion years no doubt hid many details of how translation emerged. What was overwritten and/or lost from view has, however, recently become experimentally accessible through structural biology—experimental dissection and characterization of the two aaRS superfamilies (Carter 2014, 2017; Carter et al. 2014), protein design and biochemistry (Martinez et al. 2015) new methods of phylogenetic analysis (Chandrasekaran et al. 2013), and characterization of the amino acid side chain phase transfer behavior (Carter and Wolfenden 2015, 2016; Wolfenden et al. 2015). The answer to the question: "why study the origin of translation—now?" is therefore twofold: first, because we can and second, because the emergence of coded protein synthesis is the quintessential transition necessary for biology.

Independently of the experimental work, computational study of the requirements for introducing genetic coding into emergent biology (Wills et al. 2015; Wills 1993, 1994, 2001, 2004, 2009, 2014a, b; Nieselt-Struwe and Wills 1997; Füchslin and McCaskill 2001; Markowitz et al. 2006; Hordijk et al. 2014) confirmed and elaborated on the conclusions of earlier work (Eigen and Schuster 1977, 1978; Eigen 1971a, b; Eigen et al. 1988, 1989). That work, together with the study of problems posed by molecular parasites (Hickinbotham and Hogeweg 2016; Colizzi and Hogeweg 2016a, b; Takeuchi and Hogeweg 2012; Hogeweg and Takeuchi 2003) and the increasing clarity with which error thresholds and their influence must be understood (Takeuchi et al. 2017), has characterized to an unprecedented extent the nature of the problems that must have been overcome by nature in order for genetic coding to emerge.

We are fortunate to have been able to assemble the experimental work into the context of the theoretical studies. What we have found is that what initially appeared as curiosities—evidence for ancestral bidirectional coding of the two synthetase classes, the extended inversion symmetries in higher-order structure and functionality resulting from in the experimental work, and the strong correlations between amino acid physical chemistry and both protein folding and the tRNA identity elements used by synthetases to recognize cognate tRNAs—fit like a hand in a glove into the landscape painted by the theoretical studies.

The resulting answers shape our understanding of the proteome, and more! Thus, no matter how much remains unknown, sufficient knowledge has emerged about the evolution of the aaRS to shift the focus of the origin of translation from a highly improbable system of ribozymal translatases to an eminently natural system of rudimentary peptide aaRS. That shift doubtless increases the probability that the origin of genetics could have overcome both of Eigen's obstacles within a terrestrial time frame:

i. Bidirectional aaRS coding decisively partitioned sequence space, enabling differentiation and creating hypercycle-like interdependence to stabilize and assure survival of multiple genetic units.

ii. As amino acid physical chemistry is essential for protein folding, a nanosensing feedback mechanism was necessary to find an optimal genetic code rapidly enough to account for a terrestrial origin of translation.

iii. Reflexivity, necessary and sufficient for the growth of complexity (Hofstadter 1979), emerges only if protein folding is incorporated into the feedback loop driving selection of molecular species. Reflexivity enabled bootstrapping to accelerate discovery of an optimal code via self-organization.

iv. Differential equations for hybrid ribozyme-protein enzyme translation behavior confirm the intuitive notion that inferior peptide-based aaRS or polymerases could never have "taken over" from a more sophisticated, preexisting ribozyme-based systems (Wills and Carter 2018).

v. (i–iv) imply that the RNA World is neither necessary nor epistemologically useful. Coded peptides have always been more potent, diverse catalysts than ribozymes.

vi. The need to balance replication/translation error rates to minimize free energy and information dissipation strongly implies that the genetic code grew by incorporating small numbers of new letters and that evolution of inheritance, and gene expression had to have proceeded such that the inevitable error rates of one process were matched to those of the other.

References

Andino R, Domingo E (2015) Viral quasispecies. Virology 479–480:46–51. https://doi.org/10.101 6/j.virol.2015.03.022

Bedian V (1982) The possible role of assignment catalysts in the origin of the genetic code. Orig Life 12(2):181–204

Bedian V (2001) Self-description and the origin of the genetic code. BioSystems 60:39–47

Bernhardt HS (2012) The RNA world hypothesis: the worst theory of the early evolution of life (except for all the others). Biol Direct 7:23

Bowman JC, Hud NV, Williams LD (2015) The ribosome challenge to the RNA world. J Mol Evol 80:143–161. https://doi.org/10.1007/s00239-015-9669-9

Cammer S, Carter CW Jr (2010) Six Rossmannoid folds, including the Class I aminoacyl-tRNA synthetases, share a partial core with the anticodon-binding domain of a Class II aminoacyl-tRNA synthetase. Bioinformatics 26(6):709–714. https://doi.org/10.1093/bioinformatics/btq039

Carter CW Jr (1975) Cradles for molecular evolution. New Sci March 27:784–787

Carter CW Jr (1993) Cognition mechanism and evolutionary relationships in aminoacyl-tRNA synthetases. Annu Rev Biochem 62:715–748

Carter CW Jr (2014) Urzymology: experimental access to a key transition in the appearance of enzymes. J Biol Chem 289(44):30213–30220. https://doi.org/10.1047/jbcR114.576495

Carter CW Jr (2015) What RNA world? Why a peptide/RNA partnership merits renewed experimental attention. Life 5:294–320. https://doi.org/10.3390/life5010294

Carter CW Jr (2016) An alternative to the RNA world. Nat Hist 125(1):28–33

Carter CW Jr (2017) Coding of Class I and II aminoacyl-tRNA synthetases. Adv Exp Med Biol: Protein Rev 18:103–148. https://doi.org/10.1007/5584_2017_93

Carter CW Jr, Kraut J (1974a) A proposed model for interaction of polypeptides with RNA. Proc Natl Acad Sci USA 71(2):283–287

Carter CW Jr, Kraut J (1974b) A proposed model for interaction of polypeptides with RNA. Proc Natl Acad Sci USA 71:283–287

Carter CW Jr, Wills PR (2018a) Identifying how acceptor-stem bases determine tRNA groove recognition by Class I and II aminoacyl-tRNA synthetases fulfills the operational RNA code. Nucleic Acids Res (In revision)

Carter CW Jr, Wills PR (2018b) Interdependence, reflexivity, fidelity, and impedance matching, and the evolution of genetic coding. Mol Biol Evol 35(2):269–286. https://doi.org/10.1101/139139

Carter CW Jr, Wolfenden R (2015) tRNA acceptor-stem and anticodon bases form independent codes related to protein folding. Proc Nat Acad Sci USA 112(24):7489–7494. http://www.pnas.org/cgi/doi/10.1073/pnas.1507569112

Carter CW Jr, Wolfenden R (2016) Acceptor-stem and anticodon bases embed amino acid chemistry into tRNA. RNA Biol 13(2):145–151. https://doi.org/10.1080/15476286.2015.1112488

Carter CW Jr, Li L, Weinreb V, Collier M, Gonzales-Rivera K, Jimenez-Rodriguez M, Erdogan O, Chandrasekharan SN (2014) The Rodin-Ohno hypothesis that two enzyme superfamilies descended from one ancestral gene: an unlikely scenario for the origins of translation that will not be dismissed. Biol Direct 9:11

Chandrasekaran SN, Yardimci G, Erdogan O, Roach JM, Carter CW Jr (2013) Statistical evaluation of the Rodin-Ohno hypothesis: sense/antisense coding of ancestral Class I and II aminoacyl-tRNA synthetases. Mol Biol Evol 30(7):1588–1604. https://doi.org/10.1093/molbev/mst070

Colizzi ES, Hogeweg P (2016a) High cost enhances cooperation through the interplay between evolution and self-organisation. BMC Evol Biol 16:31. https://doi.org/10.1186/s12862-016-060 0-9

Colizzi ES, Hogeweg P (2016b) Parasites sustain and enhance RNA-like replicators through spatial self-organisation. PLoS Comput Biol 12(4):e1004902. https://doi.org/10.1371/journal.pcbi.100 4902

Crick FHC (1955) On degenerate templates and the adaptor hypothesis. Unpublished. https://profi les.nlm.nih.gov/ps/retrieve/Narrative/SC/p-nid/153

Crick FHC (1970) Central dogma of molecular biology. Nature 227(August 8):561–563

Cullis PM, Wolfenden R (1981) Affinities of nucleic acid bases for solvent water? Biochemistry 20:3024–3028

Di Giulio M (1992) On the origin of the transfer RNA molecule. J Theor Biol 159:199–214

Dill K, Chan HS (1997) From Levinthal to pathways to funnels. Nat Struct Biol 4:10–19

Dutta S, Choudhury K, Banik SD, Nandi N (2014) Active site nanospace of aminoacyl tRNA synthetase: difference between the Class I and Class II synthetases. J Nanosci Nanotechnol 14:2280–2298

Eigen M (1971a) Selforganization of matter and the evolution of biological macromolecules. Naturwissenschaften 58:465–523

Eigen M (1971b) Molecular self-organisation and the early stages of evolution. Quart Rev Biophys 4:149–212

Eigen M, Schuster P (1977) The hypercyde: a principle of natural self-organization Part A: Emergence of the hypercycle. Naturwissenschaften 64:541–565

Eigen M, Schuster P (1978) The hypercycle: a principle of natural self-organization Part C: The realistic hypercycle. Die Naturwissenschaften 65:341–369

Eigen M, McCaskill JS, Schuster P (1988) Molecular quasi-species. J Phys Chem 92:6881–6891

Eigen M, Lindemann BF, Tietze M, Winkler-Oswatitsch R, Dress A, von Haesler A (1989) How old is the genetic code? statistical geometry of tRNA provides an answer. Science 244(12 May 1989):673–679

Eriani G, Delarue M, Poch O, Gangloff J, Moras D (1990) Partition of tRNA synthetases into two classes based on mutually exclusive sets of sequence motifs. Nature 347:203–206

Freeland SJ, Hurst LD (1998) The genetic code is one in a million. J Mol Evol 47:238–248

Füchslin RM, McCaskill JS (2001) Evolutionary self-organization of cell-free genetic coding. Proc Natl Acad Sci USA 98:9185–9190

Giegé R, Sissler M, Florentz C (1998) Universal rules and idiosyncratic features in tRNA identity. Nucleic Acids Res 26(22):5017–5035

Gladstone E (2016) Error in information diffusion processes. Cornell University, Ithaca, NY

Hickinbotham S, Hogeweg P (2016) Evolution towards extinction in replicase models: inevitable unless. In: Beslon G, Elena S, Hogeweg P, Schneider D, Stepney S (eds) Evolution of evolution

Hofstadter DR (1979) Gödel, Escher, Bach: an eternal golden braid. Basic Books Inc., New York

Hogeweg P, Takeuchi N (2003) Multilevel selection in models of prebiotic evolution: compartments and spatial self-organization. Orig Life Evol Biosph 33:375–403

Hordijk W, Wills PR, Steel M (2014) Autocatalytic sets and biological specificity. Bull Math Biol 76:201–224. https://doi.org/10.1007/s11538-013-9916-4

Kaila VRI, Annila A (2008) Natural selection for least action. Proc R Soc A 464:3055–3070. https://doi.org/10.1098/rspa.2008.0178

Katsnelson MI, Wolf YI, Koonin EV (2018) Towards physical principles of biological evolution. arXiv orgabs1709.00284

Koonin EV (2011) The logic of chance: the nature and origin of biological evolution. Pearson Education; FT Press Science, Upper Saddle River, NJ

Koonin EV, Novozhilov AS (2009) Origin and evolution of the genetic code: the universal enigma. IUBMB Life 61((2) February):99–111. https://doi.org/10.1002/iub.146

Li L, Carter CW Jr (2013) Full implementation of the genetic code by tryptophanyl-trna synthetase requires intermodular coupling. J Biol Chem 288(29 November):34736–34745. https://doi.org/1 0.1074/jbc.m113.510958

Li L, Weinreb V, Francklyn C, Carter CW Jr (2011) Histidyl-tRNA synthetase Urzymes: Class I and II aminoacyl-tRNA synthetase Urzymes have comparable catalytic activities for cognate amino acid activation. J Biol Chem 286:10387–10395. https://doi.org/10.1074/jbc.M110.198929

Li L, Francklyn C, Carter CW Jr (2013) Aminoacylating Urzymes challenge the RNA world hypothesis. J Biol Chem 288:26856–26863. https://doi.org/10.1074/jbc.M113.496125

Markowitz S, Drummond A, Nieselt K, Wills PR (2006) Simulation model of prebiotic evolution of genetic coding. In: Rocha LM, Yaeger LS, Bedau MA, Floreano D, Goldstone RL, Vespignani A (eds) Artificial Life, vol 10. MIT Press. Cambridge, MA, pp 152–157

Martin P (2005) Spatial interpolation in other dimensions. Oregon State University, Corvallis, OR

Martinez L, Jimenez-Rodriguez M, Gonzalez-Rivera K, Williams T, Li L, Weinreb V, Niranj Chandrasekaran S, Collier M, Ambroggio X, Kuhlman B, Erdogan O, Carter CW Jr (2015a) Functional Class I and II amino acid activating enzymes can be coded by opposite strands of the same gene. J Biol Chem 290(32):19710–19725. https://doi.org/10.1074/jbc.M115.642876

Martinez L, Jimenez-Rodriguez M, Gonzalez-Rivera K, Williams T, Li L, Weinreb V, Chandrasekaran SN, Collier M, Ambroggio X, Kuhlman B, Erdogan O, Carter CW Jr (2015b) Functional Class I and II amino acid activating enzymes can be coded by opposite strands of the same gene. J Biol Chem 290(32):19710–19725. https://doi.org/10.1074/jbc.M115.642876

Nieselt-Struwe K, Wills PR (1997) The emergence of genetic coding in physical systems. J Theor Biol 187:1–14

Pham Y, Li L, Kim A, Weinreb V, Butterfoss G, Kuhlman B, Carter CW Jr (2007) A minimal TrpRS catalytic domain supports sense/antisense ancestry of Class I and II aminoacyl-tRNA synthetases. Mol Cell 25:851–862

Pham Y, Kuhlman B, Butterfoss GL, Hu H, Weinreb V, Carter CW Jr (2010) Tryptophanyl-tRNA synthetase Urzyme: a model to recapitulate molecular evolution and investigate intramolecular complementation. J Biol Chem 285:38590–38601. https://doi.org/10.1074/jbc.M110.136911

Prigogine I, Nicolis G (1971) Biological order, structure and instabilities. Q Rev Biophys 4(2&3):107–148. https://doi.org/10.1017/S0033583500000615

Rodin SN, Ohno S (1995) Two types of aminoacyl-tRNA synthetases could be originally encoded by complementary strands of the same nucleic acid. Orig Life Evol Biosph 25:565–589

Schimmel P, Giegé R, Moras D, Yokoyama S (1993) An operational RNA code for amino acids and possible relationship to genetic code. Proc Natl Acad Sci USA 90:8763–8768

Schneider TD (2010) A brief review of molecular information theory. Nano Commun Netw 1:173–180

Schroeder GK, Wolfenden R (2007) The rate enhancement produced by the ribosome: an improved model. Biochemistry 46:4037–4044

Sharma V, Annila A (2007) Natural process—natural selection. Biophys Chem 127:123–128. https://doi.org/10.1016/j.bpc.2007.01.005

Sonnino G, Evslin J, Sonnino A (2015) Minimum dissipation principle in nonlinear transport. Entropy 17:7567–7583. https://doi.org/10.3390/e17117567

Takeuchi N, Hogeweg P (2012) Evolutionary dynamics of RNA-like replicator systems: a bioinformatic approach to the origin of life. Phys Life Rev 9:219–263

Takeuchi N, Hogeweg P, Kaneko K (2017) Conceptualizing the origin of life in terms of evolution. Philos Trans R Soc A 375:20160346. https://doi.org/10.1098/rsta.2016.0346

Weinreb V, Li L, Chandrasekaran SN, Koehl P, Delarue M, Carter CW Jr (2014) Enhanced amino acid selection in fully-evolved tryptophanyl-tRNA synthetase, relative to its Urzyme, requires domain movement sensed by the D1 switch, a remote. Dyn Pack Motif J Biol Chem 289:4367–4376. https://doi.org/10.1074/jbc.M113.538660

Wills PR (1993) Self-organization of genetic coding. J Theor Biol 162:267–287

Wills PR (1994) Does information acquire meaning naturally? Berichte der Bunsengesellschaft für Physikaliche Chemie 98:1129–1134

Wills PR (2001) Autocatalysis, information, and coding. BioSystems 50:49–57

Wills PR (2004) Stepwise evolution of molecular biological coding. In: Pollack J, Bedau M, Husbands P, Ikegami T, Watson RA (eds) Artificial life IX. MIT Press, Cambridge, pp 51–56

Wills PR (2009) Informed generation: physical origin and biological evolution of genetic codescript interpreters. J Theor Biol 257:345–358

Wills PR (2014a) Genetic information, physical interpreters and thermodynamics; the material-informatic basis of biosemiosis. Biosemiotics 7:141–165. https://doi.org/10.1007/s12304-013-9 196-2

Wills PR (2014b) Spontaneous mutual ordering of nucleic acids and proteins. Orig Life Evol Biosph 44:293–298. https://doi.org/10.1007/s11084-014-9396-z

Wills PR (2016) The generation of meaningful information in molecular systems. Philos Trans R Soc A A 374:20150016. https://doi.org/10.1098/rsta.20150066

Wills PR, Carter CW Jr (2018a) Insuperable problems of an initial genetic code emerging from an RNA world. BioSystems 164:155–166. https://doi.org/10.1101/140657

Wills PR, Carter CW Jr (2018b) Insuperable problems of an initial genetic code emerging from an RNA world. Biosystems 164:155–166. https://doi.org/10.1101/140657

Wills PR, Nieselt K, McCaskill JS (2015) Emergence of coding and its specificity as a physico-informatic problem. Orig Life Evol Biosph 45(June):249–255. https://doi.org/10.1007/s11084-0 15-9434-5

Woese CR (1965a) On the origin of the genetic code. Proc Natl Acad Sci USA 54:1546–1552

Woese CR (1965b) Order in the origin of the genetic code. Proc Natl Acad Sci USA 54:71–75

Woese CR, Dugre DH, Saxinger WC, Dugre SA (1966) The molecular basis for the genetic code. Proc Natl Acad Sci USA 55:966–974

Wolfenden R (2011) Benchmark reaction rates, the stability of biological molecules in water, and the evolution of catalytic power in enzymes. Ann Rev Biochem 80:645–667

Wolfenden R, Lewis CA, Yuan Y, Carter CW Jr (2015) Temperature dependence of amino acid hydrophobicities. Proc Nat Acad Sci USA 112(24):7484–7488. https://doi.org/10.1073/pnas.15 07565112

Xia Y, Levitt M (2002) Roles of mutation and recombination in the evolution of protein thermody-namics. Proc Natl Acad Sci USA 99(16 August 6):10382–10387. https://doi.org/10.1073/pnas.1 62097799

Zull JE, Smith SK (1990) Is genetic code redundancy related to retention of structural information in both DNA strands? TIBS 15:257–261

Part V
Concepts

Biological Dogmas in Relation to the Origin of Evolutionary Novelties

Patricia Tatemoto and Carlos Guerrero-Bosagna

Abstract Current evolutionary knowledge emerges influenced by two main currents of thinking. While some call for more variables to be considered in evolution, others consider that evolutionary outcomes are mainly associated with the genomic component. In spite of each individual point of view, however, current knowledge from many biological disciplines is constantly telling us of new advancements in biological processes. The aim of the present book chapter is to evaluate what historically in evolutionary thinking has been assumed as true and what is currently being contested by current biological knowledge. Importantly, no matter what the current knowledge about these topics is, they are still assumed as true in evolutionary thinking, making them factual dogmas. In the present chapter we describe and discuss three 'evolutionary dogmas', present in scientific and/or pedagogical writing, for which we think that there is plenty of biological evidence demonstrating the contrary: (i) 'Natural selection is the main/only guiding force in evolution'; (ii) 'Evolution proceeds driven by what is beneficial'; (iii) 'Lamarck was wrong about environmental effects and heritability'. Many of our arguments are based on recent epigenetic knowledge showing that epigenetic mechanism can on one hand induce genomic changes, and on the other hand respond to environmental factors. Through epigenetic mechanisms, the environment is able to modulate individual phenotypes even before conception, with consequences that can transcend generations. We propose that recent epigenetic knowledge can help to change some of these described dogmas to advance scientific knowledge in evolution.

P. Tatemoto · C. Guerrero-Bosagna (✉)
Avian Behavioural Genomics and Physiology Group, IFM, Linköping University, 58 183 Linköping, Sweden
e-mail: carbo@ifm.liu.se

P. Tatemoto
FMVZ, Department of Veterinary Medicine and Animal Health, School of Veterinary Medicine and Animal Science, Center for Comparative Studies in Sustainability, Health and Welfare, University of São Paulo - USP - São Paulo State, Pirassununga, SP 13635-900, Brazil

© Springer International Publishing AG, part of Springer Nature 2018
P. Pontarotti (ed.), *Origin and Evolution of Biodiversity*,
https://doi.org/10.1007/978-3-319-95954-2_17

317

A growing number of biologists are claiming for change in how evolution is conceptualized (Laland et al. 2014). The scientific community is increasingly recognizing that a systemic approach is needed in evolution, in which developmental biology, genomics, epigenetics, ecology and behavioural sciences have to be considered for a thorough understanding of evolutionary processes. Under this approach, the relation between cause and effect is more dynamic than in the traditional evolutionary thinking, in which evolutionary biology has mainly been associated with ultimate causes (Mayr 1961). It is fair to say that in recent years proximate causes have had an increasingly predominant role in evolutionary thinking. Recent evidence is showing that evolutionary processes are not solely shaped by changes in allele frequencies, as previously thought (Dobzhansky 1977), but by the synergistic interaction of many variables, including developmental, physiological and environmental, which play a fundamental role (Noble 2013).

It is nowadays patent in evolutionary biology that biological knowledge emerges influenced by two main currents of thinking. On the one side, there are those calling for more variables to be considered in evolution (Mattick 2012; Danchin 2013; Noble 2013), while on the other side there are those that consider that evolutionary outcomes are mainly associated with the genomic component (Kalinowski et al. 2010), and/or that organism are passive entities that respond to the will of their genes (Dawkins 1978). In spite of each individual point of view, however, current scientific knowledge from many biological disciplines is constantly telling us what is actually happening in biological processes. The aim of the present chapter is to evaluate what historically in evolutionary thinking has been assumed as true and what is currently being contested by current biological knowledge. Importantly, no matter what the current knowledge about these topics is, they are still assumed as true in evolutionary thinking, making them true dogmas. In the present chapter, we describe three 'evolutionary dogmas', present in scientific and/or pedagogical writing, for which we think that there is plenty of biological evidence demonstrating the contrary: (i) *'Natural selection is the main/only guiding force in evolution'*; (ii) *'Evolution proceeds driven by what is beneficial'*; (iii) *'Lamarck was wrong about environmental effects and heritability'*.

1   "*Natural Selection Is the Main/Only Guiding Force in Evolution*"

Selective forces are certainly an important factor involved in evolution. While this does not mean that natural selection is the only relevant pathway, most of evolutionary knowledge divulgation is focused on natural selection as being the **only** acting force in evolution, and thus based on the premise that the survivors in evolution have only done so at the detriment of less fit individuals. Although these two concepts, namely *'natural selection is the only/main force in evolution'* and *'evolution proceeds towards beneficial outcomes'*, relate, we have separated them as two dogmas based on the thinking that the first relates more to how the general public perceives the concept of evolution in general, while the second relates more to how scientists and non-scientists alike understand the causes of evolutionary change. A(n) (incomplete) list of other mechanisms playing a relevant role in evolution is described here.

Common evolutionary interpretations portrait changes in allele frequencies as the main drivers of evolution. Under this interpretation, mutations are random (stochastic) events that emerge independent of environmental influences while the environment acts solely through sorting alleles to one evolutionary branch or another (Dobzhansky 1977). Additionally, common knowledge overestimates the consideration of the DNA as a 'code' and underestimates its consideration as a chemical entity who as such, reacts with other chemical entities. Recent evidence, however, shows that the genome is very reactive to the environment (Gregoire et al. 2013). In some sense, these effects that the environment has on the genome are related to certain aspects of what is known as 'Lamarckian evolution' (described in detail later in the text). One of the ways this connection between environment and genetic changes occurs is through epigenetic mechanisms. Environmental factors can alter epigenetic mechanisms such as DNA methylation, which is in turn related to biased mutations (Guerrero-Bosagna 2017). Recent evidence shows that epigenetic changes could be the first step of speciation, since differing environmental conditions can generate diverging epigenomes that persist for generations (Johannes et al. 2009; Cortijo et al. 2014). Thus, this persistence of epigenetic changes provides many opportunities in which biased mutations can occur to create genomic evolutionary novelties. Epigenetic mechanisms, however, are only one of the many factors that may induce mutations. Genetic changes are known to be influenced by biased CG conversions (Galtier and Duret 2007), hypermutability of transitions compared to transversions (Collins and Jukes 1994), activity of retrotransposons (Adelson et al. 2015), or microsatellite instability (Sequeira et al. 2015), among others.

One of the most basic factors playing a role in genomic evolutionary change is genetic drift. In simple words, genetic drift describes changes in the frequency of alleles in a population that occur by chance through time (Necsulea and Kaessmann 2014). Based on the fact that alleles in the offspring are a sample of the alleles in the parents, fluctuation in allele frequency will occur from one generation to the next just driven by which alleles will be transmitted from parents to offspring, even if all individuals have the same fitness. Interestingly, Darwin himself recognized the conceptual existence of genetic drift (Duret 2008). Genetic drift in connection to neutral evolution will be described later in the text.

Another mechanism of great relevance for evolution and speciation is sexual selection. Sexual selection is a powerful evolutionary mechanism that contributes to the origin of new species by acting directly on traits related to mate preference, thereby promoting pre-mating reproductive isolation. Sexual selection arises from competition (commonly between males) for attracting mates. The success will depend on male–male competition, female mate choice or both (McCullough et al. 2016). It has been shown that female preference is a powerful driver of speciation (Kraaijeveld et al. 2011; Rodríguez et al. 2013), while male–male competition would be unrelated to reproductive isolation (Qvarnström et al. 2012). A recent meta-analysis demonstrated a correlation between sexual selection and speciation rate (Kraaijeveld et al.

2011). Learning is also shown to affect mate preference and traits under mate selection, thereby also playing a role on sexual selection and speciation (see Svensson et al. 2012).

Geographic speciation has also been described as a powerful force in evolution. Classified according a spatial isolation gradient, geographic speciation includes: (i) allopatric speciation, which is the gradual divergence of populations with completely separate geographic ranges; (ii) parapatric speciation, which is the divergence of populations with adjacent geographic ranges that experience occasional genes exchange; and (iii) sympatric speciation, which occurs between populations with close physical contact and thus, experiencing high levels of gene flow (Kocher 2004; Thorpe et al. 2010; Friedman et al. 2013; Kautt et al. 2016).

A less common evolutionary mechanism is the 'founder effect', which occurs when a few individuals become isolated from a larger population, thereby starting a new population whose gene pool does not match one of the original population. The founder effect is an extreme example of genetic drift and is the alteration of gene frequencies resulting from the establishment of new population generated by individuals isolated from the parental population (Ichinose and Arita 2008). Similarly, the 'bottleneck effect' occurs when a population becomes so reduced in size due to a particular environmental event that the surviving population does not genetically represent the original. Genetic bottleneck is commonly associated with biological invasions (Prentis et al. 2008).

It is becoming increasingly evident that the origin of species occurs mainly through a synergetic effect of more than one biological mechanism. Divergence of genomes has been shaped by cycles of population expansion, fragmentation, pluvial regimes, climate changes, environmental catastrophes, competition by niches, sexual selection, among others (Kraaijeveld et al. 2011; Svensson et al. 2012; Leijs et al. 2012; Rodríguez et al. 2013; Kautt et al. 2016). Also, occupancy of ecological niche, social behaviour and motivational variation can affect, or be affected by, brain function, thereby influencing diversity (Buss 1983; Dukas 2013; Fernald 2015; Mesoudi et al. 2016; Wood and Komarova 2018). Thus, many biological mechanisms are shown to be relevant for evolution, which are unrelated to the action of natural selection.

Among the evolutionary mechanisms described above, many have been incorporated into the framework of the modern synthesis, such as sexual selection, geographic speciation, and founder and bottleneck effects. Other mechanisms, however, involve the study of processes that challenge deeply rooted neo-Darwinian ideas such as the role of random mutations in originating evolutionary novelty, and the invulnerability of the Weissmann barrier (Haig 2011). Such biological processes include transgenerational epigenetic inheritance, environmentally induced mutations, niche construction, developmental constraints and bias, and eco-evo-devo. Because of the unease of these processes with neo-Darwinian ideas, there have been calls in recent years for an 'Extended Evolutionary Synthesis' that would fully integrate these concepts into evolutionary theory (Laland et al. 2014).

2 "*Evolution Proceeds Driven by What Is Beneficial*"

One of the most widespread dogmas in evolution is assuming that the majority of the traits we observe in organisms exist because they confer, or have conferred, advantage to some individuals over others (Guerrero-Bosagna 2012). Too often adaptation is assumed as the explanation for evolutionary processes without appropriate experimental testing (Lynch 2007; Levasseur et al. 2007). This assumption is deeply rooted in evolutionary thinking, both in scientists and non-scientists, and has previously been the focus of criticism by the now famous article '*The Spandrels of San Marco and the Panglossian Paradigm: A Critique of the Adaptationist Programme*' by Gould and Lewonting (1979). The consideration of 'benefit' as the main driving force for evolution originated within the realm of neo-Darwinian evolution, which assumes that mutational changes accumulate in a species exclusively by the action of positive selection. In other words, there has to be fitness advantages associated with the mutations (i.e. base replacements or genomic rearrangements) in order for them to be fixed in the population. Although some mutations are associated with increased fitness (Johnston et al. 2013), and others with negative outcomes (Mohan and Mellick 2017; Stojković and Fujimori 2017; Srinivas et al. 2018) that include the production of defective proteins and uncontrolled cell division related to cancer (Johnson 2017), the reality is that mutations in most cases will not associate with detrimental or positive outcomes (Eyre-Walker and Keightley 2007). Estimations show that the vast majority of mutations that are maintained in evolution are neutral (unrelated to fitness), while a small fraction is detrimental, and those positively associated with fitness are indeed rare (Ponting and Lunter 2006).

The view that mutations with neutral fitness consequences are the main ones maintained in evolution is not new. In 1968, scientist Motoo Kimura developed what is known as 'The Neutral Theory of Evolution' (Kimura 1968) which states that the majority of evolutionary changes at the molecular level are caused not by natural selection acting in the advantageous mutants, rather by random fixation, selectively neutral or close-to-neutral mutants under continued mutation pressure (Gojobori et al. 1990). The neutral theory adopts the thinking that most of the intraspecific variability at the molecular level is selectively neutral from a fitness standpoint, so that polymorphisms are maintained based on a balance between mutational input and random extinction (Gojobori et al. 1990). Beneficial mutations can occur, but the neutral theory assumes (correctly) that they are so rare that is not the main mechanism through which evolution works. The neutral theory challenges natural selection in the sense of the evolutionary relevance of the association between fitness and mutations, while in neo-Darwinism, the mutations associated with increased fitness are the most relevant; in the neutral theory, mutations with no fitness associations are the most important.

A recent example of the relevance of neutrality in evolution has been described in cichlid fish, which represent one of the most striking examples of adaptive radiation, with the highest sustained rates of speciation known to date in vertebrates (Kocher 2004). The wide variation in lineages and rates of speciation of cichlid fish has intrigued scientists for many years. Interestingly, neutral evolution was key in producing the genetic variability related to the variety of phenotypes registered in cichlids, with neutral processes being the main responsible for the accumulation of the observed genomic variation (Brawand et al. 2015). Thus, it seems that neutral evolution has a great role in the generation of genetic variation, which involves the emergence of single nucleotide polymorphisms or major genomic rearrangements.

It is postulated that neutrality facilitates molecular innovation by enabling the emergence of robust molecules that can tolerate the inclusion of neutral mutations (Wagner 2008). These mutations would ultimately constitute a 'reservoir' of new gene functions. In an attempt to reconcile neo-Darwinism and neutral theory, Wagner (Wagner 2008) suggests recognizing the importance of neutral mutations for evolutionary innovation, while considering selection a key factor in explaining patterns of genetic variation.

"The Neutral Theory of Evolution" has recently been used as the base for 'The Neutral Theory of Biodiversity' or 'The Unified Neutral Theory of Biodiversity and Biogeography'. This theory is presented as a model for testing the mechanisms of species coexistence and biodiversity maintenance in ecological communities. It also offers a mechanistic explanation of the species abundance distributions, by stating that differences between members among similar species within the same trophic level are neutral, or irrelevant for their success (Hubbell 2005, 2014).

3 *"Lamarck Was Wrong on His Assumption About Environmental Effects and Heritability"*

One of the most debated issues in biology is related to the relative importance of genes or environment in shaping the phenotype of living organisms. The discussion between nature and nurture has crossed millennia, beginning with philosophers René Descartes (nature) and John Locke (nurture), continuing with Darwin and Lamarck, and reaching contemporary scientists. Darwin and Lamarck certainly had quite different, but not uncompromising, views about this issue. While Darwin emphasized the role of environment as restricting inherent traits of living organisms (nature), Lamarck emphasizes that environmental conditions can exert direct influence on the development and inheritance of organisms (Guerrero-Bosagna 2012). Although both perspectives were theoretically sound and based on observations, the historical perception of Lamarck and Darwin has been quite dissimilar (Noble 2013); while Darwin is considered the main name related to evolutionary thinking, Lamarck is often the object of discredit and mock. In the words of Waddington (Waddington 1954) '*Lamarck is the only major figure in the history of biology whose name has become to all extents and purposes, a term of abuse. Most scientists' contributions are fated to be outgrown, but very few authors have written works which, two centuries later, are still rejected with an indignation so intense that the skeptic may suspect something akin to an uneasy conscience*'. This is a sad outcome to the legacy of the scientist that, among other contributions, put forward the first comprehensive theory of organic evolution (Burkhardt Jr. 2013). In spite of this, in the classroom, from high school to undergraduate teaching, as well as in documentaries, what is still being taught is that Darwin was right and Lamarck wrong. In this section, we do not intend to say the opposite by portraying that Lamarck got it right and Darwin wrong. Our only aim is to contribute to the view that both Darwin and Lamarck had relevant things to say from an evolutionary standpoint and that current science should rescue the value of both in order to fully understand how evolution proceeds. Evolutionary processes will not be fully understood without a serious consideration of both Darwinian and Lamarckian theoretical postulates, observations, and ideas, since they both have described complementary and relevant evolutionary mechanism. With this spirit, we attempt to summarize which aspects of Lamarck's contribution are still pertinent to biology, particularly to evolutionary biology.

Lamarck's ideas were focused on finding the causes underlying the expansion of organic complexity (Burkhardt Jr. 2013), thereby initiating the conceptual trend of inquiring into the role of the environment in producing heritable phenotypic variability. Currently, many disciplines have taken into consideration these (initially) Lamarckian views. Evo-devo, for example, has focused on developmental processes involved in the formation of structures, as well as in the emergence of evolutionary novelty in organisms (Hall 2000). Eco-evo-devo has recently emerged to focus on environmental factors that influence developmental processes, phenotype formation, and ultimately evolution (Gilbert et al. 2015). Epigenetics, in turn, is the discipline that bridges the gap between environmental influences and long-term regulation of gene expression, which also influences the phenotype (Richards 2006).

Thus, studies on the 'nurture' part of the 'nature versus nurture' dilemma are flourishing in recent years. It is becoming increasingly accepted that the 'nature' contribution is influenced by 'nurture' even before fertilization, and that the nature–nurture dynamics keep changing during development. Lifestyle is currently shown to affect health and lifespan at individual level (Govindaraju et al. 2015; Grey 2017; Rizzuto et al. 2017). Moreover, recent data provide evidence that lifestyle in the parents can also affect gametes and produce effects across generations (Jacobs et al. 2017). In this sense, the environment is modulating individual phenotypical variability even before conception, with consequences that can transcend generations.

Multiple time windows exist in early postnatal life, known as experience-dependent events, when developing systems become especially sensitive to environmental cues (Poletto et al. 2006; Teicher et al. 2016). The environment in which an animal is maintained during gestation may result in changes in several parameters in the offspring (Braastad 159AD; Urakubo et al. 2001; Darnaudéry and Maccari 2008; Meyer et al. 2009; Rutherford et al. 2014; Baxter et al. 2016). Factors such as emotional reactivity, responsiveness to stressors and cognition can be modulated by challenges in the prenatal and neonatal periods (Poletto et al. 2006; Weinstock 2008; Rutherford et al. 2014). For example, in farm, animals' studies have shown that exposure to stressors during gestation, such as negative interactions with handlers (Coulon et al. 2013; Rutherford et al. 2014; Baxter et al. 2016), social stress (Rutherford et al. 2014) or hunger (Bernardino et al. 2016), influence factors in the offspring such as emotional reactivity, social behaviour, responsiveness to stressors, cognition and memory.

The recent boom in epigenetic research has given Lamarck's ideas new prominence (Gissis and Jablonka 2011; Skinner 2014). Epigenetic mechanisms involve chemical modifications of the DNA that can regulate gene expression and be maintained after cell divisions (Skinner et al. 2010). Epigenetic mechanisms are fundamental players in the development of phenotypes and are sensitive to environmental influence (Jirtle and Skinner 2007). The most studied epigenetic mechanisms are DNA methylation, post-translational histone modifications, the action of noncoding RNAs, chromatin loops, and long-range chromatin interactions (Kim et al. 2009; Dudley et al. 2011; Kramer et al. 2011). Epigenetic mechanisms are involved in a dynamic and constant interaction between nature and nurture, which include strong and rapid effects of nurture over nature that produce sustained consequences in the ontogeny of animals (Hrvatin et al. 2018).

Epigenetics, however, is involved not only in correlating environmental exposures to phenotype formation. Epigenetics has recently been shown to have a relevant role in the inheritance of environmentally influenced traits, in intimate correlation to the Lamarckian concept of 'inheritance of acquired traits'. Epigenetic mechanisms are involved both in mitotic and meiotic inheritance. In mitotic inheritance, epigenetic mechanisms are involved in the potential of the genome to produce multiple distinct, and at some point stable, phenotypes, such as the various differentiated cell types present in organisms (Sharma and Rando 2017).

Since epigenomic conformations in cells influence biological regulation, epigenomic disruption leads to a wide variety of disease phenotypes such as diabetes mellitus (Johnson and Evans-Molina 2015), autoimmune diseases (Sun et al. 2016; Carnero-Montoro and Alarcón-Riquelme 2018), obesity (Craig et al. 2017), or cancer (Khan et al. 2018; Ramadoss and Mahadevan 2018). Complex dynamic links exist between metabolism and epigenetic modifications, both in somatic cells and in the germline (Sharma and Rando 2017). Studies show, for example, that the etiology of autoimmune diseases involves aberrant DNA methylation in the genome of cells that can be hypomethylated, demethylated or hypermethylated (see review Sun et al. 2016). Similarly, obesity and diabetes have also been intimately related with epigenetic disruption (Lind et al. 2016; Alonso-Magdalena et al. 2016).

Besides the potential of the environment to alter the epigenome of exposed individuals, environmentally induced epigenetic changes are also related to meiotic inheritance when the gametic epigenome is affected, and this can impact the phenotype and epigenome (germ line and somatic) of unexposed future generations (Guerrero-Bosagna and Jensen 2015). Disrupted epigenetic patterns in the sperm, for example, are responsible for changes in the embryonic development and the phenotype of the generated offspring (Donkin and Barrès 2018). This process is known as 'Transgenerational Epigenetic Inheritance'. Transgenerational effects have been reported in a variety of organisms, including laboratory rodents (Anway et al. 2005; Guerrero-Bosagna et al. 2010, 2012), fish (Baker et al. 2014a, b; Bhandari et al. 2015), quails (Leroux et al. 2017), ducks (Brun et al. 2015) and chicken (Goerlich et al. 2012). Overall, it is becoming increasingly evident that, as predicted by Lamarck, environmental exposures can have an important role in inducing heritable phenotypic variability. Again, our aim is not to oversell Lamarck but to rescue what in his ideas is still pertinent nowadays, knowing for example, that many of Lamarckian ideas were indeed teleological. For example, Lamarck adhered to the '*adaptation towards benefit*' argument to justify evolutionary novelties (Guerrero-Bosagna 2012) and attributed organisms with the faculty of recognition of what could be beneficial for them in novel environments (Bowler 1983).

4 Conclusion

In this chapter, we described and discussed three 'evolutionary dogmas' that we consider highly relevant in both the scientific and pedagogical contexts. Certainly, the concepts that are needed to be discussed and rethought in evolution, based on the current status of biological knowledge, are not limited to these three aspects discussed here. However, with this chapter, we want to offer a small contribution of what is currently changing in terms of biological knowledge, mainly from the genomic perspective that could be highly relevant for evolutionary thinking. It is not our aim to say, for example, that neo-Darwinian evolution should be considered wrong, as it has been the attitude towards Lamarckian ideas. Rather, we expect that the new knowledge emerging from biology will help to discover the limits that apply for the most known evolutionary theories to date. In physics, for example, the theory of relativity by Einstein helped to set limits of predictability of Newtonian physics. Similarly, in biology, we should now give current evidence serious consideration to understand what the real predictability of Darwinian evolution is, and also acknowledge the contributions of Lamarckism. This should not only concern scientific literature, but it should permeate to the general public.

The beauty of science is that change in perspective is not the exception but the rule, and such a change is based on the emergence of new findings. The way that current science operates initiated from the Latin assumption *ignoramus*, which is basically assuming that '*we don't know*'. Under this concept, we must navigate carefully to not stop scientific knowledge at dogmatic points that might even move us to the opposite direction: '*thinking that we know everything*'. Science needs constant rethinking of paradigms, and evolutionary biology is not the exception. Enough evidence exists already in biology to change our current evolutionary paradigms and thus, to move the knowledge and science forward.

References

Adelson DL, Buckley RM, Ivancevic AM et al (2015) Retrotransposons: genomic and trans-genomic agents of change. Evolutionary biology: biodiversification from genotype to phenotype. Springer International Publishing, Cham, pp 55–75

Alonso-Magdalena P, Rivera FJ, Guerrero-Bosagna C (2016) Bisphenol-A and metabolic diseases: epigenetic, developmental and transgenerational basis. Environ Epigenet 2. https://doi.org/10.10 93/eep/dvw022

Anway MD, Cupp AS, Uzumcu M, Skinner MK (2005) Epigenetic transgenerational actions of endocrine disruptors and male fertility. Science (80–) 308:1466–1469. https://doi.org/10.1126/s cience.1108190

Baker TR, King-Heiden TC, Peterson RE, Heideman W (2014a) Dioxin induction of transgenerational inheritance of disease in zebrafish. Mol Cell Endocrinol 398:36–41. https://doi.org/10.10 16/j.mce.2014.08.011

Baker TR, Peterson RE, Heideman W (2014b) Using zebrafish as a model system for studying the transgenerational effects of dioxin. Toxicol Sci 138:403–411. https://doi.org/10.1093/toxsci/kfu 006

Baxter EM, Mulligan J, Hall SA et al (2016) Positive and negative gestational handling influences placental traits and mother-offspring behavior in dairy goats. Physiol Behav 157:129–138. https:// doi.org/10.1016/j.physbeh.2016.02.001

Bernardino T, Tatemoto P, Morrone B et al (2016) Piglets born from sows fed high fibre diets during pregnancy are less aggressive prior to weaning. PLoS ONE 11:1–11. https://doi.org/10.1371/jou rnal.pone.0167363

Bhandari RK, vom Saal FS, Tillitt DE (2015) Transgenerational effects from early developmental exposures to bisphenol A or 17α-ethinylestradiol in medaka, Oryzias latipes. Sci Rep 5:9303. https://doi.org/10.1038/srep09303

Bowler PJ (1983) The eclipse of Darwinism: anti-Darwinian evolution theories in the decades around 1900. Johns Hopkins University Press, Baltimore

Braastad BO (159AD) Effects of prenatal stress on behaviour of offspring\nof laboratory and farmed mammals. Appl Anim Behav Sci 61:1998. https://doi.org/10.1016/s0168-1591(98)00188-9

Brawand D, Wagner CE, Li YI et al (2015) The genomic substrate for adaptive radiation in African cichlid fish. Nature 513:375–381. https://doi.org/10.1038/nature13726

Brun J-M, Bernadet M-D, Cornuez A et al (2015) Influence of grand-mother diet on offspring performances through the male line in Muscovy duck. BMC Genet 16:145. https://doi.org/10.11 86/s12863-015-0303-z

Burkhardt RW Jr (2013) Lamarck, evolution, and the inheritance of acquired characters. Genetics 194:793–805. https://doi.org/10.1534/genetics.113.151852

Buss DM (1983) Evolutionary biology and personality psychology: implications of genetic variability. Pers Individ Dif 4:51–63. https://doi.org/10.1016/0191-8869(83)90052-1

Carnero-Montoro E, Alarcón-Riquelme ME (2018) Epigenome-wide association studies for systemic autoimmune diseases: the road behind and the road ahead. Clin Immunol. https://doi.org/ 10.1016/j.clim.2018.03.014

Collins DW, Jukes TH (1994) Rates of transition and transversion in coding sequences since the human-rodent divergence. Genomics 20:386–396. https://doi.org/10.1006/geno.1994.1192

Cortijo S, Wardenaar R, Colome-Tatche M et al (2014) Mapping the epigenetic basis of complex traits. Science (80–) 343:1145–1148. https://doi.org/10.1126/science.1248127

Coulon M, Wellman CL, Marjara IS et al (2013) Early adverse experience alters dendritic spine density and gene expression in prefrontal cortex and hippocampus in lambs. Psychoneuroendocrinology 38:1112–1121. https://doi.org/10.1016/j.psyneuen.2012.10.018

Craig JR, Jenkins TG, Carrell DT, Hotaling JM (2017) Obesity, male infertility, and the sperm epigenome. Fertil Steril 107:848–859. https://doi.org/10.1016/j.fertnstert.2017.02.115

Danchin E (2013) Avatars of information: towards an inclusive evolutionary synthesis. Trends Ecol Evol 28(6):351–358

Darnaudéry M, Maccari S (2008) Epigenetic programming of the stress response in male and female rats by prenatal restraint stress. Brain Res Rev 57:571–585. https://doi.org/10.1016/j.brainresre v.2007.11.004

Dawkins R (1978) The selfish gene. Oxford University Press

Dobzhansky T (1977) Evolution. Freeman, W.H

Donkin I, Barrès R (2018) Sperm epigenetics and influence of environmental factors. Mol Metab 1–11. https://doi.org/10.1016/j.molmet.2018.02.006

Dudley KJ, Li X, Kobor MS et al (2011) Epigenetic mechanisms mediating vulnerability and resilience to psychiatric disorders. Neurosci Biobehav Rev 35:1544–1551. https://doi.org/10.10 16/j.neubiorev.2010.12.016

Dukas R (2013) Effects of learning on evolution: robustness, innovation and speciation. Anim Behav 85:1023–1030. https://doi.org/10.1016/J.ANBEHAV.2012.12.030

Duret L (2008) Neutral theory: the null hypothesis of molecular evolution. Nat Educ 1:803–806

Eyre-Walker A, Keightley PD (2007) The distribution of fitness effects of new mutations. Nat Rev Genet 8:610–618. https://doi.org/10.1038/nrg2146

Fernald RD (2015) Social behaviour: can it change the brain? Anim Behav 103:259–265. https://d oi.org/10.1016/j.anbehav.2015.01.019

Friedman J, Alm EJ, Shapiro BJ (2013) Sympatric speciation: when is it possible in bacteria? PLoS ONE 8:e53539. https://doi.org/10.1371/journal.pone.0053539

Galtier N, Duret L (2007) Adaptation or biased gene conversion? Extending the null hypothesis of molecular evolution. Trends Genet 23:273–277. https://doi.org/10.1016/j.tig.2007.03.011

Gilbert SF, Bosch TCG, Ledón-Rettig C (2015) Eco-Evo-Devo: developmental symbiosis and developmental plasticity as evolutionary agents. Nat Rev Genet 16:611–622. https://doi.org/10.1038/ nrg3982

Gissis S, Jablonka E (2011) Transformations of lamarckism: from subtle fluids to molecular biology. MIT Press

Goerlich VC, Nätt D, Elfwing M et al (2012) Transgenerational effects of early experience on behavioral, hormonal and gene expression responses to acute stress in the precocial chicken. Horm Behav 61:711–718. https://doi.org/10.1016/J.YHBEH.2012.03.006

Gojobori T, Moriyama EN, Kimura M (1990) Molecular clock of viral evolution, and the neutral theory. Proc Natl Acad Sci USA 87:10015–10018. https://doi.org/10.1073/pnas.87.24.10015

Gould SJ, Lewontin RC (1979) The spandrels of San Marco and the Panglossian paradigm: a critique of the adaptationist programme. Proc R Soc London Ser B Biol Sci 205:581 LP-598

Govindaraju D, Atzmon G, Barzilai N (2015) Genetics, lifestyle and longevity: lessons from centenarians. Appl Transl Genomics 4:23–32. https://doi.org/10.1016/j.atg.2015.01.001

Gregoire M-C, Massonneau J, Simard O et al (2013) Male-driven de novo mutations in haploid germ cells. Mol Hum Reprod 19:495–499. https://doi.org/10.1093/molehr/gat022

Grey M (2017) Lifestyle determinants of health: isn't it all about genetics and environment? Nurs Outlook 65:501–505. https://doi.org/10.1016/J.OUTLOOK.2017.04.011

Guerrero-Bosagna C (2012) Finalism in Darwinian and Lamarckian evolution: lessons from epigenetics and developmental biology. Evol Biol 39:283–300. https://doi.org/10.1007/s11692-012-9 163-x

Guerrero-Bosagna C (2017) Evolution with no reason: a neutral view on epigenetic changes, genomic variability, and evolutionary novelty. Bioscience 67:469–476. https://doi.org/10.1093/b iosci/bix021

Guerrero-Bosagna C, Jensen P (2015) Globalization, climate change, and transgenerational epigenetic inheritance: will our descendants be at risk? Clin Epigenet 7:8. https://doi.org/10.1186/s13 148-014-0043-3

Guerrero-Bosagna C, Settles M, Lucker B, Skinner MK (2010) Epigenetic transgenerational actions of vinclozolin on promoter regions of the sperm epigenome. PLoS ONE 5:e13100. https://doi.or g/10.1371/journal.pone.0013100

Guerrero-Bosagna C, Covert TR, Haque MM et al (2012) Epigenetic transgenerational inheritance of vinclozolin induced mouse adult onset disease and associated sperm epigenome biomarkers. Reprod Toxicol 34:694–707. https://doi.org/10.1016/j.reprotox.2012.09.005

Haig D (2011) Lamarck ascending! Philos Theory Biol 3. https://doi.org/10.3998/ptb.6959004.00 03.004

Hall BK (2000) Guest editorial: Evo-devo or devo-evo—does it matter? Evol Dev 2:177–178. https://doi.org/10.1046/j.1525-142x.2000.00003e.x

Hrvatin S, Hochbaum DR, Nagy MA et al (2018) Single-cell analysis of experience-dependent transcriptomic states in the mouse visual cortex. Nat Neurosci 21:120–129. https://doi.org/10.1 038/s41593-017-0029-5

Hubbell S (2005) Neutral theory in community ecology and the hypothesis of functional equivalence. Funct Ecol 19:166–172. https://doi.org/10.1111/j.0269-8463.2005.00965.x

Hubbell SP (2014) Neutral theory and the evolution of ecological equivalence. Ecology 87:1387–1398. https://doi.org/10.1890/0012-9658(2006)87[1387:NTATEO]2.0.CO;2

Ichinose G, Arita T (2008) The role of migration and founder effect for the evolution of cooperation in a multilevel selection context. Ecol Model 210:221–230. https://doi.org/10.1016/J.ECOLMO DEL.2007.07.025

Jacobs MN, Marczylo EL, Guerrero-Bosagna C, Rüegg J (2017) Marked for life: epigenetic effects of endocrine disrupting chemicals. Annu Rev Environ Resour 42:105–160. https://doi.org/10.11 46/annurev-environ-102016-061111

Jirtle RL, Skinner MK (2007) Environmental epigenomics and disease susceptibility. Nat Rev Genet 8:253–262. https://doi.org/10.1038/nrg2045

Johannes F, Porcher E, Teixeira FK et al (2009) Assessing the impact of transgenerational epigenetic variation on complex traits. PLoS Genet 5:e1000530. https://doi.org/10.1371/journal.pgen.1000 530

Johnson AV (2017) An update: genetic mutations and childhood cancers. J Nurse Pract 14:230–237. https://doi.org/10.1016/j.nurpra.2017.08.016

Johnson JS, Evans-Molina C (2015) Translational implications of the β-cell epigenome in diabetes mellitus. Transl Res 165:91–101. https://doi.org/10.1016/j.trsl.2014.03.002

Johnston SE, Gratten J, Berenos C et al (2013) Life history trade-offs at a single locus maintain sexually selected genetic variation. Nature 502:93–95. https://doi.org/10.1038/nature12489

Kalinowski ST, Leonard MJ, Andrews TM (2010) Nothing in evolution makes sense except in the light of DNA. CBE Life Sci Educ 9:87–97. https://doi.org/10.1187/cbe.09-12-0088

Kautt AF, Machado-Schiaffino G, Meyer A (2016) Multispecies outcomes of sympatric speciation after admixture with the source population in two radiations of Nicaraguan Crater Lake Cichlids. PLoS Genet 12:e1006157. https://doi.org/10.1371/journal.pgen.1006157

Khan MI, Rath S, Adhami VM, Mukhtar H (2018) Targeting epigenome with dietary nutrients in cancer: current advances and future challenges. Pharmacol Res 129:375–387. https://doi.org/10. 1016/j.phrs.2017.12.008

Kim JK, Samaranayake M, Pradhan S (2009) Epigenetic mechanisms in mammals. Cell Mol Life Sci 66:596–612. https://doi.org/10.1007/s00018-008-8432-4

Kimura M (1968) Evolutionary rate at the molecular level. Nature 217:624–626. https://doi.org/1 0.1038/217624a0

Kocher TD (2004) Adaptive evolution and explosive speciation: the cichlid fish model. Nat Rev Genet 5:288–298. https://doi.org/10.1038/nrg1316

Kraaijeveld K, Kraaijeveld-Smit FJL, Maan ME (2011) Sexual selection and speciation: the comparative evidence revisited. Biol Rev 86:367–377. https://doi.org/10.1111/j.1469-185X.2010.00 150.x

Kramer JM, Kochinke K, Oortveld MAW et al (2011) Epigenetic regulation of learning and memory by Drosophila EHMT/G9a. PLoS Biol 9:e1000569. https://doi.org/10.1371/journal.pbio.100056 9

Laland K, Uller T, Feldman M et al (2014) Does evolutionary theory need a rethink? Nature 514:161–164. https://doi.org/10.1038/514161a

Leijs R, van Nes EH, Watts CH et al (2012) Evolution of blind beetles in isolated aquifers: a test of alternative modes of speciation. PLoS ONE 7:e34260. https://doi.org/10.1371/journal.pone.003 4260

Leroux S, Gourichon D, Leterrier C et al (2017) Embryonic environment and transgenerational effects in quail. Genet Sel Evol 49:14. https://doi.org/10.1186/s12711-017-0292-7

Levasseur A, Orlando L, Bailly X et al (2007) Conceptual bases for quantifying the role of the environment on gene evolution: the participation of positive selection and neutral evolution. Biol Rev 82:551–572. https://doi.org/10.1111/j.1469-185X.2007.00024.x

Lind L, Lind PM, Lejonklou MH et al (2016) Uppsala consensus statement on environmental contaminants and the global obesity epidemic. Environ Health Perspect 124:A81–A83. https://doi.org/10.1289/ehp.1511115

Lynch M (2007) The frailty of adaptive hypotheses for the origins of organismal complexity. Proc Natl Acad Sci 104:8597 LP-8604

Mattick JS (2012) Rocking the foundations of molecular genetics. Proc Natl Acad Sci 109:16400–16401

Mayr E (1961) Cause and effect in biology. Science 134:1501–1506. https://doi.org/10.1126/SCIENCE.134.3489.1501

McCullough EL, Miller CW, Emlen DJ (2016) Why sexually selected weapons are not ornaments. Trends Ecol Evol 31:742–751. https://doi.org/10.1016/j.tree.2016.07.004

Mesoudi A, Chang L, Dall SRX, Thornton A (2016) The evolution of individual and cultural variation in social learning. Trends Ecol Evol 31:215–225. https://doi.org/10.1016/j.tree.2015.12.012

Meyer U, Feldon J, Fatemi SH (2009) In-vivo rodent models for the experimental investigation of prenatal immune activation effects in neurodevelopmental brain disorders. Neurosci Biobehav Rev 33:1061–1079. https://doi.org/10.1016/j.neubiorev.2009.05.001

Mohan M, Mellick GD (2017) Role of the VPS35 D620N mutation in Parkinson's disease. Park Relat Disord 36:10–18. https://doi.org/10.1016/j.parkreldis.2016.12.001

Necsulea A, Kaessmann H (2014) Evolutionary dynamics of coding and non-coding transcriptomes. Nat Rev Genet 15.734–748. https://doi.org/10.1038/nrg3802

Noble D (2013) Physiology is rocking the foundations of evolutionary biology. Exp Physiol 98:1235–1243. https://doi.org/10.1113/expphysiol.2012.071134

Poletto R, Steibel JP, Siegford JM, Zanella AJ (2006) Effects of early weaning and social isolation on the expression of glucocorticoid and mineralocorticoid receptor and 11beta-hydroxysteroid dehydrogenase 1 and 2 mRNAs in the frontal cortex and hippocampus of piglets. Brain Res 1067:36–42. https://doi.org/10.1016/j.brainres.2005.10.001

Ponting CP, Lunter G (2006) Signatures of adaptive evolution within human non-coding sequence. Hum Mol Genet 15:R170–R175. https://doi.org/10.1093/hmg/ddl182

Prentis PJ, Wilson JRU, Dormontt EE et al (2008) Adaptive evolution in invasive species. Trends Plant Sci 13:288–294. https://doi.org/10.1016/J.TPLANTS.2008.03.004

Qvarnström A, Vallin N, Rudh A (2012) The role of male contest competition over mates in speciation. Curr Zool 58:493–509. https://doi.org/10.1093/czoolo/58.3.493

Ramadoss M, Mahadevan V (2018) Targeting the cancer epigenome: synergistic therapy with bromodomain inhibitors. Drug Discov Today 23:76–89. https://doi.org/10.1016/j.drudis.2017.09.011

Richards EJ (2006) Inherited epigenetic variation—revisiting soft inheritance. Nat Rev Genet 7:395–401. https://doi.org/10.1038/nrg1834

Rizzuto D, Mossello E, Fratiglioni L et al (2017) Personality and survival in older age: the role of lifestyle behaviors and health status. Am J Geriatr Psychiatry 25:1363–1372. https://doi.org/10.1016/J.JAGP.2017.06.008

Rodríguez RL, Boughman JW, Gray DA et al (2013) Diversification under sexual selection: the relative roles of mate preference strength and the degree of divergence in mate preferences. Ecol Lett 16:964–974. https://doi.org/10.1111/ele.12142

Rutherford KMD, Piastowska-Ciesielska A, Donald RD et al (2014) Prenatal stress produces anxiety prone female offspring and impaired maternal behaviour in the domestic pig. Physiol Behav 129:255–264. https://doi.org/10.1016/j.physbeh.2014.02.052

Sequeira P, Chen Y-S, Weiss MA (2015) Mutation-driven evolution: microsatellite instability drives speciation in a mammalian taxon. Evolutionary biology: biodiversification from genotype to phenotype. Springer International Publishing, Cham, pp 141–164

Sharma U, Rando OJ (2017) Metabolic inputs into the epigenome. Cell Metab 25:544–558. https://doi.org/10.1016/j.cmet.2017.02.003

Skinner MK (2014) Endocrine disruptor induction of epigenetic transgenerational inheritance of disease. Mol Cell Endocrinol 398:4–12. https://doi.org/10.1016/j.mce.2014.07.019

Skinner MK, Manikkam M, Guerrero-Bosagna C (2010) Epigenetic transgenerational actions of environmental factors in disease etiology. Trends Endocrinol Metab 21:214–222. https://doi.org/10.1016/j.tem.2009.12.007

Srinivas M, Verselis VK, White TW (2018) Human diseases associated with connexin mutations. Biochim Biophys Acta Biomembr 1860:192–201. https://doi.org/10.1016/j.bbamem.2017.04.024

Stojković V, Fujimori DG (2017) Mutations in RNA methylating enzymes in disease. Curr Opin Chem Biol 41:20–27. https://doi.org/10.1016/j.cbpa.2017.10.002

Sun B, Hu L, Luo ZY et al (2016) DNA methylation perspectives in the pathogenesis of autoimmune diseases. Clin Immunol 164:21–27. https://doi.org/10.1016/j.clim.2016.01.011

Svensson EI, Verzijden MN, Cate C et al (2012) The impact of learning on sexual selection and speciation. The impact of learning on sexual selection and speciation 27:511–519. https://doi.org/10.1016/j.tree.2012.05.007

Teicher MH, Samson JA, Anderson CM, Ohashi K (2016) The effects of childhood maltreatment on brain structure, function and connectivity. Nat Rev Neurosci 17:652–666. https://doi.org/10.1038/nrn.2016.111

Thorpe RS, Surget-Groba Y, Johansson H (2010) Genetic tests for ecological and allopatric speciation in Anoles on an Island Archipelago. PLoS Genet 6:e1000929. https://doi.org/10.1371/journal.pgen.1000929

Urakubo A, Jarskog LF, Lieberman JA, Gilmore JH (2001) Prenatal exposure to maternal infection alters cytokine expression in the placenta, amniotic fluid, and fetal brain. Schizophr Res 47:27–36. https://doi.org/10.1016/S0920-9964(00)00032-3

Waddington CH (1954) Evolution and epistemology. Nature 173:880–880. https://doi.org/10.1038/173880a0

Wagner A (2008) Neutralism and selectionism: a network-based reconciliation. Nat Rev Genet 9:965–974. https://doi.org/10.1038/nrg2473

Weinstock M (2008) The long-term behavioural consequences of prenatal stress. Neurosci Biobehav Rev 32:1073–1086. https://doi.org/10.1016/j.neubiorev.2008.03.002

Wood KE, Komarova NL (2018) Cooperation-based branching as a mechanism of evolutionary speciation. J Theor Biol 445:166–186. https://doi.org/10.1016/J.JTBI.2018.02.033

A Proposed Terminology of Convergent Evolution

George R. McGhee, Isabelle Hue, Justine Dardaillon and Pierre Pontarotti

Abstract In this paper, we propose a unified terminology of convergence, in which the phenomenon of convergent evolution is the *set*, and the types of convergent evolution are *subsets* of that phenomenon. The set of the convergence phenomenon contains all of the same or very similar traits that have evolved independently in different lineages of organisms. This set contains three subsets that specify the three pathways by which evolution may produce convergence: allo-convergent evolution, iso-convergent evolution, and retro-convergent evolution. Allo-convergent evolution is the independent evolution of the same or very similar new trait from *different precursor traits* in different lineages, iso-convergent evolution is the independent evolution of the same or very similar new trait from the *same precursor trait* in different lineages, and retro-convergent evolution is the independent *re-evolution* of the same or very similar trait to an *ancestral trait* in different lineages. This unified convergence terminology is proposed to replace the older and confusing terms parallel evolution, reverse evolution, and convergent evolution sensu stricto.

G. R. McGhee (✉)
Wright-Rieman Laboratories, Department of Earth and Planetary Sciences,
Rutgers University, Piscataway, NJ 08854, USA
e-mail: mcghee@eps.rutgers.edu

I. Hue
UMR BDR INRA ENVA, Université Paris Saclay, 78350 Jouy en Josas, France

J. Dardaillon
CNRS Centrale Marseille, 12M UMR 7373 équipe EBM (Evolution Biologique Modélisation),
Aix Marseille Université, 13453 Marseille Cedex, France

P. Pontarotti
Evolutionary Biology Team, Aix Marseille Université, IRD, APHM, Microbe, Evolution,
Phylogénie, Infection, IHU Méditerranée Infection, Marseille, France

P. Pontarotti
CNRS, Marseille, France

© Springer International Publishing AG, part of Springer Nature 2018
P. Pontarotti (ed.), *Origin and Evolution of Biodiversity*,
https://doi.org/10.1007/978-3-319-95954-2_18

1 Introduction

Convergence is the evolution of the same or very similar traits independently in different lineages of organisms. The analysis of the evolutionary phenomenon of convergence has been hampered in the past by a confusing terminology concerning the mechanisms and directionality of that phenomenon, specifically usage of the terms parallel evolution, reverse evolution, and convergent evolution sensu stricto. "Parallel evolution" is a particularly unfortunate term in that this evolutionary process in the past has been considered not only to be different from convergent evolution, but in some cases has been considered to be the actual disproof of convergent evolution (see discussion in McGhee 2011, pp. 5–7). Likewise, the term "reverse evolution" has been used to describe both an evolutionary direction and/or an evolutionary mechanism such as devolution.

In this paper, we propose a unified terminology of convergence, in which the phenomenon of convergent evolution is the *set*, and the types of convergent evolution are *subsets* of that phenomenon. In this terminology, the phenomenon of convergence is the *suffix* of the term, and the type of convergent evolution is the *prefix* of the term. This terminology was first proposed by McGhee et al. (2017) at the 21st Evolutionary Biology Meeting at Marseilles (France), discussed in a review of convergence in McGhee (2018), and now proposed in full form in this paper.

2 Proposed Terminology

Our proposed convergence terminology is outlined in Table 1. The *set* of the convergence phenomenon contains all of the same or very similar traits that have evolved independently in different lineages of organisms. This set contains three *subsets* that specify the three pathways by which evolution may produce convergence: allo-convergent evolution, iso-convergent evolution, and retro-convergent evolution (Table 1).

Allo-convergent evolution is the independent evolution of the same or very similar new trait from *different precursor traits* in different lineages (the prefix "allo" for different). Iso-convergent evolution is the independent evolution of the same or very similar new trait from the *same precursor trait* in different lineages (the prefix "iso" for same). Retro-convergent evolution is the independent *re-evolution* of the same or very similar trait to an *ancestral trait* in different lineages (the prefix "retro" for reverse).

The subset of retro-convergent evolution itself contains two subsets that specify the type of retro-convergence: retro-alloconvergent evolution and retro-isoconvergent evolution. Retro-alloconvergent evolution is the independent *re-evolution* of the same or very similar trait to an ancestral trait from *different precursor traits* in different lineages. Retro-isoconvergent evolution is the independent *re-evolution* of the same or very similar trait to an ancestral trait from the *same precursor trait* in different lineages.

Table 1 Proposed terminology of convergent evolution

SET: CONVERGENT EVOLUTION: The evolution of the same or very similar traits independently in different lineages of organisms *SUBSETS*: 1. Allo-convergent evolution: The independent evolution of the same or very similar new trait from different precursor traits in different lineages 2. Iso-convergent evolution: The independent evolution of the same or very similar new trait from the same precursor trait in different lineages 3. Retro-convergent evolution: The independent re-evolution of the same or very similar trait to an ancestral trait in different lineages *Subsets*: 3a. Retro-alloconvergent evolution: The independent re-evolution of the same or very similar trait to an ancestral trait from different precursor traits in different lineages 3b. Retro-isoconvergent evolution: The independent re-evolution of the same or very similar trait to an ancestral trait from the same precursor trait in different lineages

3 Examples of Allo-convergent, Iso-convergent, and Retro-convergent Evolution

A classic example of allo-convergence is the independent evolution of wings in bats and dragonflies (Table 2). The evolutionary lineage of a bat (Bilateria: Deuterostomia: Chordata: Vertebrata) is very different and divergent from that of a dragonfly (Bilateria: Protostomia: Ecdysozoa: Arthropoda), yet both have evolved wings—but from radically *different precursor traits*. The bat's wings are modified vertebrate forelimbs, whereas the dragonfly's wings are modifications of gill branches originally present in the larval stage of development of the animal (the pleural origin hypothesis; Averof and Cohen 1997), or a combination of gill branch segments and dorsal body-wall segments (the dual origin hypothesis; Linz and Tomoyasu 2018).

A classic example of iso-convergence is the independent evolution of wings in bats, birds, and the extinct pterodactyls of the Mesozoic (Table 2). A bat is a member of the clade of the mammals, a bird is a member of the clade of the dinosaurs, and a pterodactyl is a member of the clade of the pterosaurs—yet all three animals are members of the larger inclusive clade of the vertebrates. And the *same precursor trait*—the vertebrate forelimb—was independently modified to form wing structures iso-convergently in bats, birds, and pterodactyls.

A classic example of retro-convergence is the independent *re-evolution* of ancestral fusiform-bodied fish forms in porpoises and the extinct ichthyosaurs (Table 3). Fast-swimming, fusiform-bodied fish forms were independently, iso-convergently evolved both in the clade of the chondrichthyan cartilaginous fishes, such as the great white shark, and the osteichthyan bony fishes, such as the swordfish (Table 3,

Fig. 1), from the same ancestral swimming-fish forms. However, the porpoises and the ichthyosaurs are descendants of land-dwelling mammals and reptiles, respectively, that once possessed four walking limbs, no fins, and no fish-like fusiform bodies. Both the mammalian porpoise and the reptilian ichthyosaur independently, retro-convergently, *re-evolved* fast-swimming, fusiform-bodied fish forms like their osteichthyan ancestors (Fig. 1) when species in their respective lineages made the evolutionary habitat shift from the land back into the oceans.

A porpoise fin is not inherited from a fish ancestor but is an independently derived retro-convergent mammalian fin that is similar in form to an ancestral sarcopterygian-osteichthyan fish fin. Likewise for the ichthyosaur fin—it is an independently derived retro-convergent reptilian fin, not a fish fin. When we examine the retro-convergent evolution of fins in these two lineages, we see that the evolution of mammalian and reptilian fins followed both iso-convergent and allo-convergent pathways (Table 3). The pectoral fins of both the porpoises and ichthyosaurs are retro-isoconvergently evolved from the *same precursor structure*: tetrapod pentadactylus forelimbs (Benton 2015). In contrast, the dorsal fins in the porpoises and ichthyosaurs were retro-alloconvergently evolved from two *different precursor tissue structures* to produce a rope-like structure of fine filaments in the dorsal fin of the extinct ichthyosaurs (Smithwick et al. 2017), and a dorsal fin of soft tissue with core placements of the blood vessels in living porpoises (Felts 1966).

Table 2 Iso-convergent and allo-convergent evolution of wings in animals

1. *Iso-convergent evolution of wings in vertebrates*:
1.1. Hairy devil pterosaur (Metazoa: Bilateria: Deuterostomia: Chordata: Vertebrata: Gnathostomata: Osteichthyes: Sarcopterygii: Tetrapoda: Amniota: Reptilia: Diapsida: Archosauromorpha: Avemetatarsalia: Pterosauria: Rhamphorhynchidae; *Sordes pilosus* †Jurassic)
1.2. Great blue heron (Avemetatarsalia: Dinosauria: Saurischia: Theropoda: Maniraptora: Aves: Neognathae: Neoaves: Ciconiiformes: Ardeidae; *Ardea herodias*)
1.3. Mouse-eared bat (Amniota: Synapsida: Therapsida: Mammalia: Eutheria: Laurasiatheria: Chiroptera: Microchiroptera: Vespertilionidae; *Myotis myotis*)
2. *Allo-convergent evolution of wings in insects*:
2.1. Green darner dragonfly (Bilateria: Protostomia: Ecdysozoa: Arthropoda: Mandibulata: Hexapoda: Odonata: Aeschnidae; *Anax junius*)

Note Extinct taxa are marked with a dagger (†), and their geologic age is given

Table 3 Iso-convergent and retro-convergent evolution of marine, fast-swimming fish forms in the vertebrates

1. *Iso-convergent evolution of fast-swimming, fusiform-bodied fish forms*:
1.1. Great white shark (Metazoa: Bilateria: Deuterostomia: Chordata: Vertebrata: Gnathostomata: Chondrichthyes: Elasmobranchii: Lamnidae; *Carcharodon carcharias*)
1.2. Swordfish (Vertebrata: Gnathostomata: Osteichthyes: Actinopterygii, Teleostei: Xiphiidae; *Xiphias gladius*)
2. *Retro-convergent evolution of fast-swimming, fusiform-bodied fish forms*:
2.1. Ichthyosaur (Vertebrata: Gnathostomata: Osteichthyes: Sarcopterygii: Tetrapoda: Amniota: Reptilia: Diapsida: Lepidosauromorpha: Ichthyosauria: Ichthyosauridae; *Ichthyosaurus platyodon* †Jurassic)
2.2. Harbor porpoise (Tetrapoda: Amniota: Synapsida: Therapsida: Mammalia: Eutheria: Laurasiatheria: Cetartiodactyla: Cetacea: Odontoceti: Phocaenidae; *Phocaena phocaena*)
3. *Retro-convergent evolution of fins in ichthyosaurs and porpoises*:
3.1. *Retro-isoconvergent evolution of* PECTORAL FINS (pectoral fins retro-convergently evolved from the same precursor structure: tetrapod pentadactylus forelimbs)
3.2. *Retro-alloconvergent evolution of* DORSAL FINS (dorsal fins retro-convergently evolved from different precursor tissues to produce a rope-like structure of fine filaments in the dorsal fin of the extinct ichthyosaurs, and a dorsal fin of soft tissue with core placements of the blood vessels in living porpoises)

Note Extinct taxa are marked with a dagger (†), and their geologic age is given

4   Replacement of Confusing Older Terminology by the New Terminology of Convergence

In the older terminology of evolutionary convergence, iso-convergent evolution replaces the term parallel evolution, retro-convergent evolution replaces the term reverse evolution, and allo-convergent evolution is equivalent to the term convergent evolution sensu stricto. The problem with the term "parallel" evolution is that it generated confusion in that it is often misunderstood not only to be different from convergent evolution but in some cases the actual disproof of convergent evolution. For example, the molecular discovery that the same regulatory gene (*Pax-6*) was independently, in parallel, modified in the convergent evolution of eyes was mistakenly taken by Rey et al. (1998, p. 6212) to be a disproof of convergent evolution: "An oft invoked example of convergent evolution has been compound eyes of insects versus singular eyes of vertebrates. This proved to be a wrong example, however, because development of all metazoan eyes recently has been shown to be under control of the same regulatory gene that encodes Pax-6 protein." In contrast, phylogenetic systematists consider parallel evolution to be a phenomenon that is "a special

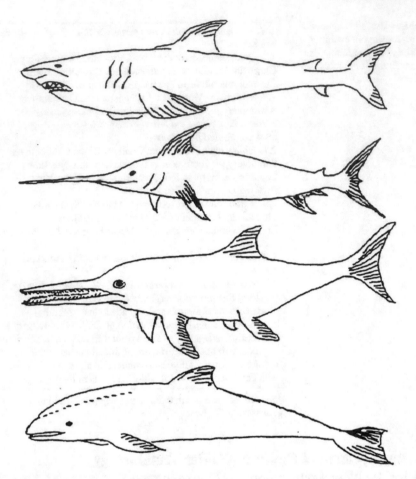

Fig. 1 Iso-convergently evolved fast-swimming, fusiform-bodied fish forms in the chondrichthyan great white shark and osteichthyan swordfish (top two figures), and retro-convergently evolved fast-swimming, fusiform-bodied fish forms in the reptilian ichthyosaur and mammalian harbor porpoise (bottom two figures). From McGhee (2007)

case of convergence" in which the same trait has independently evolved "from the same ancestral character in different taxa" (Lecointre and Le Guyader 2006, p. 541).

The term "parallel" evolution has created even greater confusion in that this same term has been used to describe different evolutionary phenomena. For example, in some molecular analyses Pontarotti and Hue (2016, p. 4) point out that the term parallel evolution has been used to describe molecular convergence from the same ancestral amino acid in different taxa and convergent evolution only used to describe molecular convergence from different ancestral amino acids in different taxa, whereas in other molecular analyses "the distinction between parallel and convergent evolution is not based on the evolutionary history of the characters, but on the similarity of the genetic mechanisms that are involved in the repeated phenotype" such that "if the molecular mechanisms are the same, the evolution is said to be parallel; if the genetic mechanisms are different, the evolution is said to be convergent." Even more confusing, in phenotypic analyses the concept of "phylogenetic proximity" (however defined) has been used to differentiate parallel and convergent evolution, where the independent evolution of the same trait in phylogenetically close species is described as parallel evolution, whereas the independent evolution of the same trait in phylogenetically distant species is described as convergent evolution (Pontarotti and Hue 2016).

Likewise, the term "reverse" evolution has generated confusion as it has been used to describe both an evolutionary direction and/or an evolutionary mechanism such as devolution. A term "retroconvergence" has been introduced by Krassilov (1995) but it is not the same as the term "retro-convergence" used in this paper. Krassilov (1995, 2004) proposed the terms "cyclic evolution (retroconvergence)" (Krassilov 2004, p. 265) for a process in which "retroconvergent morphological trends seem to indicate gradual reversal of a modified genetic program to the original Gestalt inbuilt in the genomic memory" (Krassilov 1995, p. 70). Krassilov (2004, p. 280) further stated that "[a]n adaptive meaning of such cyclic transformations is scarcely conceivable. Rather, the meaning is related to the nature of genomic memory retaining images of lost basic structures, which can be conferred upon their derivative organs." Thus Krassilov's proposed "cyclic evolution (retroconvergence)" is a process of devolution of a descendant structure backwards into an ancestral structure.

In contrast, our term retro-convergent extends the same logic as the term iso-convergence for "parallel" convergent evolution to the case of "reverse" convergent evolution, where the evolutionary pathway producing that convergence involves the re-evolution of a trait very similar to an ancestral trait. That process involves modifying different precursor traits (retro alloconvergence) or the same precursor traits (retro-isoconvergence) in independent lineages to produce a descendant trait that has the same or very similar form to that of an ancestral trait. That process is not Krassilov's (1995, 2004) devolution or backwards evolution. Using our terminology, Krassilov's (1995, 2004) devolutionary "cyclic evolution (retroconvergence)" process would represent an extreme case in which a descendant species is retro-isoconvergent on its own ancestor.

For example, the pectoral fins of the ichthyosaurs and porpoises (Table 3) both resemble the ancestral pectoral fins of their sarcopterygian-osteichthyan ancestors, both in form and function. Yet the pectoral fin of an ichthyosaur is a reptile fin and the pectoral fin of a porpoise is a mammalian fin—neither is a sarcopterygian fish fin. The pentadactylus-manus bone structures of land-dwelling tetrapod forelimbs are still present, although highly modified, in the pectoral fins of both the ichthyosaurs and porpoises (Benton 2015). In contrast, in the "cyclic evolution (retroconvergence)" scenario of Krassilov (1995, 2004), the pectoral fins of both the ichthyosaurs and porpoises would have to have been produced by a devolutionary process that would have reversed the development of pentadactyly in the manus of the forelimbs of the reptilian and mammalian ancestors of the ichthyosaurs and porpoises, backwards to the original non-pentadactylus bone arrangement in the pectoral fins of sarcopterygian fish.

The new terminology is based on the analysis of phylogeny (Pontarotti and Hue 2016; McGhee 2018) and is strictly empirical, avoiding any terminological linkage with the causation or development of the convergent phenomena described (contra Gould 2002; Hall 2003). The analysis of the underlying causation of the empirical phenomena of allo-convergent, iso-convergent, retro-alloconvergent, and retro-isoconvergent evolution is a separate process from the demonstration of the existence of the phenomena. Once convergent evolution at the phenotypic level has been identified, co-convergent evolution at other levels can be tested: from the coding level (DNA, genes) to gene-regulatory network levels, to the non-coding (epigenetic, transcriptional) and molecular levels (proteins, enzymes), to the physiological (biochemical, organ systems) and phenotypic levels (single trait versus multivariate convergence).

The new terminology also avoids any linkage with the interminable debate as to what Gould (1989, 2002) meant by the term "contingent," and how it relates to "convergent" evolution versus "parallel" evolution (Pearce 2012; Powell 2012). Beatty (2006, 2016), Sapp (2012), and McGhee (2016) have argued that Gould (1989, 2002) misused the term "contingent" throughout his work, confusing the distinction between contingent per se and contingent *upon* (Beatty 2006, 2016) and creating a false dichotomy between truly contingent evolution and teleological evolution that continues to generate confusion in the analysis of convergence (Sapp 2012; McGhee 2016).

5 Summary and Proposed Future Research

In this paper, we have proposed a unified terminology of convergence, in which the phenomenon of convergent evolution is the *set*, and the types of convergent evolution are *subsets* of that phenomenon. In this terminology, the phenomenon of convergence is the *suffix* of the term, and the type of convergent evolution is the *prefix* of the term. The unified terminology of allo-convergence, iso-convergence, retro-alloconvergence, and retro-isoconvergence covers the conceptual spectrum of the types of convergent evolution.

In conclusion, we propose the following future research pathway in the analysis of the evolutionary phenomenon of convergence: First, the re-analysis of all cases of convergent evolution that are described in the scientific literature at the phenotypic level. The goal here is to sort out those that are iso-convergent and retro-isoconvergent. Iso-convergence and retro-isoconvergence at the phenotypic level could be linked to similar mechanisms at other biological levels (see detailed discussion in Pontarotti and Hue 2016, pp. 10–17), thus the causative mechanism for the known cases of phenotypic convergences may be analyzed at this research step. Second, the documentation of new and previously unrecognized cases of iso-convergent and retro-isoconvergent evolution, particularly those that may have been reported as parallel or reverse and not convergent. And third, the creation of an internationally accessible database that integrates the different cases of iso-convergent and retro-isoconvergent evolution at all different biological levels that were documented in the first and second steps of the proposed research pathway.

Acknowledgments McGhee would like to acknowledge the support of the Konrad Lorenz Institute for Evolution and Cognition Research (the KLI) and the Santa Fe Institute (the SFI) for his research into the phenomenon of convergent evolution. Hue would like to thank her INRA and CNRS collaborators in the EvolUCIA proposal as, even though the proposal was not granted, it sparked her interest in evolutionary concepts. Pontarotti would like to acknowledge that this work was supported by the French Government under the "investissements d'avenir" (Investments for the Future) program managed by the Agence Nationale de la Recherche (ANR, National Agency for Research) (reference: Méditerranée Infection 10-IAHU-03).

References

Averof M, Cohen SM (1997) Evolutionary origin of insect wings from ancestral gills. Nature 385:627–630

Beatty J (2006) Replaying life's tape. J Philos 103:336–362

Beatty J (2016) What are local narratives good for? Stud Hist Philos Biol Biomed Sci 58:33–40

Benton MJ (2015) Vertebrate palaeontology, 4th edn. Wiley Blackwell, Chichester, England

Felts WJL (1966) Some functional and structural characteristics of cetacean flippers and flukes. In: Norris KS (ed) Whales, dolphins, and porpoises. University of California Press, Berkeley, pp 255–276

Gould SJ (1989) Wonderful life: the Burgess Shale and the nature of history. WW Norton, New York

Gould SJ (2002) The structure of evolutionary theory. Belknap Press of Harvard University Press, Cambridge MA, USA

Hall BK (2003) Descent with modification: the unity underlying homology and homoplasy as seen through an analysis of development and evolution. Biol Rev 78:409–433

Krassilov VA (1995) *Scytophyllum* and the origin of angiosperm leaf characters. Paleontol J 29(1A):63–73

Krassilov VA (2004) Macroevolutionary events and the origin of higher taxa. In: Wasser SP (ed) Evolutionary theory and processes: modern horizons, papers in honor of Eviatar Nevo. Kluwer Academic Publishers, The Netherlands, pp 265–289

Lecointre G, Le Guyader H (2006) The tree of life: a phylogenetic classification. Belknap Press of Harvard University Press, Cambridge MA, USA

Linz DM, Tomoyasu Y (2018) Dual evolutionary origin of insect wings supported by an investigation of the abdominal wing serial homologs in *Tribolium*. Proc Nat Acad Sci USA. www.pnas.org/cgi/doi/10.1073/pnas.1711128115

McGhee GR (2007) Adaptive landscapes and theoretical morphospaces. Cambridge University Press, Cambridge, England

McGhee GR (2011) Convergent evolution: limited forms most beautiful. The MIT Press, Cambridge MA, USA

McGhee GR (2016) Can evolution be directional without being teleological? Stud Hist Philos Biol Biomed Sci 58:93–99

McGhee GR (2018) Convergence. In: Nuño de la Rosa L, Müller G (eds) Evolutionary developmental biology. Springer International Publishing, Berlin. https://doi.org/10.1007/978-3-319-33038-9_124-1

McGhee GR, Hue I, Pontarotti P (2017) A proposed terminology of convergent evolution. In: A.E.E.B.: 21st evolutionary biology meeting at Marseilles, France. http://aeeb.fr/program-2/

Pearce T (2012) Convergence and parallelism in evolution: a neo-Gouldian account. Br J Philos Sci 63:429–448

Pontarotti P, Hue I (2016) Road map to study convergent evolution: a proposition for evolutionary systems biology approaches. In: Pontarotti P (ed) Evolutionary biology: convergent evolution, evolution of complex traits, concepts and methods. Springer, Berlin, pp 3–21

Powell R (2012) Convergent evolution and the limits of natural selection. Eur J Philos Sci 2:355–373

Rey M, Ohno S, Pinter-Toro JA, Llobel A, Benitez T (1998) Unexpected homology between inducible cell wall protein QID74 of filamentous fungi and BR3 salivary protein of the insect *Chironomus*. Proc Nat Acad Sci USA 95:6212–6216

Sapp J (2012) Evolution replayed. BioScience 62:693–694

Smithwick FM, Mayr G, Saitta ET, Benton MJ, Vinther V (2017) On the purported presence of fossilized collagen fibres in an ichthyosaur and a theropod dinosaur. Palaeontology 60(3):409–422

Natura Fecit Saltum: Punctuationalism Pervades the Natural Sciences

Graham P. Wallis

Abstract Despite parallel revolutions of uniformitarianism in geology and gradualism in biology c200 years ago, it is now clear that rare events of large effect prevail in the natural world. While astronomical and geological phenomena are more overt (e.g. asteroid impact, supervolcanoes, earthquakes, tsunami), the general and widespread prevalence of major genetic, ecological and climatic events, and their long-term effects on life on Earth is becoming more widely appreciated. Biogeography abounds with examples of rare dispersal events shaping the biota of volcanic islands; ecological studies are showing us how adaptation can happen rapidly in association with habitat change; genomic studies show that major adaptations and speciation can happen rapidly through selection on just a few genetic variants; ecology shows us how tipping points can lead to major and irreversible shifts in ecosystems; climate change is exacerbating the frequency and degree of extreme weather events all over the globe. An unfortunate corollary of climate change is that the long-term integrated effects of global warming are experienced most strongly in connection with the most extreme events.

1 Introduction—Huxley's Dissent

> 'And Mr. Darwin's position might, we think, have been even stronger than it is if he had not embarrassed himself with the aphorism, "*Natura non facit saltum*," which turns up so often in his pages.' Westminster Review, 1860 (Huxley 1906)

Change in nature, be it physical or biotic, is typically characterized as the result of continuous, gradual processes, immeasurable over our lifespan, accreted over eons of time. This concept was central to the thinking of Darwin (1859), building on the work of Hutton and Lyell (1830–1833). Their great insight was that large-scale past change could be explained in terms of small-scale ongoing processes (Penny and Phillips 2004), upending the orthodoxies of biblical creation in biology and

G. P. Wallis (✉)
Department of Zoology, University of Otago, PO Box 56, Dunedin 9054, New Zealand
e-mail: g.wallis@otago.ac.nz

© Springer International Publishing AG, part of Springer Nature 2018
P. Pontarotti (ed.), *Origin and Evolution of Biodiversity*,
https://doi.org/10.1007/978-3-319-95954-2_19

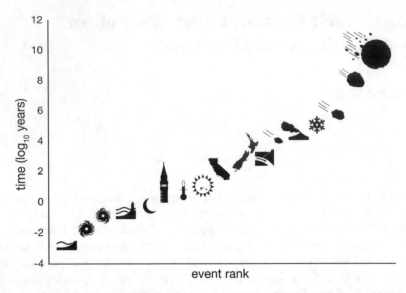

Fig. 1 Frequency of punctuational natural events: rank order plotted against $\log_{10}t$. From left to right: tides, Pacific cyclones, Atlantic hurricanes, king tides, lunar perigee, Acqua alta Venezia, El Niño, sunspots, San Andreas Fault, Alpine Fault, Cascadia Subduction Zone, 200 m asteroids, supervolcanoes, Milankovitch eccentricity cycle, 1-km asteroids, 10-km asteroids, Heavy Bombardment Era

catastrophism in geology [e.g. Cuvier and Whewell (Penny 2017), but see Gould (2002)]. Darwin portrayed the formation of adaptations and new species as the result of gradual change by natural selection; Hutton and Lyell saw mountain-building, erosion and formation of geological layers resulting from the imperceptibly slow processes of uplift, weathering and sedimentation. Thus, adaptation and speciation were effected by microevolution, and geological change by uniformitarianism, both over a timescale of millions of years.

Despite his own identification as 'Darwin's bulldog', when first reading *The Origin of Species*, Thomas Huxley famously remarked: 'you have loaded yourself with an unnecessary difficulty in adopting *Natura non facit saltum* so unreservedly', and (with reference to organic chemistry as an example): '... I have a sort of notion that similar laws of definite combination rule over the modifications of organic bodies, and that in passing from species to species *"Natura fecit saltum"*.' (Provine 1971).

The modern synthesis of evolution famously melded genetics (Dobzhansky 1937), systematics (Mayr 1942) and palaeontology (Simpson 1944). The first two of these evolutionists and their subsequent followers largely adhered to a more classical Darwinian gradualism, not attributing much evolutionary importance to macromutations ('sports'), favoured by Goldschmidt (1940) as material to evolutionary change. Simpson was more pluralist, seeing a continuum of ground between gradualism and punctuationalism, perhaps reflecting his palaeontological background, which Gould was later to promote. Ongoing developments in these areas, as well as in climatology, ecology and molecular genetics, increasingly suggest that natural processes are indeed, as Huxley perceived, actually rather 'lumpy', with the majority of change occurring during relatively brief, extreme events, punctuating periods of relative calm or stasis. Here, I review the widespread significance of these punctuational events across all natural sciences (Fig. 1).

2 Extraterrestrial Events

Conceived after findings from the Apollo moon mission, and what would be the most cataclysmic event in Earth's history if it were to happen again, is that of the Late Heavy Bombardment nearly 4 billion years ago (yBP). Although the narrowness of this event is now under question (Mann 2018), there was at least a period in early Earth history when our planet was pounded with asteroids, perhaps delaying, if not re-setting, the origin of life. Since then, at least some of the largest mass extinctions were probably the result of bolide impacts (Chapman and Morrison 1994). These include the famous Cretaceous–Paleogene (K–Pg) boundary dinosaur-extinction event 66 million years ago (Ma), linked to the Chicxulub crater and atmospheric soot (Bardeen et al. 2017). There is a strong inverse relationship between size of asteroid and impact rate; the biggest the events are the rarest, with 1-km asteroids hitting at $c10^8$ year intervals. The Earth's eccentric orbit, and obliquity and precession of its axis led to a period of Pleistocene glaciations, ongoing since 2.6 Ma. These have occurred on a cycle of $\sim 10^5$ years over the last million years, and about 2.5× that frequency before (Hodell 2016; Tzedakis et al. 2017), with both destructive and creative effects on the planet's flora and fauna (Hewitt 2000, 2004; Jansson and Dynesius 2002; Wallis et al. 2016; Weir et al. 2016).

3 Geology

Geology provides many dramatic episodic events, epitomized by massive subduction zone earthquakes and ensuing tsunami of the Pacific Rim. Tsunami have been shown to be responsible for deposition of deep sand beds in sea caves as well as further inland, making up the large part sedimentation in these places (Rubin et al. 2017). The huge tsunami generated by the 2011 East Japan earthquake transported a documented 289 species (16 phyla) of coastal species to the Hawaiian islands and western coast of North America, mostly by rafting (Carlton et al. 2017). The recent Kaikoura earthquake in New Zealand lifted the foreshore a few metres in places (Hamling et al. 2017), temporarily eradicating local intertidal and subtidal communities of kelp, molluscs and echinoderms. Major landslips, like the Usoy rock-slide of 1911, can create new lakes; Green Lake in Fiordland, New Zealand, was created 12,000–13,000 years ago when a 9-km block of mountains slipped down 700 m and along 2.5 km (Graham 2008), and more recently the Dart River was dammed by an enormous rockfall, which locals felt as an earthquake.

Evidence strongly points towards volcanism (in Siberia in both cases) as the cause of two mass extinctions: the biggest of all at the Permian–Triassic (252 Ma) leading to global anoxia (Stordal et al. 2017), and the Triassic–Jurassic (201.5 Ma) (Percival et al. 2017). Fifteen planet-altering mega-eruptions have been identified over the globe, likely associated with extinctions of varying degrees (Witze 2017). Supervolcanoes such as Yellowstone, Toba and Taupo have the potential to be 3–4 orders of magnitude larger than the Mount St Helens eruption in 1980, with far-reaching effects on Earth's climate and biota (McDowall 1996). Super-eruptions (>1000 gigatonnes) have recently been estimated to occur as frequently as every 17 kyr on average (Rougier et al. 2018).

When the Straits of Gibraltar closed 6 Ma, the Mediterranean dried, leaving a basin 3–5 km deep in places, probably with a few hypersaline lakes and pans, until filling again some 700 kyr later (Hsü et al. 1977). Filling of the Black Sea through the Bosphorus probably occurred in a monumental flood some 7–8 kyr ago, turning a freshwater lake into marine habitat (Ryan et al. 1997). Erosion and tectonic uplift can lead to headwater capture (river piracy), transferring biota from one system to another, an important mode of range expansion and speciation in freshwater fishes (Craw et al. 2016; Tagliacollo et al. 2015; Waters et al. 2001).

In Derek Ager's words: 'the history of any one part of the earth, like the life of a soldier, consists of long periods of boredom and short periods of terror' (Ager 1973).

4 Biogeography

The colonization of new landmasses, particularly islands, is dominated by chance events, often happening soon after condition become habitable. Lack's 'first arrival' hypothesis is apparent from island radiations of groups that just happened to get there first and thus be able to adapt to multiple niches in the absence of other occupants (Lack 1947), and rapidly spread to nearby islands (Wilmshurst et al. 2011). Thus, finches radiated in the Galápagos (Schluter 2000) and became the honeycreepers of Hawaii (Freed et al. 1987), while ancestors of Californian tarweeds became the Hawaiian silverswords (Carlquist et al. 2003). Monophyletic radiations can dominate the fauna, such as *Anolis* lizards in the Caribbean, *Nesotes* beetles on the Canaries and picture-winged *Drosophila* on Hawaii (Emerson 2002). Lemurs and chameleons dominate the Madagascan fauna whereas New Zealand is known for radiations of ratites, reptiles, land-snails, earthworms, cicadas, moths, ensiferans, *Veronica*, *Ranunculus* and *Celmisia*, though depauperate in many other groups (Gibbs 2016; Wallis and Trewick 2009). New Zealand's freshwater fish fauna is dominated by galaxiids and gobiids (McDowall 2000), and Hawaii by gobiids (McDowall 2003), which happened to get there first, reflecting their diadromous life history (Burridge et al. 2012). Thus, island faunas can reflect a few unlikely chance events early in their colonization. Once settled, plants and animals undergo rapid adaptive radiations, filling niches normally occupied by other groups on continents. Intriguingly, recent work on city park populations shows evidence of rapid morphological and genetical divergence in a lizard species, over the last century or more that cannot be attributed to drift alone (Littleford-Colquhoun et al. 2017).

5 Adaptation

Ironically, Darwin's own metaphor for natural selection-artificial selection by humans (Darwin 1859)—is evidence that morphological change can be effected rapidly in a wide range of species (DeLong et al. 2016; Ellner et al. 2011). Many classic natural examples of adaptation are also rapid and represent responses to human habitat modification (Alberti 2015; Bosse et al. 2017; Bradshaw and Holzapfel 2008; Carroll et al. 2007; Franks and Weis 2008; Palumbi 2001a, b; Rudman et al. 2017), including domestication (Zeng et al. 2017), physiological responses to pollution (van't Hof et al. 2016; Wright et al. 2013) and over-harvesting (Borrell 2013), resistance to drugs (Brown and Wright 2016), herbicides (Délye et al. 2013) and pesticides (Crossley et al. 2017), and introductions to new environments (Cattau et al. 2017; Gordon et al. 2015; Lee et al. 2011; Novy et al. 2013). These have sweeping local effects such that original populations can be completely replaced by genetically distinct ones in a few generations, or more widespread in the case of spread of viruses or plasmids carrying resistance factors. Recent work using *Tribolium* suggests that rapid evolution can be the cause of range expansion rather than merely a result (Szücs et al. 2017).

Change can be rapid when there is plentiful standing genetic variation for a quantitative trait, or 'soft sweep' (Crossley et al. 2017; Jain and Stephan 2017) and can be effected by change in gene expression as well (Becks et al. 2012; Huang et al. 2017; Passow et al. 2017; Rollins et al. 2015; Uusi-Heikkila et al. 2017). Indeed, altering gene expression through quantitative trait selection can be a rapid mode of adaptive evolution, since the protein is already present (Goncalves et al. 2017; Margres et al. 2017; Mathieson et al. 2015; Nandamuri et al. 2017). Strictly speaking, the rapid sorting of existing polymorphism may not strictly constitute 'evolution' in its entirety, but it is certainly an important component and can happen on an ecological timescale (Lallensack 2018). A recent review detailed numerous examples of evolutionary changes associated with extreme climatic events (Grant et al. 2017).

Adaptation on islands can be particularly rapid due to intense selection caused by novel food, habitat and competitors. When the large ground finch, *Geospiza magnirostris*, established breeding on the small island of Daphne in the Galápagos, the resident medium-sized *G. fortis* underwent a sharp reduction in bill size over the course of two seasons, coinciding with a drought the previous year (Grant and Grant 2006). In contrast, termite-eating geckos on five islands created by flooding behind a dam showed parallel increase in head-size within 15 years, reflecting inclusion of larger prey items in the absence of competition following extinction of four larger species (de Amorim et al. 2017). Another reptile showed a rapid change in perching behaviour when an invading congeneric competitor was introduced, resulting in a concomitant increase in toe-pad size as a result (Stuart et al. 2014). Rapid size increase on an even shorter timeframe has been documented in lake *Daphnia* in response to invasion by predatory zooplankton (Gillis and Walsh 2017).

Application of genomics is providing examples where rapid adaptation has been achieved through selection acting on a small number of genes (Lamichhaney et al. 2015, 2016; Rogers et al. 2017), in some cases caused by transposons (Brawand et al. 2014; van't Hof et al. 2016), or genes within inversions (see Speciation section). An example in steelhead trout (*Oncorhynchus mykiss*) shows rapid adaptation to land-locked lake conditions through change in transcription rates of four osmoregulatory genes within 14 generations (Aykanat et al. 2011). Such examples are particularly compelling when they can be shown to have happened repeatedly, in cases of convergent or parallel evolution (Arendt and Reznick 2008; Martin and Orgogozo 2013). In three-spine stickleback (*Gasterosteus aculeatus*), for example, repeated deletions in *Pitx1* have caused reduction in pelvic girdle structure associated with adaptation to freshwater habitat (Chan et al. 2010) by a single mutational leap. Different cave populations of tetras (*Astyanax*) show different 'regressive' loss of function mutations in the same pigmentation gene (*Mc1r*) in separate cave systems, one of them, remarkably, homologous to the substitution leading to red hair in humans (Gross et al. 2009), as well as independent deletions in another gene (*Oca*) causing complete albinism (Protas et al. 2006). Populations of rock pocket mice (*Chaetodipus intermedius*) on dark substrates show different mutations at *Mc1r* modulating adaptive melanism, while some populations utilise different loci altogether (Nachman et al. 2003). A recent case in *Anolis* lizards showed physiological, transcriptomic and genomic shifts after a single extreme winter (Campbell-Staton et al. 2017), recapitulating the classic work of Bumpus (1898). Repeated rapid evolution can happen on a seasonal basis (Becks et al. 2012) and has been shown for innate immune genes in *Drosophila* (Behrman et al. 2017).

As well as these examples of parallelism, a beneficial mutation in one species can be 'transported' to another through hybridization, as in the case of genes modulating bill size and shape in Darwin's finches (Lamichhaney et al. 2015, 2016), colour polymorphism in *Heliconius* butterflies (Dasmahapatra et al. 2012; Kronforst and Papa 2015), warfarin resistance in *Mus* (Song et al. 2011) and enhanced vectorial capacity in *Anopheles* mosquitoes (Fontaine et al. 2015). Repeated transport of low-armour *Ectodysplasin* (*Eda*) and low pigmentation (*Kitlg*) alleles, present at low frequency in the marine stock, has allowed rapid parallel evolution of the freshwater benthic form of *G. aculeatus* (Colosimo et al. 2005; Schluter and Conte 2009). There are even examples of horizontal gene transfer from bacteria and fungi into their phytophagous insect hosts (McKenna et al. 2016). Thus, a single mutational event can have far-reaching effects beyond the species in which it arose.

Major adaptations can often lead to incipient, or indeed, complete speciation (Wright et al. 2013). For example, movement of vertebrates onto land was preceded by a rapid threefold increase in eye size at 385–395 Ma, affording a million-fold increase in the volume of space in which objects could be seen (MacIver et al. 2017), and launching several major new lineages. Developmental genetics has shown that major changes in body plan often derive from tinkering with cis-regulation of a small number of otherwise conserved gene families (Carroll 2008).

6 Speciation

Eldredge and Gould (1972) fervently championed (Gould 1980) the theory of 'punctuated equilibrium'—the idea that phenotypic change tends to occur in bursts associated with cladogenesis, separated by long periods of stasis (Gould and Eldredge 1977). On a geological timescale, the punctuated evolutionary pattern is frequently upheld in morphologies preserved in fossils (Gould and Eldredge 1993; Jackson and Cheetham 1999; Stanley 1979; Vrba 1980). While some major geological boundaries (i.e. mass extinctions) are attributed to extraterrestrial and terrestrial geological causes (above), other links include marine transgression, atmospheric shifts, selenium depletion, and biotic changes, including the current human-induced event (Ceballos et al. 2017), subsequently promoting major new radiations (Lamsdell and Selden 2017), probably rapid ones (Ksepka et al. 2017), with new body plans exploiting new niches. The Cambrian explosion at 570 Ma led to a host of multicellular lineages in a very short space of time on a geological scale, most of which subsequently went extinct. These mass extinctions were followed by rapid phases of lineage diversification within certain groups (Simpson 1944), before slowing as the available niches became filled.

Punctuated equilibrium is undergoing something of a renaissance, given the recent availability of extensive molecular phylogenies and development of new analytical approaches (Hopkins and Lidgard 2012). A recent review (Pennell et al. 2014) distinguished two separate issues relating to evolutionary rate: (i) gradual versus pulsed evolution, and (ii) anagenetic versus cladogenetic evolution. Several examples supporting both concepts are presented (Uyeda et al. 2011), though that second is deemed hard to show on the basis of current data, particularly given the difficulty of morphologically cryptic species in fossils. This 'quantum evolution' of George Simpson has since been demonstrated in bird bill shape evolution (Cooney et al. 2017). Asteroid impact event at the K-Pg boundary led to sudden rapid diversification of bill shape (Simpsonian), followed by more gradual (Darwinian) evolution within lineages. Rapid radiations within three major superfamilies of frogs are similarly aligned with this event (Feng et al. 2017), though its implication in radiation of placental mammals is a matter of some debate (Liu et al. 2017; Penny and Phillips 2004). A recent analysis of the evolution of vertebrate body size used a maximum likelihood method to fit Lévy processes to comparative phylogenetic data and showed a pulsed mode of evolution with intervening stasis (Landis and Schraiber 2017). The oligochaete genus *Mesenchytraeus* appears to have undergone an explosive Pliocene radiation following uplift in the Pacific northwest, driven by habitat type (Lang et al. 2017). A pattern of rapid initial cladogenesis followed by slower lineage diversification could be a general observation in evolution of new groups (Rabosky and Lovette

2008), but may be caused by a correlation between the rate of phenotypic evolution and the rate of speciation, giving the impression that speciation is driving phenotypic divergence in a punctuated manner (Rabosky 2012). For example, ecological complexity and opportunity can drive rates of both phenotypic diversification and speciation. If these co-vary with each other across different lineages across a tree, it can lead to the perception that more change is associated with speciation.

Claims that Eldredge and Gould had uncovered a new evolutionary process were quickly repudiated: population genetics can easily explain geologically 'rapid' change over a few thousand generations (Charlesworth et al. 1982; Winsor et al. 2017; Wright 1931). Even Mayr's seminal book made several mentions of 'genetic revolutions' (Mayr 1963). He took care, however, to clearly distinguish his view of post-founder effect genetic change, which results from re-organisation of variation already present, from Goldschmidt's suggestion of systemic macromutations (Goldschmidt 1940).

Examples of comparatively rapid speciation in extant genera include dwarf planktivorous whitefish (*Coregonus*) repeatedly evolved from the normal benthic form (Dion-Côté et al. 2017), benthic and limnetic stickleback (*Gasterosteus*) from the anadromous form (Peichel and Marques 2017), and various divergent sympatric populations of trout (*Salmo*), charr (*Salvelinus*) and smelt (*Osmerus*), all formed in post-glacial lakes over the last several thousand years (Schluter 2000). The large species flocks of cichlid fishes in the African Rift Valley lakes show very little genetic divergence, implying rapid formation (Brawand et al. 2014; Meyer et al. 1990), despite an impressive array of feeding adaptations, patterns and colours. The 13 neotropical Midas cichlid species are only a few thousand years old, yet show attendant rapid evolution of opsins (Torres-Dowdall et al. 2017). One Nicaraguan crater-lake cichlid (*Amphilophus zaliosus*) probably arose from a more widespread species in the last 10,000 years (Barluenga et al. 2006). One of the fastest cases of speciation in vertebrates appears to have occurred in Baltic Sea flounder, where a demersal breeding form has evolved in low salinity conditions over the last 2,400 generations (Momigliano et al. 2017). A genetically homogeneous, selfing species of brooding sea star is hypothesized to have arisen in ecological time as a result of disruptive selection (Puritz et al. 2012). Phytophagous insects have the ability to switch host plants and speciate rapidly as a result (Coyne and Orr 2004). The tephritid fruit fly *Rhagoletis* switched from its native hawthorn (*Crataegus*) to apple (*Malus*) crops in the eastern USA in the 1850s, and has now switched back to novel endemic host hawthorn species with the introduction of infested apples into the Pacific northwest (Mattsson et al. 2015).

These examples of rapid divergence are often driven by adaptation to different ecological niches, but sexual selection is another likely driver, particularly in birds and fishes (Nosil 2012; Schluter 2000). More rapid still are potential examples of genetic isolation via *Wolbachia* infection in *Drosophila* and other insects (Charlat et al. 2003). Application of genomics to studies of speciation is also providing many examples of rapid speciation by 'magic traits', achieved either through pleiotropic effects of a few selected loci (Ferris et al. 2017), or hitchhiking of postzygotic isolation genes with a strongly selected locus (Wright et al. 2013). Some studies find punctuation at the molecular level causing departures from a molecular clock (Pagel et al. 2006), echoing an early suggestion that the molecular clock may be episodic and driven at least partly by selection (Gillespie 1984). Ongoing in vitro studies using *E. coli* shows sustained molecular evolution, often with very rapid fixation of positively selected mutations, even in the face of environmental constancy (Good et al. 2017).

Gould was criticized for his dalliance with the extreme saltationist 'chromosome repatterning' ideas of Richard Goldschmidt (Goldschmidt 1940; Gould 2002), but there is good evidence that rapid speciation has happened: we should maintain a pluralist view on rates and mechanisms of speciation. Chromosomal rearrangements appear to play a big role, since they can have the effect of reducing recombination and maintaining linkage of desirable combinations of genes in the face of gene flow (Ortiz-Barrientos et al. 2016). Saltational speciation through structural rearrangements (Lewis 1966) and polyploidy are clear examples in plants (Soltis and Soltis 1999); as many as 80% of angiosperms may be polyploids (Soltis and Soltis 2012). More recently, inversions have been shown to play a role in rapid ecological speciation in animals too (Hoffman and Rieseberg 2008), including *Gasterosteus* (Jones et al. 2012), *Rhagoletis* (Powell et al. 2013) and butterflies (*Heliconius, Papilio*) (Joron et al. 2011). Although it has often been thought that polyploids might represent an evolutionary 'dead end', evidence is building that suggests that it can be of long-term significance (Van de Peer et al. 2017). Genomics has confirmed that tetraploidisation has happened at least four times in major chordate lineages: twice at the beginning of vertebrate evolution, again in the teleosts, and again in salmoniform fishes. Each of these fundamental events spanned a single generation, becoming fixed in the descendant lineages, giving these groups vast scope for innovation and evolution through redundancy in gene copy number (Ohno 1970).

Recent work is revealing the hybrid nature of many species (Mallet 2007), including our own, where there has been widespread introgression of genes from archaic hominids conferring physiological adaptations (Evans et al. 2006; Huerta-Sánchez et al. 2014). Merging of genomes can derive from a small number of matings, providing traits with far-reaching consequences, such as range expansion (Pfennig et al. 2016), or 'transgressive segregation' allowing exploitation of novel habitat (Rieseberg et al. 2003), both greatly reducing risk of extinction. Hybridization followed by recombination is an appealing and plausible way to explain the rapid emergence of new forms (Dittrich-Reed and Fitzpatrick 2013). Stabilization of hybridity can be rapid, leading to new species in a few tens of generations, in animals as well as plants (Grant and Grant 2014; Lamichhaney et al. 2018; Mallet 2007; Marques et al. 2017; Meier et al. 2017; Soltis and Soltis 1999; Stemshorn et al. 2011; Ungerer et al. 1998).

7 Ecology

Although ecological theory is largely concerned with gradual processes and theory based on equilibria, disturbance (e.g. fire, invasions, overgrazing, disease, drought, anthropogenic effects) (Darimont et al. 2015) has been crucial to shaping ecosystems and communities. These disturbances can compound to produce cataclysmic events, such as loss of forest to scrub or grassland, or over-running of coral reefs by algae, sponges or grazers (Scheffer et al. 2001). Temporal environmental variability reduces the value of predictions based on competition theory (Wiens 1977), just as genetic drift can scatter population gene frequencies away from expected equilibrium conditions. Ecological systems have limited resilience; perturbations can cause a threshold to be crossed, leading to a new ecological regime. Habitat modification and over-harvesting can lead to nonlinear population dynamics and feedback loops when systems are perturbed beyond a tipping point (Scheffer et al. 2012), beyond which they may never recover, as is often the case with overfishing (Hsieh et al. 2005; Travis et al. 2014). Over the last 15 years, southwestern Australia has seen a rapid loss of kelp forest as its coastal ecosystem has become 'tropicalized' by warming of the Indian Ocean (Wernberg et al. 2016). In combination, multiple stressors can have larger and less predictable effects. Overfishing of spiny lobster (*Jasus*) in Tasmania has allowed invasion of long-spined sea urchin (*Centrostephanus*) into the now warmer waters, whose overgrazing of kelp in turn threatens a catastrophic phase shift (Ling et al. 2009)

As in social and financial systems, 'black swan' events (unlikely events with profound consequences) happen more frequently than expected in natural systems: they are the 'heavy tails' of normal distributions of events (Anderson et al. 2017). They are usually (86% of the time) destructive rather than constructive, probably because there are fewer limits to the rate of population decline compared with growth. Disease has the potential for local extirpation, and pandemic epizootics can cause more widespread extinction. Nowhere is this more obvious than with anthropogenic introductions of pathogens into new areas, such as the effect of chytrid fungus on endemic amphibian populations (Stegen et al. 2017).

8 Climate

On a geological timescale, cyclical onset and ending of glaciation is very rapid, with each current glacial cycle representing some 0.002% of the Earth's history. Yet on our own timeframe, such transitions would be imperceptibly slow: *Homo sapiens* has only been in existence through two such epochs. Some of the effects of post-glacial warming were, however, very immediate. Glacial lake outburst floods have devastating downstream effects, such as occurred in the Columbia River system at the end of the last glaciation, leaving signatures over a millennial timescale (Baynes et al. 2015; Larsen and Lamb 2016). The Straits of Dover were likely created by draining of a lake from Doggerland at the end of the previous glaciation. Heinrich events, involving rapid glacial decay (counter-intuitively during cold periods), led to sudden discharge of huge quantities of ice from the Laurentide Ice Sheet into the North Atlantic. Recent modelling suggests that these events were triggered by very small climatic perturbations, with implications for our understanding of current melting of the Greenland and Antarctic ice sheets (Bassis et al. 2017), which also brings a threat of megadroughts to the Sahel region of Africa (Defrance et al. 2017). Current melting is itself exacerbated by the positive feedback effects of decreasing albedo-darkening of surface ice caused by residual particulate matter and growth of microbes and algae (Kintisch 2017). Receding glaciers can also cause abrupt re-routing of rivers, affecting landscape evolution (Shugar et al. 2017).

Our coastal cities and installations may not appear to be immediately threatened by a monthly sea-level rise of about 0.3 mm, with diurnal variation on a scale of several metres. Even monthly spring/king tides can differ by a metre from each other, swamping tiny annual increases. But hurricanes like Katrina, Sandy, Harvey and Irma are becoming more frequent, as are disastrous coincidences of tides, winds and low pressure (Garner et al. 2017), such as the North Sea Flood of 1953, which inundated low-lying parts of UK, Belgium and Netherlands. Current assessment practices tend not to take into account the multiplicative effects of coinciding high sea level and heavy precipitation (Moftakhari et al. 2017), the effects of which were only too evident in Harvey's impact on Houston.

A Sandy-like event had a return frequency of about 400 years in the year 2000, but that will fall to ~90 yrs by 2100 (Lin et al. 2016). Keeping time constant, a 100-year event will be 0.65–1.7 m higher in 2100 than in 2000, depending on the global climate change model used. This is already higher by about the same amount compared to AD 850. Thus, global warming may add a metre or more in sea level to an extreme event that cost $50 billion the last time it occurred (Reed et al. 2015). Recent modelling shows a 1/500 yr event before 1800 to be a 1/5 yr event after 2030 (Garner et al. 2017). Ironically, the rarer and more extreme the event, the greater the added integrated effect of increased sea level. Likewise, modelling of precipitation accumulation (Neelin et al. 2017), convective available potential energy (Singh et al. 2018) and direct satellite observation (Taylor et al. 2017) show that extreme storm events get more frequent, and the biggest events get even bigger. El Niño events have caused repeated (1998, 2002, 2016) coral bleaching events on the Great Barrier Reef, with the most recent being the most extreme. As the frequency and intensity of these extreme events increases, the reef will likely reach a point beyond which recovery is not possible (Hughes et al. 2017).

As with our own mortality, demise usually follows a single major event (heart attack, stroke, disease) rather than old age alone. So it will be for coastal cities like New Orleans and Venice, and low-lying regions such as the Netherlands, Vietnam and Pacific island nations. Thinking about how we might mitigate future major events is of immediate importance to humanity. Although we must act to halt the slow, inexorable change caused by greenhouse gases, much of it will happen anyway because of inertia in the system.

9 Conclusion

Nature plays out as a series of major events, be it climatic or ecological, evolutionary or geological. We live in a jumpy natural world of serial catastrophism as opposed to true gradualism. The length and frequency of the events depends upon the process concerned; the common feature is that the duration is relatively very brief, and effects profound, compared to the long intervals of stasis or gradual change in between events. New adaptations can arise and become the new wild type over a few generations, yet remain in place for hundreds of thousands more; rapid speciation can happen on a scale of years or decades, though species typically exist for millions of years; dispersal events typically happen over periods of weeks or months, but resulting adaptive radiations may endure for millions of years; extreme weather events are measured by hours, but are separated by years or decades.

These extreme events are major purveyors of change through time. When we consider change through deep time, we are often looking at the results of numerous catastrophes or relatively sudden events telescoped together, rather than gradual change. This pattern extends to macroeconomic cycles of booms and crashes; even language (Atkinson et al. 2008) and cultural change (Kolodny et al. 2015) have been characterized as following a punctuated mode. There is a parallel here with Kuhn's portrayal of paradigm shifts in science (Kuhn 1996) and Marxist philosophy on the necessity for revolution to effect political transition. Although Darwin, Hutton and Lyell debunked biblical catastrophism, evidence for neocatastrophism on a shorter timeframe abounds. Although we live in a world of both gradual and punctuated processes, it is arguably the latter, as Huxley implied, that is responsible for most of the change in nature.

References

Ager DV (1973) The nature of the stratigraphical record. Halsted (Wiley) Press, New York

Alberti M (2015) Eco-evolutionary dynamics in an urbanizing planet. Trends Ecol Evol 30:115–126

Anderson SC, Branch TA, Cooper AB, Dulvy NK (2017) Black-swan events in animal populations. Proc Natl Acad Sci USA 114:3252–3257

Arendt J, Reznick D (2008) Convergence and parallelism reconsidered: what have we learned about the genetics of adaptation? Trends Ecol Evol 23:26–32

Atkinson QD, Meade A, Venditti C, Greenhill SJ, Pagel M (2008) Languages evolve in punctuational bursts. Science 319:588

Aykanat T, Thrower FP, Heath DD (2011) Rapid evolution of osmoregulatory function by modification of gene transcription in steelhead trout. Genetica 139:233–242

Bardeen CG, Garcia RR, Toon OB, Conley AJ (2017) On transient climate change at the Cretaceous-Paleogene boundary due to atmospheric soot injections. Proc Natl Acad Sci USA 114:E7415–E7424

Barluenga M, Stölting KN, Salzburger W, Muschick M, Meyer A (2006) Sympatric speciation in Nicaraguan crater lake cichlid fish. Nature 439:719–723

Bassis JN, Petersen SV, Cathles LM (2017) Heinrich events triggered by ocean forcing and modulated by isostatic adjustment. Nature 542:332–334

Baynes ERC, Attal M, Niedermann S, Kirstein LA, Dugmore AJ, Naylor M (2015) Erosion during extreme flood events dominates Holocene canyon evolution in northeast Iceland. Proc Natl Acad Sci USA 112:2355–2360

Becks L, Ellner SP, Jones LE, Hairston NG (2012) The functional genomics of an eco-evolutionary feedback loop: linking gene expression, trait evolution, and community dynamics. Ecol Lett 15:492–501

Behrman EL, Howick VM, Kapun M, Staubach F, Bergland AO, Petrov DA, Lazzaro BP, Schmidt PS (2017) Rapid seasonal evolution in innate immunity of wild *Drosophila melanogaster*. Proc R Soc B 285:20172599

Borrell B (2013) A big fight over little fish. Nature 493:597–598

Bosse M, Spurgin LG, Laine VN, Cole EF, Firth JA, Gienapp P, Gosler AG, McMahon K, Poissant J, Verhagen I, Groenen MAM, van Oers K, Sheldon BC, Visser ME, Slate J (2017) Recent natural selection causes adaptive evolution of an avian polygenic trait. Science 358:365–368

Bradshaw WE, Holzapfel CM (2008) Genetic response to rapid climate change: it's seasonal timing that matters. Mol Ecol 17:157–166

Brawand D, Wagner CE, Li YI, Malinsky M, Keller I, Fan S, Simakov O, Ng AY, Lim ZW, Bezault E, Turner-Maier J, Johnson J, Alcazar R, Noh HJ, Russell P, Aken B, Alfoldi J, Amemiya C, Azzouzi N, Baroiller J-F, Barloy-Hubler F, Berlin A, Bloomquist R, Carleton KL, Conte MA, D'Cotta H, Eshel O, Gaffney L, Galibert F, Gante HF, Gnerre S, Greuter L, Guyon R, Haddad NS, Haerty W, Harris RM, Hofmann HA, Hourlier T, Hulata G, Jaffe DB, Lara M, Lee AP, MacCallum I, Mwaiko S, Nikaido M, Nishihara H, Ozouf-Costaz C, Penman DJ, Przybylski D, Rakotomanga M, Renn SCP, Ribeiro FJ, Ron M, Salzburger W, Sanchez-Pulido L, Santos ME, Searle S, Sharpe T, Swofford R, Tan FJ, Williams L, Young S, Yin S, Okada N, Kocher TD, Miska EA, Lander ES, Venkatesh B, Fernald RD, Meyer A, Ponting CP, Streelman JT, Lindblad-Toh K, Seehausen O, Di Palma F (2014) The genomic substrate for adaptive radiation in African cichlid fish. Nature 513:375–381

Brown ED, Wright GD (2016) Antibacterial drug discovery in the resistance era. Nature 529:336–343

Bumpus HC (1898) The elimination of the unfit as illustrated by the introduced sparrow, *Passer domesticus* (A fourth contribution to the study of variation.). Biol Lect Woods Hole Mar Biol Lab 1898:209–225

Burridge CP, McDowall RM, Craw D, Wilson MVH, Waters JM (2012) Marine dispersal as a pre-requisite for Gondwanan vicariance among elements of the galaxiid fish fauna. J Biogeogr 39:306–321

Campbell-Staton SC, Cheviron ZA, Rochette N, Catchen J, Losos JB, Edwards SV (2017) Winter storms drive rapid phenotypic, regulatory, and genomic shifts in the green anole lizard. Science 357:495–498

Carlquist S, Baldwin BG, Carr GD (2003) Tarweeds and silverswords: evolution of the Madiinae (Asteraceae). Missouri Botanical Gardens Press, St. Louis, MO, p 293

Carlton JT, Chapman JW, Geller JB, Miller JA, Carlton DA, McCuller MI, Treneman NC, Steves BP, Ruiz GM (2017) Tsunami-driven rafting: transoceanic species dispersal and implications for marine biogeography. Science 357:1402–1406

Carroll SB (2008) Evo-devo and an expanding evolutionary synthesis: a genetic theory of morphological evolution. Cell 134:25–36

Carroll SP, Hendry AP, Reznick DN, Fox CW (2007) Evolution on ecological time-scales. Funct Ecol 21:387–393

Cattau CE, Fletcher RJ, Kimball RT, Miller CW, Kitchens WM (2017) Rapid morphological change of a top predator with the invasion of a novel prey. Nat Ecol Evolut 2

Ceballos G, Ehrlich PR, Dirzo R (2017) Biological annihilation via the ongoing sixth mass extinction signaled by vertebrate population losses and declines. Proc Natl Acad Sci USA 114:E6089–E6096

Chan YF, Marks ME, Jones FC, Villarreal G, Shapiro MD, Brady SD, Southwick AM, Absher DM, Grimwood J, Schmutz J, Myers RM, Petrov D, Jónsson B, Schluter D, Bell MA, Kingsley DM

(2010) Adaptive evolution of pelvic reduction in sticklebacks by recurrent deletion of a *Pitx1* enhancer. Science 327:302–305

Chapman CR, Morrison D (1994) Impacts on the Earth by asteroids and comets: assessing the hazard. Nature 367:33–40

Charlat S, Hurst GDD, Merçot H (2003) Evolutionary consequences of *Wolbachia* infections. Trends Genet 19:217–223

Charlesworth B, Lande R, Slatkin M (1982) A neo-darwinian commentary on macroevolution. Evolution 36:474–498

Colosimo PF, Hosemann KE, Balabhadra S, Villarreal G, Dickson M, Grimwood J, Schmutz J, Myers RM, Schluter D, Kingsley DM (2005) Widespread parallel evolution in sticklebacks by repeated fixation of ectodysplasin alleles. Science 307:1928–1933

Cooney CR, Bright JA, Capp EJR, Chira AM, Hughes EC, Moody CJA, Nouri LO, Varley ZK, Thomas GA (2017) Mega-evolutionary dynamics of the adaptive radiation of birds. Nature 542:344–347

Coyne JA, Orr HA (2004) Speciation. Sinauer Associates Inc., Sunderland (MA)

Craw D, Upton P, Burridge CP, Wallis GP, Waters JM (2016) Rapid biological speciation driven by tectonic evolution in New Zealand. Nat Geosci 9:140–145

Crossley MS, Chen YH, Groves RL, Schoville SD (2017) Landscape genomics of Colorado potato beetle provides evidence of polygenic adaptation to insecticides. Mol Ecol 26:6284–6300

Darimont CT, Fox CH, Bryan HM, Reimchen TE (2015) The unique ecology of human predators. Science 349:858–860

Darwin C (1859) On the origin of species by means of natural selection. John Murray, London

Dasmahapatra KK, Walters JR, Briscoe AD, Davey JW, Whibley A, Nadeau NJ, Zimin AV, Hughes DST, Ferguson LC, Martin SH, Salazar C, Lewis JJ, Adler S, Ahn S-J, Baker DA, Baxter SW, Chamberlain NL, Chauhan R, Counterman BA, Dalmay T, Gilbert LE, Gordon K, Heckel DG, Hines HM, Hoff KJ, Holland PWH, Jacquin-Joly E, Jiggins FM, Jones RT, Kapan DD, Kersey P, Lamas G, Lawson D, Mapleson D, Maroja LS, Martin A, Moxon S, Palmer WJ, Papa R, Papanicolaou A, Pauchet Y, Ray DA, Rosser N, Salzberg SL, Supple MA, Surridge A, Tenger-Trolander A, Vogel H, Wilkinson PA, Wilson D, Yorke JA, Yuan F, Balmuth AL, Eland C, Gharbi K, Thomson M, Gibbs RA, Han Y, Jayaseelan JC, Kovar C, Mathew T, Muzny DM, Ongeri F, Pu L-L, Qu J, Thornton RL, Worley KC, Wu Y-Q, Linares M, Blaxter ML, ffrench-Constant RH, Joron M, Kronforst MR, Mullen SP, Reed RD, Scherer SE, Richards S, Mallet J, McMillan WO, Jiggins CD (2012) Butterfly genome reveals promiscuous exchange of mimicry adaptations among species. Nature 487:94–98

de Amorim ME, Schoener TW, Santoro GRCC, Lins ACR, Piovia-Scott J, Brandão RA (2017) Lizards on newly created islands independently and rapidly adapt in morphology and diet. Proc Natl Acad Sci USA 114:8812–8816

Defrance D, Ramstein G, Charbit S, Vrac M, Famien A, Sultan B, Swingedouw D, Dumas C, Gemenne F, Alvarez-Solas J, Vanderlinden J-P (2017) Consequences of rapid ice sheet melting on the Sahelian population vulnerability. Proc Natl Acad Sci USA 114:6533–6538

DeLong JP, Forbes VE, Galic N, Gibert JP, Laport RG, Phillips JS, Vavra JM (2016) How fast is fast? Eco-evolutionary dynamics and rates of change in populations and phenotypes. Ecol Evol 6:573–581

Délye C, Jasieniuk M, Le Corre V (2013) Deciphering the evolution of herbicide resistance in weeds. Trends Genet 29:649–658

Dion-Côté A-M, Symonová R, Lamaze FC, Pelikánová S, Ráb P, Bernatchez L (2017) Standing chromosomal variation in Lake Whitefish species pairs: the role of historical contingency and relevance for speciation. Mol Ecol 26:178–192

Dittrich-Reed DR, Fitzpatrick BM (2013) Transgressive hybrids as hopeful monsters. Evol Biol 40:310–315

Dobzhansky T (1937) Genetics and the origin of species. Columbia University Press, New York

Eldredge N, Gould SJ (1972) Punctuated equilibria: an alternative to phyletic gradualism. In: Schopf TJM (ed) Models in paleobiology. Freeman, Cooper and Co, San Francisco, pp 82–115

Ellner SP, Geber MA, Hairston NG (2011) Does rapid evolution matter? Measuring the rate of contemporary evolution and its impacts on ecological dynamics. Ecol Lett 14:603–614

Emerson BC (2002) Evolution on oceanic islands: molecular phylogenetic approaches to understanding pattern and process. Mol Ecol 11:951–966

Evans PD, Mekel-Bobrov N, Vallender EJ, Hudson RR, Lahn BT (2006) Evidence that the adaptive allele of the brain size gene *microcephalin* introgressed into *Homo sapiens* from an archaic *Homo* lineage. Proc Natl Acad Sci USA 103:18178–18183

Feng Y-J, Blackburn DC, Liang D, Hillis DM, Wake DB, Cannatella DC, Zhang P (2017) Phylogenomics reveals rapid, simultaneous diversification of three major clades of Gondwanan frogs at the Cretaceous-Paleogene boundary. Proc Natl Acad Sci USA 114:E5864–5870

Ferris KG, Barnett LL, Blackman BK, Willis JH (2017) The genetic architecture of local adaptation and reproductive isolation in sympatry within the *Mimulus guttatus* species complex. Mol Ecol 26:208–224

Fontaine MC, Pease JB, Steele A, Waterhouse RM, Neafsey DE, Sharakhov IV, Jiang X, Hall AB, Catteruccia F, Kakani E, Mitchell SN, Wu Y-C, Smith HA, Love RR, Lawniczak MK, Slotman MA, Emrich SJ, Hahn MW, Besansky NJ (2015) Extensive introgression in a malaria vector species complex revealed by phylogenomics. Science 347:42

Franks SJ, Weis AE (2008) A change in climate causes rapid evolution of multiple life-history traits and their interactions in an annual plant. J Evol Biol 21:1321–1334

Freed LA, Conant S, Fleischer RC (1987) Evolutionary ecology and radiation of Hawaiian passerine birds. Trends Ecol Evol 2:196–203

Garner AJ, Mann ME, Emanuel KA, Kopp RE, Lin N, Alley RB, Horton BP, DeConto RM, Donnelly JP, Pollard D (2017) Impact of climate change on New York City's coastal flood hazard: increasing flood heights from the preindustrial to 2300 CE. Proc Natl Acad Sci USA 114:11861–11866

Gibbs G (2016) Ghosts of Gondwana: the history of life in New Zealand. Potton & Burton Ltd, Nelson

Gillespie JH (1984) The molecular clock may be an episodic clock. Proc Natl Acad Sci USA 81:8009–8013

Gillis MK, Walsh MR (2017) Rapid evolution mitigates the ecological consequences of an invasive species (*Bythotrephes longimanus*) in lakes in Wisconsin. Proc R Soc B 284:20170814

Goldschmidt R (1940) The material basis of evolution. Yale University Press, New Haven, CT

Goncalves P, Jones DB, Thompson EL, Parker LM, Ross PM, Raftos DA (2017) Transcriptomic profiling of adaptive responses to ocean acidification. Mol Ecol 26:5974–5988

Good BH, McDonald MJ, Barrick JE, Lenski RE, Desai MM (2017) The dynamics of molecular evolution over 60,000 generations. Nature 551:45–50

Gordon SP, Reznick D, Arendt JD, Roughton A, Ontiveros Hernandez MN, Bentzen P, López-Sepulcre A (2015) Selection analysis on the rapid evolution of a secondary sexual trait. Proc R Soc B 282:20151244

Gould SJ (1980) Is a new and general theory of evolution emerging? Paleobiology 61:119–130

Gould SJ (2002) The structure of evolutionary theory. The Belknap Press of Harvard University Press, Cambridge, MA

Gould SJ, Eldredge N (1977) Punctuated equilibria: the tempo and mode of evolution reconsidered. Paleobiology 3:115–151

Gould SJ, Eldredge N (1993) Punctuated equilibrium comes of age. Nature 366:223–227

Graham IJ (2008) A continent on the move: New Zealand geoscience into the 21st century. The Geological Society of New Zealand in Association with GNS Science, Wellington, p 388

Grant PR, Grant BR (2006) Evolution of character displacement in Darwin's finches. Science 313:224–226

Grant PR, Grant BR (2014) 40 years of evolution. Darwin's finches on Daphne Major Island. Princeton University Press, Princeton, NJ

Grant PR, Grant BR, Huey RB, Johnson MTJ, Knoll AH, Schmitt J (2017) Evolution caused by extreme events. Philos Trans R Soc B 372:20160146

Gross JB, Borowsky R, Tabin CJ (2009) A novel role for *Mc1r* in the parallel evolution of depigmentation in independent populations of the cavefish Astyanax mexicanus. PLoS Genet 5:e1000326

Hamling IJ, Hreinsdóttir S, Clark K, Elliott J, Liang C, Fielding E, Litchfield N, Villamor P, Wallace L, Wright TJ, D'Anastasio E, Bannister S, Burbidge D, Denys P, Gentle P, Howarth J, Mueller C, Palmer N, Pearson C, Power W, Barnes P, Barrell DJA, Van Dissen R, Langridge R, Little TA, Nicol A, Pettinga J, Rowland J, Stirling M (2017) Complex multifault rupture during the 2016 *M*w 7.8 Kaikōura earthquake, New Zealand. Science 356:154

Hewitt GM (2000) The genetic legacy of the Quaternary ice ages. Nature 405:907–913

Hewitt GM (2004) Genetic consequences of climatic oscillations in the Quaternary. Philos Trans R Soc Lond B Biol Sci 359:183–195

Hodell DA (2016) The smoking gun of the ice ages. Science 354:1235–1236

Hoffman AA, Rieseberg LH (2008) Revisiting the impact of inversions in evolution: from population genetic markers to drivers of adaptive shifts and speciation? Annu Rev Ecol Syst 39:21–42

Hopkins MJ, Lidgard S (2012) Evolutionary mode routinely varies among morphological traits within fossil species lineages. Proc Natl Acad Sci USA 109:20520–20525

Hsieh C-H, Glaser SM, Lucas AJ, Sugihara G (2005) Distinguishing random environmental fluctuations from ecological catastrophes for the North Pacific Ocean. Nature 435:336–340

Hsü KJ, Garrison RE, Montadert L, B KR, Bernoulli D, Mèlierés F, Cita MB, Müller C, Erickson A, Wright R (1977) History of the Mediterranean salinity crisis. Nature 267:399–403

Huang X, Li S, Ni P, Gao Y, Jiang B, Zhou Z, Zhan A (2017) Rapid response to changing environments during biological invasions: DNA methylation perspectives. Mol Ecol 26:6621–6633

Huerta-Sánchez E, Jin X, Asan Bianba Z, Peter BM, Vinckenbosch N, Liang Y, Yi X, He M, Somel M, Ni P, Wang B, Ou X, Huasang Luosang J, Cuo ZXP, Li K, Gao G, Yin Y, Wang W, Zhang X-C, Xu X, Yang H, Li Y, Wang J, Wang J, Nielsen R (2014) Altitude adaptation in Tibetans caused by introgression of Denisovan-like DNA. Nature 512:194–197

Hughes TP, Kerry JT, Álvarez-Noriega M, Álvarez-Romero JG, Anderson KD, Baird AH, Babcock RC, Beger M, Bellwood DR, Berkelmans R, Bridge TC, Butler IR, Byrne M, Cantin NE, Comeau S, Connolly SR, Cumming GS, Dalton SJ, Diaz-Pulido G, Eakin CM, Figueira WF, Gilmour JP, Harrison HB, Heron SF, Hoey AS, Hobbs J-PA, Hoogenboom MO, Kennedy EV, C-y Kuo, Lough JM, Lowe RJ, Liu G, McCulloch MT, Malcolm HA, McWilliam MJ, Pandolfi JM, Pears RJ, Pratchett MS, Schoepf V, Simpson T, Skirving WJ, Sommer B, Torda G, Wachenfeld DR, Willis BT, Wilson SK (2017) Global warming and recurrent mass bleaching of corals. Nature 543:373–377

Huxley TH (1906) Darwin on the origin of species Man's place in nature and other essays. J M Dent & Sons Ltd., London, pp 299–336

Jackson JBC, Cheetham AH (1999) Tempo and mode of speciation in the sea. Trends Ecol Evol 14:72–77

Jain K, Stephan W (2017) Modes of rapid polygenic adaptation. Mol Biol Evol 34:3169–3317

Jansson R, Dynesius M (2002) The fate of clades in a world of recurrent climatic change: Milankovitch oscillations and evolution. Annu Rev Ecol Syst 33:741–777

Jones FC, Grabherr MG, Chan YF, Russell P, Mauceli E, Johnson J, Swofford R, Pirun M, Zody MC, White S, Birney E, Searle S, Schmutz J, Grimwood J, Dickson MC, Myers RM, Miller CT, Summers BR, Knecht AK, Brady SD, Zhang H, Pollen AA, Howes T, Amemiya C, Platform BIGS, Team WGA, Lander ES, Di Palma F, Lindblad-Toh K, Kingsley DM (2012) The genomic basis of adaptive evolution in threespine sticklebacks. Nature 484:55–61

Joron M, Frezal L, Jones RT, Chamberlain NL, Lee SF, Haag CR, Whibley A, Becuwe M, Baxter SW, Ferguson L, Wilkinson PA, Salazar C, Davidson C, Clark R, Quail MA, Beasley H, Glithero R, Lloyd C, Sims S, Jones MC, Rogers J, Jiggins CD, French-Constant RH (2011) Chromosomal rearrangements maintain a polymorphic supergene controlling butterfly mimicry. Nature 477:203–206

Kintisch E (2017) Meltdown. Science 355:788–791

Kolodny O, Creanza N, Feldman MW (2015) Evolution in leaps: the punctuated accumulation and loss of cultural innovations. Proc Natl Acad Sci USA 112:E6762–E6769

Kronforst MR, Papa R (2015) The functional basis of wing patterning in *Heliconius* butterflies: the molecules behind mimicry. Genetics 200:1–19

Ksepka DT, Stidham TA, Williamson TE (2017) Early Paleocene landbird supports rapid phylogenetic and morphological diversification of crown birds after the K-Pg mass extinction. Proc Natl Acad Sci USA 114:8047–8052

Kuhn TS (1996) The structure of scientific revolutions. University of Chicago Press, Chicago

Lack DL (1947) Darwin's finches. Cambridge University Press, Cambridge

Lallensack R (2018) Evology. Nature 554:1–40

Lamichhaney S, Bergland J, Almén MS, Maqbool K, Grabherr M, Martinez-Barrio A, Promerová M, Rubin C-J, Wang C, Zamani N, Grant BR, Grant PR, Webster MT, Andersson L (2015) Evolution of Darwin's finches and their beaks revealed by genome sequencing. Nature 518:371–375

Lamichhaney S, Han F, Bergland J, Wang C, Almén MS, Webster MT, Grant BR, Grant PR, Andersson L (2016) A beak size locus in Darwin's finches facilitated character displacement during a drought. Science 352:470–474

Lamichhaney S, Han F, Webster MT, Andersson L, Grant BR, Grant PR (2018) Rapid speciation in Darwin's finches. Science 359:224–228

Lamsdell JC, Selden PA (2017) From success to persistence: identifying an evolutionary regime shift in the diverse Paleozoic aquatic arthropod group Eurypterida, driven by the Devonian biotic crisis. Evolution 71:95–110

Landis MJ, Schraiber JG (2017) Pulsed evolution shaped modern vertebrate body sizes. Proc Natl Acad Sci USA 114:13224–13229

Lang SA, Saglam N, Kawash J, Shain DH (2017) Punctuated invasion of water, ice, snow and terrestrial ecozones by segmented worms (Oligochaeta: Enchytraeidae: *Mesenchytraeus*). Proc R Soc B 284:20171081

Larsen IJ, Lamb MP (2016) Progressive incision of the Channeled Scablands by outburst floods. Nature 538:229–232

Lee CE, Kiergaard M, Gelembiuk GW, Eads BD, Posavi M (2011) Pumping ions: rapid parallel evolution of ionic regulation following habitat invasions. Evolution 65:2229–2244

Lewis H (1966) Speciation in flowering plants. Science 152:167–172

Lin N, Kopp RE, Horton BP, Donnelly JP (2016) Hurricane Sandy's flood frequency increasing from year 1800 to 2100. Proc Natl Acad Sci USA 113:12071–12075

Ling SD, Johnson CR, Frusher SD, Ridgway KR (2009) Overfishing reduces resilience of kelp beds to climate-driven catastrophic phase shift. Proc Natl Acad Sci USA 106:22341–22345

Littleford-Colquhoun BL, Clemente C, Whiting MJ, Ortiz-Barrientos D, Frère CH (2017) Archipelagos of the Anthropocene: rapid and extensive differentiation of native terrestrial vertebrates in a single metropolis. Mol Ecol 26:2466–2481

Liu L, Zhang J, Rheindt FE, Lei F, Qu Y, Wang Y, Zhang Y, Sullivan C, Nie W, Wang J, Yang F, Chen J, Edwards SV, Meng J, Wu S (2017) Genomic evidence reveals a radiation of placental mammals uninterrupted by the KPg boundary. Proc Natl Acad Sci USA E7282–E7290

Lyell C (1830–1833) Principles of geology, being an attempt to explain the former changes of the Earth's surface by reference to causes now in operation. John Murray, London

MacIver MA, Schmitz L, Mugan U, Murphey TD, Mobley CD (2017) Massive increase in visual range preceded the origin of terrestrial vertebrates. Proc Natl Acad Sci USA 114:E2375–E2384

Mallet J (2007) Hybrid speciation. Nature 446:279–283

Mann A (2018) Cataclysm's end. Nature 553:393–395

Margres MJ, Wray KP, Hassinger ATB, Ward MJ, McGivern JJ, Lemmon EM, Lemmon AR, Rokyta DR (2017) Quantity, not quality: rapid adaptation in a polygenic trait proceeded exclusively through expression differentiation. Mol Biol Evol 34:3099–3110

Marques DA, Lucek K, Haesler MP, Feller AF, Meier JI, Wagner CE, Excoffier L, Seehausen O (2017) Genomic landscape of early ecological speciation initiated by selection on nuptial colour. Mol Ecol 26:7–24

Martin A, Orgogozo V (2013) The loci of repeated evolution: a catalog of genetic hotspots of phenotypic variation. Evolution 67:1235–1250

Mathieson I, Lazaridis I, Rohland N, Mallick S, Patterson N, Roodenberg SA, Harney E, Stewardson K, Fernandes D, Novak M, Sirak K, Gamba C, Jones ER, Llamas B, Dryomov S, Pickrel J, Arsuaga JL, Bermúdez de Castro JM, Carbonell E, Gerritsen F, Khokhlov A, Kuznetsov P, Lozano M, Meller H, Mochalov O, Moiseyev V, Rojo Guerra MA, Roodenberg J, Vergès JM, Krause J, Cooper A, Alt KW, Brown D, Anthony D, Lalueza-Fox C, Haak W, Pinhasi R, Reich D (2015) Genome-wide patterns of selection in 230 ancient Eurasians. Nature 528:499–503

Mattsson M, Hood GR, Feder JL, Ruedas LA (2015) Rapid and repeatable shifts in life-history timing of *Rhagoletis pomonella* (Diptera: Tephritidae) following colonization of novel host plants in the Pacific Northwestern United States. Ecol Evol 5:5823–5837

Mayr E (1942) Systematics and the origin of species. Columbia University Press, New York

Mayr E (1963) Animal species and evolution. Belknap Press, Cambridge, MA

McDowall RM (1996) Volcanism and freshwater fish biogeography in the northeastern North Island of New Zealand. J Biogeogr 23:139–148

McDowall RM (2000) The reed field guide to New Zealand freshwater fishes. Reed Publishing, Auckland

McDowall RM (2003) Hawaiian biogeography and the islands' freshwater fish fauna. J Biogeogr 30:703–710

McKenna DD, Scully ED, Pauchet Y, Hoover K, Kirsch R, Geib SM, Mitchell RF, Waterhouse RM, Ahn S-J, Arsala D, Benoit JB, Blackmon H, Bledsoe T, Bowsher JH, Busch A, Calla B, Chao H, Childers AK, Childers C, Clarke DJ, Cohen L, Demuth JP, Dinh H, Doddapaneni H, Dolan A, Duan JJ, Dugan S, Friedrich M, Glastad KM, Goodisman MAD, Haddad S, Han Y, Hughes DST, Ioannidis P, Johnston JS, Jones JW, Kuhn LA, Lance DR, Lee C-Y, Lee SL, Lin H, Lynch JA, Moczek AP, Murali SC, Muzny DM, Nelson DR, Palli SR, Panfilio KA, Pers D, Poelchau MF, Quan H, Qu J, Ray AM, Rinehart JP, Robertson HM, Roehrdanz R, Rosendale AJ, Shin S, Silva C, Torson AS, Vargas Jentzsch IM, Werren JH, Worley KC, Yocum G, Zdobnov EM, Gibbs RA, Richards S (2016) Genome of the Asian longhorned beetle (*Anoplophora glabripennis*), a globally significant invasive species, reveals key functional and evolutionary innovations at the beetle–plant interface. Genome Biol 17:227

Meier JI, Sousa VC, Marques DA, Selz OM, Wagner CE, Excoffier L, Seehausen O (2017) Demographic modelling with whole-genome data reveals parallel origin of similar *Pundamilia* cichlid species after hybridization. Mol Ecol 26:123–141

Meyer A, Kocher TD, Basasibwaki P, Wilson AC (1990) Monophyletic origin of Lake Victoria cichlid fishes suggested by mitochondrial DNA sequences. Nature 347:550–553

Moftakhari HR, Salvadori G, AghaKouchak A, Sanders BF, Matthew RA (2017) Compounding effects of sea level rise and fluvial flooding. Proc Natl Acad Sci USA 114:9785–9790

Momigliano P, Jokinen H, Fraimout A, Florin A-B, Norkko A, Merilä J (2017) Extraordinarily rapid speciation in a marine fish. Proc Natl Acad Sci USA 114:6074–6079

Nachman MW, Hoekstra HE, D'Agostino SL (2003) The genetic basis of adaptive melanism in pocket mice. Proc Natl Acad Sci USA 100:5268–5273

Nandamuri SP, Yourick MR, Carleton KL (2017) Adult plasticity in African cichlids: Rapid changes in opsin expression in response to environmental light differences. Mol Ecol 26:6036–6052

Neelin JD, Sahany S, Stechmann SN, Bernstein DN (2017) Global warming precipitation accumulation increases above the current-climate cutoff scale. Proc Natl Acad Sci USA 114:1258–1263

Nosil P (2012) Ecological speciation. Oxford University Press, Oxford

Novy A, Flory SL, Hartman JM (2013) Evidence for rapid evolution of phenology in an invasive grass. J Evol Biol 26:443–450

Ohno S (1970) Evolution by gene duplication. Springer, Berlin

Ortiz-Barrientos D, Engelstädter J, Rieseberg LH (2016) Recombination rate evolution and the origin of species. Trends Ecol Evol 31:226–236

Pagel M, Venditti C, Meade A (2006) Large punctuational contribution of speciation to evolutionary divergence at the molecular level. Science 314:119–121

Palumbi SR (2001a) The evolution explosion: how humans cause rapid evolutionary change. W W Norton & Co, New York

Palumbi SR (2001b) Humans as the world's greatest evolutionary force. Science 293:1786–1790

Passow CN, Henpita C, Shaw JH, Quackenbush CR, Warren WC, Schartl M, Arias-Rodriguez L, Kelley JL, Tobler M (2017) The roles of plasticity and evolutionary change in shaping gene expression variation in natural populations of extremophile fish. Mol Ecol 26:6384–6399

Peichel CL, Marques DA (2017) The genetic and molecular architecture of phenotypic diversity in sticklebacks. Philos Trans R Soc B 372:20150486

Pennell MW, Harmon LJ, Uyeda JC (2014) Is there room for punctuated equilibrium in macroevolution? Trends Ecol Evol 29:23–32

Penny D (2017) Evolution now. Xlibris

Penny D, Phillips MJ (2004) The rise of birds and mammals: are microevolutionary processes sufficient for macroevolution? Trends Ecol Evol 19:516–522

Percival LME, Ruhl M, Hesselbo SP, Jenkyns HC, Mather TA, Whiteside JH (2017) Mercury evidence for pulsed volcanism during the end-Triassic mass extinction. Proc Natl Acad Sci USA 114:7929–7934

Pfennig KS, Kelly AL, Pierce AA (2016) Hybridization as a facilitator of species range expansion. Proc R Soc B 283

Powell THQ, Hood GR, Murphy MO, Heilveil JS, Berlocher SH, Nosil P, Feder JL (2013) Genetic divergence along the speciation continuum: the transition from host race to species in *Rhagoletis* (Diptera: Tephritidae). Evolution 67:2561–2576

Protas ME, Hersey C, Kochanek D, Zhou Y, Wilkens H, Jeffery WR, Zon LI, Borowsky R, Tabin CJ (2006) Genetic analysis of cavefish reveals molecular convergence in the evolution of albinism. Nat Genet 38:107–111

Provine WB (1971) The origins of theoretical population genetics. The University of Chicago Press, Chicago

Puritz JB, Keever CC, Addison JA, Byrne M, Hart MW, Grosberg RK, Toonen RJ (2012) Extraordinarily rapid life-history divergence between *Cryptasterina* sea star species. Proc R Soc B 279:3914–3922

Rabosky DL (2012) Positive correlation between diversification rates and phenotypic evolvability can mimic punctuated equilibrium on molecular phylogenies. Evolution 66:2622–2627

Rabosky DL, Lovette IJ (2008) Explosive evolutionary radiations: decreasing speciation or increasing extinction through time? Evolution 62:1866–1875

Reed AJ, Mann ME, Emanuel KA, Lin N, Horton BP, Kemp AC, Donnelly JP (2015) Increased threat of tropical cyclones and coastal flooding to New York City during the anthropogenic era. Proc Natl Acad Sci USA 112:12610–12615

Rieseberg LH, Widmer A, Arntz AM, Burke JM (2003) The genetic architecture necessary for transgressive segregation is common in both natural and domesticated populations. Philos Trans R Soc B 358:1141–1147

Rogers SM, Xu S, Schluter PM (2017) Introduction: integrative molecular ecology is rapidly advancing the study of adaptation and speciation. Mol Ecol 26:1–6

Rollins LA, Richardson MF, Shine R (2015) A genetic perspective on rapid evolution in cane toads (*Rhinella marina*). Mol Ecol 24:2264–2276

Rougier J, Sparks RSJ, Cashman KV, Brown SK (2018) The global magnitude-frequency relationship for large explosive volcanic eruptions. Earth Planet Sci Lett 482:621–629

Rubin CM, Horton BP, Sieh K, Pilarczyk JE, Daly P, Ismail N, Parnell AC (2017) Highly variable recurrence of tsunamis in the 7,400 years before the 2004 Indian Ocean tsunami. Nat Commun 8:16019

Rudman SM, Kreitzman M, Chan KMA, Schluter D (2017) Evosystem services: rapid evolution and the provision of ecosystem services. Trends Ecol Evol 32:403–415

Ryan WBF, Pitman WC, Major CO, Shimkus K, Moskalenko V, Jones GA, Dimitrov P, Gorür N, Sakinc M, Yüce H (1997) An abrupt drowning of the Black Sea shelf. Mar Geol 138:119–126

Scheffer M, Carpenter S, Foley JA, Folke C, Walker B (2001) Catastrophic shifts in ecosystems. Nature 413:591–596

Scheffer M, Carpenter SR, Lenton TM, Bascompte J, Brock W, Dakos V, van de Koppel J, van de Leemput IA, Levin SA, van Nes EH, Pascual M, Vandermeer J (2012) Anticipating critical transitions. Science 338:344–348

Schluter D (2000) The ecology of adaptive radiation. Oxford University Press, Oxford

Schluter D, Conte GL (2009) Genetics and ecological speciation. Proc Natl Acad Sci USA 106:9955–9962

Shugar DH, Clague JJ, Best JL, Schoof C, Willis MJ, Copland L, Roe GH (2017) River piracy and drainage basin reorganization led by climate-driven glacier retreat. Nat Geosci 10:370–375

Simpson GG (1944) Tempo and mode in evolution. Columbia University Press, New York

Singh MS, Kuang Z, Maloney ED, Hannah WM, Wolding BO (2018) Increasing potential for intense tropical and subtropical thunderstorms under global warming. Proc Natl Acad Sci USA 115:11657–11662

Soltis DE, Soltis PS (1999) Polyploidy: recurrent formation and genome evolution. Trends Ecol Evol 14:348–352

Soltis PS, Soltis DE (2012) Polyploidy and genome evolution. Springer, Heidelberg

Song Y, Endepols S, Klemann N, Richter D, Matuschka F-R, Shih C-H, Nachman MW, Kohn MH (2011) Adaptive introgression of anticoagulant rodent poison resistance by hybridization between old world mice. Curr Biol 21:1296–1301

Stanley SM (1979) Macroevolution: pattern and process. Freeman, San Francisco

Stegen G, Pasmans F, Schmidt BR, Rouffaer LO, van Praet S, Schaub M, Canessa S, Laudelout A, Kinet T, Adriaensen C, Haesebrouck F, Bert W, Bossuyt F, Martel A (2017) Drivers of salamander extirpation mediated by *Batrachochytrium salamandrivorans*. Nature 544:353–356

Stemshorn KC, Reed FA, Nolte AW, Tautz D (2011) Rapid formation of distinct hybrid lineages after secondary contact of two fish species (*Cottus* sp.). Mol Ecol 20:1475–1491

Stordal F, Svensen HH, Aarnes I, Roscher M (2017) Global temperature response to century-scale degassing from the Siberian Traps Large igneous province. Palaeogeogr Palaeoclimatol Palaeoecol 471:96–107

Stuart YE, Campbell TS, Hohenlohe PA, Reynolds RG, Revell LJ, Losos JB (2014) Rapid evolution of a native species following invasion by a congener. Science 346:463–466

Szücs M, Vahsen ML, Melbourne BA, Hoover C, Weiss-Lehman C, Hufbauer RA (2017) Rapid adaptive evolution in novel environments acts as an architect of population range expansion. Proc Natl Acad Sci USA 114:13501–13506

Tagliacollo VA, Roxo FF, Duke-Sylvester SM, Oliveira C, Albert JS (2015) Biogeographical signature of river capture events in Amazonian lowlands. J Biogeogr 42:2349–2362

Taylor CM, Belušić D, Guichard F, Parker DJ, Vischel T, Harris PP, Janicot S, Klein C, Panthou G (2017) Frequency of extreme Sahelian storms tripled since 1982 in satellite observations. Nature 544:475–478

Torres-Dowdall J, Pierotti MER, Harer A, Karagic N, Woltering JM, Henning F, Elmer KR, Meyer A (2017) Rapid and parallel adaptive evolution of the visual system of neotropical midas cichlid fishes. Mol Biol Evol 34:2469–2485

Travis J, Coleman FC, Auster PJ, Cury PM, Estes JA, Orensanz J, Peterson CH, Power ME, Steneck RS, Wootton JT (2014) Integrating the invisible fabric of nature into fisheries management. Proc Natl Acad Sci USA 111:581–584

Tzedakis PC, Crucifix M, Mitsui T, Wolff EW (2017) A simple rule to determine which insolation cycles lead to interglacials. Nature 542:427–432

Ungerer MC, Baird SJE, Pan J, Riesberg LH (1998) Rapid hybrid speciation in wild sunflowers. Proc Natl Acad Sci USA 95:11757–11762

Uusi-Heikkila S, Sävilammi T, Leder E, Arlinghaus R, Primmer CR (2017) Rapid, broad-scale gene expression evolution in experimentally harvested fish populations. Mol Ecol 26:3954–3967

Uyeda JC, Hansen TF, Arnold SJ, Pienaar J (2011) The million-year wait for macroevolutionary bursts. Proc Natl Acad Sci USA 108:15908–15913

Van de Peer Y, Mizrachi E, Marchal K (2017) The evolutionary significance of polyploidy. Nat Rev Genet 18:411–424 van't Hof AE, Campagne P, Rigden DJ, Yung CJ, Lingley J, Quail MA, Hall N, Darby AC, Saccheri IJ (2016) The industrial melanism mutation in British peppered moths is a transposable element. Nature 534:102–105

Vrba ES (1980) Evolution, species and fossils: how does life evolve? S Afr J Sci 76:61–84

Wallis GP, Trewick SA (2009) New Zealand phylogeography: evolution on a small continent. Mol Ecol 18:3548–3580

Wallis GP, Waters JM, Upton P, Craw D (2016) Transverse alpine speciation driven by glaciation. Trends Ecol Evol 31:916–926

Waters JM, Craw D, Youngson JH, Wallis GP (2001) Genes meet geology: fish phylogeographic pattern reflects ancient, rather than modern, drainage connections. Evolution 55:1844–1851

Weir JT, Haddrath O, Robertson HA, Colbourne RM, Baker AJ (2016) Explosive ice age diversification of kiwi. Proc Natl Acad Sci USA 113:E5580–E5587

Wernberg T, Bennett S, Babcock RC, de Bettignies T, Cure K, Depczynski M, Dufois F, Fromont J, Fulton CJ, Hovey RK, Harvey ES, Holmes TH, Kendrick GA, Radford B, Santana-Garcon J, Saunders BJ, Smale DA, Thomsen MS, Tuckett CA, Tuya F, Vanderklift MA, Wilson S (2016) Climate-driven regime shift of a temperate marine ecosystem. Science 353:169–172

Wiens JA (1977) On competition and variable environments. Am Sci 65:590–597

Wilmshurst JM, Hunt TL, Lipo CP, Anderson AJ (2011) High-precision radiocarbon dating shows recent and rapid initial human colonization of East Polynesia. Proc Natl Acad Sci USA 108:1815–1820

Winsor HL, Kovach RP, Allendorf FW (2017) Population genetics and demography unite ecology and evolution. Trends Ecol Evol 32:141–152

Witze A (2017) Ancient volcanoes exposed. Nature 543:295–296

Wright KM, Lloyd DG, Lowry DB, Macnair MR, Willis JII (2013) Indirect evolution of hybrid lethality due to linkage with selected locus in *Mimulus guttatus*. PLoS Biol 11:e1001497

Wright S (1931) Evolution in Mendelian populations. Genetics 16:97–159

Zeng L, Ming C, Li Y, Su L-Y, Su Y-H, Otecko NO, Liu H-Q, Wang M-S, Yao Y-G, Li H-P, Wu D-D, Zhang Y-P (2017) Rapid evolution of genes involved in learning and energy metabolism for domestication of the laboratory rat. Mol Biol Evol 34:3148–3153